Preharvest and Postharvest Factors Improving Horticultural Crops Quality and Shelf-Life

Preharvest and Postharvest Factors Improving Horticultural Crops Quality and Shelf-Life

Editors

Yang Bi
Yongcai Li
Di Gong

MDPI

Basel • Beijing • Wuhan • Barcelona • Belgrade • Novi Sad • Cluj • Manchester

Editors
Yang Bi
College of Food Science and Engineering, Gansu Agricultural University
Lanzhou
China

Yongcai Li
College of Food Science and Engineering, Gansu Agricultural University
Lanzhou
China

Di Gong
School of Public Health, Lanzhou University
Lanzhou
China

Editorial Office
MDPI
St. Alban-Anlage 66
4052 Basel, Switzerland

This is a reprint of articles from the Special Issue published online in the open access journal *Horticulturae* (ISSN 2311-7524) (available at: http://www.mdpi.com).

For citation purposes, cite each article independently as indicated on the article page online and as indicated below:

Lastname, A.A.; Lastname, B.B. Article Title. *Journal Name* **Year**, *Volume Number*, Page Range.

ISBN 978-3-7258-1053-6 (Hbk)
ISBN 978-3-7258-1054-3 (PDF)
doi.org/10.3390/books978-3-7258-1054-3

Contents

horticulturae

Editorial

Special Issue: 'Preharvest and Postharvest Factors Improving Horticultural Crops Quality and Shelf-Life'

Di Gong [1], Yang Bi [2,*] and Yongcai Li [2]

1 School of Public Health, Lanzhou University, Lanzhou 730000, China; gongd@lzu.edu.cn
2 College of Food Science and Engineering, Gansu Agricultural University, Lanzhou 730070, China; lyc@gsau.edu.cn
* Correspondence: biyang@gsau.edu.cn

1. Introduction

Although fresh horticultural crops are rich in nutrients and various biologically active chemicals, they are also perishable. During the pre- and postharvest periods, fresh horticultural products inevitably suffer from quality deterioration and decay [1]. Especially in developing countries, approximately one-third of fresh horticultural products are lost during the postharvest storage and marketing processes and do not reach consumers' tables [2]. Decay not only causes economic losses, but also produces mycotoxins that accumulate in fresh horticultural products, which is a serious threat to consumer health [3]. Effective pre- and post-harvest techniques play a key role in improving the quality of fresh horticultural products [4]. The aim of this Special Issue is to provide readers with a more in-depth understanding of preharvest and postharvest factors that affect the postharvest quality, disease, ripening and senescence, and shelf life of horticultural crops, which could better serve the horticultural industry. After a rigorous review, a total of thirteen articles (two review papers and eleven research papers) are included in this Special Issue, which examines and discusses field management and pre- and postharvest treatments, as well as proposing new methods to improve the quality and shelf life of various fresh horticultural products.

2. Overview of Published Articles

2.1. Internal Factors Affecting the Quality of Horticultural Crops

Variety and maturity are important intrinsic factors affecting the quality of fresh horticultural produce [1]. Ayuso-Yuste et al. (contribution 1) cultivated five traditional tomato varieties and one commercial variety under organic conditions. The results show that traditional varieties have higher contents of total soluble solids (TSSs), titratable acidity, vitamin C, lycopene, β-carotene, and bioactive compounds during ripening. Zou et al. (contribution 2) harvested *Akebia trifoliata* fruits at four stages of maturity and found that fruit size increased and firmness decreased during ripening. Moreover, the accumulation of TSS, fructose, glucose and ascorbic acid content and the decrease in total phenolic and total flavonoid levels in the fruit occurred during ripening. Therefore, *A. trifoliata* fruits harvested at 148 days after full bloom had the best quality and longest shelf life.

2.2. External Factors Affecting the Quality of Horticultural Products

2.2.1. Field Management

The development period is critical for the formation of postharvest quality and resistance in fresh horticultural products. Light, irrigation, soil, and fertilization affect the quality of horticultural products in terms of photosynthesis, respiration, and nutrient content [5]. Photosynthesis is the process by which plants use captured light energy for photochemical reactions and carbon assimilation, which is the basis for crop yield and quality formation [6]. Nadeem et al. (contribution 3) demonstrated that on-tree fruit bagging and 10-day cold storage maintained the postharvest quality of 'Samar Bahisht

Citation: Gong, D.; Bi, Y.; Li, Y. Special Issue: 'Preharvest and Postharvest Factors Improving Horticultural Crops Quality and Shelf-Life'. *Horticulturae* **2024**, *10*, 380. https://doi.org/10.3390/horticulturae10040380

Received: 28 March 2024
Accepted: 2 April 2024
Published: 9 April 2024

Chaunsa' mango fruit, including its firmness, TSSs, vitamin C and total phenolic content, and antioxidant activity. Major et al. (contribution 4) showed that the application of a plant cover increased the dry matter, starch, and sugar contents of 'Early' potato varieties. In addition, the potato tubers harvested from the covered plants showed a similar sensory quality after oil frying compared to potatoes without the cover. Soil moisture status also affects the quality of fresh horticultural crops, and the proper irrigation of arable land is conducive to the coordination of soil water, fertilizer, gas and heat, thereby improving the quality of fresh horticultural crops [7]. Melero-Meraz et al. (contribution 5) showed that supplemental irrigation increased fruit yield, fruit mass, and the proportion of 'Roja Lisa' cactus pears. Moreover, the irrigation treatment reduced fruit mass loss at harvest and during storage.

2.2.2. Preharvest Treatments

Preharvest chemical spraying can significantly improve the postharvest quality of horticultural products. Compared with postharvest treatment, preharvest chemical spraying during the development period is simple, which is more conducive to the application in production practice. Preharvest spraying can not only solve the problems encountered after harvest, but also play a key role in the growth and development of crops, field disease control, and postharvest quality. The spraying has a more pronounced effect [8,9]. Zhu et al. (contribution 6) showed that plants of 'Longshu No.7' potato sprayed with 3% chitosan before harvest had enhanced phenylpropanoid metabolism and an accelerated accumulation of suberin polyphenols and lignin. Moreover, periderm formed in the wounds of potato tubers, thus improving their postharvest quality. Ziogas et al. (contribution 7) demonstrated that foliar spraying with a Si-Ca-based product delayed maturation, maintained postharvest quality, and reduced decay during the storage of 'Clementine' mandarins. Gong et al. (contribution 8) found that preharvest spraying with a chemical elicitor delayed ripening and maturity and maintained the postharvest quality of fruit. Moreover, the spray improved the antioxidant activity of the harvested fruit by enhancing the antioxidant system. In addition, the spray maintained cell membrane integrity, thereby reducing the chilling injury of harvested fruit during cold storage.

2.2.3. Postharvest Treatments

After harvest, the quality of fresh horticultural crops deteriorates with ripening/ripening and senescence, and undesirable metabolic changes occur in response to storage environments, including softening, uneven color, and a loss of flavor and nutrients [10]. Although these physiological and biochemical changes cannot be stopped, appropriate postharvest measures and improved storage conditions help to delay ripening and senescence, maintain postharvest quality, and extend the shelf life of fresh horticultural crops during storage. Sortino et al. (contribution 9) showed that coating fresh-cut 'Hayward' kiwifruits with mucilage from *Opuntia ficus-indica* and *Aloe arborescens* delayed fruit ripening and maintained appearance and texture. The coating also maintained TA and TSSs, reducing weight loss and microbial activity. In addition, the coating extended the shelf life of fresh-cut kiwifruit. Vilvert et al. (contribution 10) demonstrated that chitosan–graphene–oxide-based biodegradable packaging could delay ripening and maintain the postharvest quality of 'Tommy Atkins' mangoes during cold storage. Moreover, the packaged fruit demonstrated higher levels of ascorbic acid, yellow flavonoids, phenolic compounds, and antioxidant activity. Choi et al. (contribution 11) found that chlorophyll fluorescence changes during fruit ripening and can be used to classify the ripening stages. Tomatoes stored in the dark at 20–15 °C (day–night) had delayed ripening. However, tomatoes stored at 30–20 °C (day–night) with a light intensity of 400 $\mu mol \cdot m^{-2} \cdot s^{-1}$ showed greater accumulation of sugar, phenolics and lycopene, and higher levels of antioxidant activity. Dogan et al. (contribution 12) showed that the atmospheric composition of 3% O_2 + 15% CO_2 effectively delayed the ripening of fresh 'Bursa Siyahi' fig by reducing the respiration rate and ethylene production during 28 d of storage at 0 °C. Moreover, the treatment maintained the firmness, as well as the sugar and organic acid contents, of fig fruits and prevented decay caused by microorganism growth. Subroto et al. (contribution 13) reviewed microbiological activity in the postharvest handling of cocoa beans, particularly in the fermentation process. In addition, the effect of microbial activity on the physical and chemical properties of cocoa beans was concluded. They also demonstrated that fermentation control is critical to improving the postharvest quality of cocoa beans.

3. Conclusions

These articles, which are written to a very high standard with excellent citations, were selected from a large number of submissions through rigorous evaluation and. This Special Issue strives to promote more in-depth communication among professional and technical personnel in the field of horticultural postharvest studies, improving postharvest quality, ensuring product safety, and serving the healthy development of the horticultural industry.

Author Contributions: Writing—original draft preparation, D.G.; writing—review and editing, Y.B.; visualization, Y.L. All authors have read and agreed to the published version of the manuscript.

Funding: This paper was supported by Natural Science Foundation of Gansu Province (23JRRA1082).

Conflicts of Interest: The authors declare no conflict of interest.

List of Contributions

1. Ayuso-Yuste, M.C.; González-Cebrino, F.; Lozano-Ruiz, M.; Fernández-León, A.M.; Bernalte-García, M.J. Influence of ripening stage on quality parameters of five traditional tomato varieties grown under organic conditions. *Horticulturae* **2022**, *8*, 313. https://doi.org/10.3390/horticulturae8040313.
2. Zou, S.; Gao, P.; Jia, T.; Huang, H. Physicochemical characteristics and nutritional composition during fruit ripening of *Akebia trifoliata* (Lardizabalaceae). *Horticulturae* **2022**, *8*, 326. https://doi.org/10.3390/horticulturae8040326.
3. Nadeem, A.; Ahmed, Z.F.R.; Hussain, S.B.; Omar, A.E.-D.K.; Amin, M.; Javed, S.; Ali, A.; Ullah, S.; Razzaq, K.; Rajwana, I.A.; et al. On-tree fruit bagging and cold storage maintain the postharvest quality of mango fruit. *Horticulturae* **2022**, *8*, 814. https://doi.org/10.3390/horticulturae8090814.
4. Major, N.; Goreta Ban, S.; Perković, J.; Žnidarčič, D.; Peršurić, A.S.I.; Oplanić, M.; Dumičić, G.; Urlić, B.; Ban, D. Plant cover stimulates quicker dry matter accumulation in "Early" potato cultivars without affecting nutritional or sensory quality. *Horticulturae* **2022**, *8*, 364. https://doi.org/10.3390/horticulturae8050364.
5. Melero-Meraz, V.; Zegbe, J.A.; Herrera, M.D.; Guzmán-Maldonado, S.H.; Medina-García, G.; Sánchez-Toledano, B.I.; Cruz-Bravo, R.K.; Servín-Palestina, M. On-farm supplemental irrigation of 'Roja Lisa' cactus pear: pre- and postharvest effects. *Horticulturae* **2022**, *8*, 483. https://doi.org/10.3390/horticulturae8060483.
6. Zhu, Y.; Zhang, W.; Zhao, Y.; Chen, F.; Zhu, F.; Yang, J.; Ge, X.; Gong, W.; Cui, R.; Sheng, W. Foliar application of chitosan accelerates wound periderm formation with an intensified deposition of suberin polyphenolic and lignin in the wounds of potato tubers. *Horticulturae* **2023**, *9*, 663. https://doi.org/10.3390/horticulturae9060663.
7. Ziogas, V.; Bravos, N.; Hussain, S.B. Preharvest foliar application of Si–Ca-based biostimulant affects postharvest quality and shelf-life of clementine mandarin (*Citrus clementina* Hort. Ex Tan). *Horticulturae* **2022**, *8*, 996. https://doi.org/10.3390/horticulturae8110996.
8. Gong, D.; Bi, Y.; Li, Y.; Wang, Y.; Prusky, D.; Alkan, N. Preharvest elicitors spray improves antioxidant activity, alleviates chilling injury, and maintains quality in harvested fruit. *Horticulturae* **2022**, *8*, 1208. https://doi.org/10.3390/horticulturae8121208.
9. Sortino, G.; Inglese, P.; Farina, V.; Passafiume, R.; Allegra, A. The Use of *Opuntia ficus-indica* Mucilage and *Aloe arborescens* as edible coatings to improve the physical, chemical, and microbiological properties of 'Hayward' kiwifruit slices. *Horticulturae* **2022**, *8*, 219. https://doi.org/10.3390/horticulturae8030219.

10. Vilvert, J.C.; de Freitas, S.T.; Ferreira, M.A.R.; Costa, C.d.S.R.; Leite, R.H.d.L.; dos Santos, F.K.G.; Aroucha, E.M.M. Preservation of quality and bioactive compounds in mangoes using chitosan-graphene-oxide-based biodegradable packaging. *Horticulturae* **2023**, *9*, 1145. https://doi.org/10.3390/horticulturae9101145.
11. Choi, H.G.; Park, K.S. Ripening process of tomato fruits postharvest: impact of environmental conditions on quality and chlorophyll *a* fluorescence characteristics. *Horticulturae* **2023**, *9*, 812. https://doi.org/10.3390/horticulturae9070812
12. Dogan, A.; Erkan, M. Responses of high carbon dioxide concentration on postharvest quality of fresh fig fruit during storage. *Horticulturae* **2023**, *9*, 293. https://doi.org/10.3390/horticulturae9030293.
13. Subroto, E.; Djali, M.; Indiarto, R.; Lembong, E.; Baiti, N. Microbiological activity affects post-harvest quality of cocoa (*Theobroma cacao* L.) beans. *Horticulturae* **2023**, *9*, 805.

References

1. Shipman, E.N.; Yu, J.; Zhou, J. Can gene editing reduce postharvest waste and loss of fruit, vegetables, and ornamentals? *Hortic. Res.* **2021**, *8*, 1. [CrossRef] [PubMed]
2. Porat, R.; Lichter, A.; Terry, L.A.; Harker, R.; Buzby, J. Postharvest losses of fruit and vegetables during retail and in consumers' homes: Quantifications, causes, and means of prevention. *Postharvest Biol. Technol.* **2018**, *139*, 135–149. [CrossRef]
3. Chen, T.; Ji, D.; Zhang, Z.Q.; Li, B.Q.; Qin, G.Z.; Tian, S.P. Advances and strategies for controlling the quality and safety of postharvest fruit. *Engineering* **2020**, *7*, 1177–1184. [CrossRef]
4. Ziv, C.; Fallik, E. Postharvest storage techniques and quality evaluation of fruits and vegetables for reducing food loss. *Agronomy* **2021**, *11*, 1133. [CrossRef]
5. Wang, X.; Xing, Y. Evaluation of the effects of irrigation and fertilization on tomato fruit yield and quality: A principal component analysis. *Sci. Rep.* **2017**, *7*, 350. [CrossRef] [PubMed]
6. Griñán, I.; Morales, D.; Galindo, A.; Torrecillas, A.; Pérez-López, D.; Moriana, A.; Hernández, F. Effect of preharvest fruit bagging on fruit quality characteristics and incidence of fruit physiopathies in fully irrigated and water stressed pomegranate trees. *J. Sci. Food Agric.* **2018**, *99*, 1425–1433. [CrossRef] [PubMed]
7. Mpelasoka, B.S.; Behboudian, M.H.; Mills, T.M. Effects of deficit irrigation on fruit maturity and quality of 'Braeburn' apple. *Sci. Hortic.* **2001**, *90*, 279–290. [CrossRef]
8. Xie, P.D.; Yang, Y.Y.; Gong, D.; Li, Y.; Wang, Y.; Li, Y.C.; Prusky, D.; Bi, Y. Spraying L-phenylalanine during fruit development alleviates chilling injury in harvested muskmelons by regulating membrane lipid metabolism. *Postharvest Biol. Technol.* **2024**, *211*, 112858. [CrossRef]
9. Xie, P.D.; Yang, Y.Y.; Oyom, W.; Su, T.T.; Tang, Y.B.; Wang, Y.; Li, Y.C.; Prusky, D.; Bi, Y. Chitooligosaccharide accelerated wound healing in potato tubers by promoting the deposition of suberin polyphenols and lignin at wounds. *Plant Physiol. Biochem.* **2023**, *199*, 107714. [CrossRef] [PubMed]
10. Yun, Z.; Gao, H.J.; Jiang, Y.M. Insights into metabolomics in quality attributes of postharvest fruit. *Curr. Opin. Food Sci.* **2022**, *45*, 100836. [CrossRef]

horticulturae

MDPI

Article

The Use of *Opuntia ficus-indica* Mucilage and *Aloe arborescens* as Edible Coatings to Improve the Physical, Chemical, and Microbiological Properties of 'Hayward' Kiwifruit Slices

Giuseppe Sortino , Paolo Inglese , Vittorio Farina , Roberta Passafiume and Alessio Allegra *

Department of Agricultural, Food and Forest Sciences, Università degli Studi di Palermo, Edificio 4, Ingresso H, 90128 Palermo, Italy; giuseppe.sortino@unipa.it (G.S.); paolo.inglese@unipa.it (P.I.); vittorio.farina@unipa.it (V.F.); roberta.passafiume@unipa.it (R.P.)
* Correspondence: alessio.allegra@unipa.it; Tel.: +39-09123861234

Abstract: Edible coatings (EC) with mucilage of *Opuntia ficus-indica* or *Aloe arborescens* are promising to extend the shelf life of fresh-cut fruit products by reducing weight loss and microbial spoilage. In this work, fresh-cut kiwifruits (cv. Hayward) were coated in solutions with mucilage extracted from *Opuntia ficus-indica* (MC) and *A. arborescens* (AL). We used three alternative treatments with AL or MC, MC+AL, and with distilled water as control, and stored in passive atmosphere at 5 °C for 3, 5, 7, and 9 d, respectively. For all treatments at each storage period, firmness, weight loss, color, visual quality score, respiration rate, pectin content, and microbiological characteristics were observed. The treatments with mucilage and *A. arborescens* applied on fresh-cut kiwi slices showed different significant effects until 9 days of storage, in terms of firmness and total pectin. Microbial spoilage analysis revealed the beneficial effects of this strategy after 3 d, particularly in terms of bacteria and yeast. *A. arborescens* provided a reduction of microbial activity, probably due to the higher quantity of aloin if related to Aloe species. Furthermore, the treatment with MC and AL increased the total pectin content, showing positive effects in terms weight loss and firmness. The results showed that the MC+AL treatment improved the visual score of fresh-cut kiwi fruit until 7 d of storage.

Keywords: *Actinidia deliciosa*; pectin contents; fresh-cut; post-harvest; bio-based films; visual color

Citation: Sortino, G.; Inglese, P.; Farina, V.; Passafiume, R.; Allegra, A. The Use of *Opuntia ficus-indica* Mucilage and *Aloe arborescens* as Edible Coatings to Improve the Physical, Chemical, and Microbiological Properties of 'Hayward' Kiwifruit Slices. *Horticulturae* **2022**, *8*, 219. https://doi.org/10.3390/horticulturae8030219

Academic Editors: Yang Bi, Yongcai Li and Di Gong

Received: 4 February 2022
Accepted: 28 February 2022
Published: 2 March 2022

1. Introduction

Kiwifruit (*Actinidia* spp.) is a typical climacteric fruit that, after harvest, goes through three distinct softening phases that are temporally well separated. Pectin retained in the cell wall starts to "soften" during ripening; this process clearly precedes both pectin solubilization and depolymerization (Phase 1) [1].

The degradation of solubilized pectin and loss of middle lamellae in kiwifruit (Phase 2) are processes initiated in the second softening phase, with a peak in the last softening phase, where cell wall disintegration is completed (Phase 3) [1]. After harvest, kiwifruits can be stored for a long period and sold for fresh consumption or for processing as fresh-cut. The peeling and slicing involved in minimal fruit processing can cause physical damages, and an increase of polyphenoxidase and peroxidase begins browning in the flesh. In fresh-cut fruit, the rapid softening and deterioration likely involve membrane and cell wall catabolism, accelerated or otherwise altered in response to physical wounding [2].

In recent years, limited research works have been carried out on applying various kinds of natural edible coatings to extend fresh-cut kiwifruit during storage. The use of edible coatings could be an efficient alternative to the chemical treatments applied to preserve fresh-cut kiwifruit. *Opuntia ficus-indica* (OFI) mucilage, *A. vera*, and *A. arborescen* coatings reduce physiological disorders and gas exchange [3,4]. Moreover, these coatings can maintain total soluble solids, a higher concentration of total phenols and ascorbic acid, and a better antioxidant activity when compared to the control [5–7]. The most studied gel from any *Aloe* spp. has been *A. vera* or *A. arborescens*, where the gel is found in a clear internal zone located between the abaxial and adaxial mesophyll. The chemical composition of the aloe species gel is very complex, composed mainly of polysaccarides and soluble sugars, followed by proteins; most of these are enzymes, amino acids, and vitamins [8]. Zapata et al. [9] showed that *A. arborescens* gel had more aloin concentration, total phenolics, total antioxidant activity (hydrophilic and lipophilic fractions), and a low level of putrescine and spermidine than *A. vera* gel content. However, the use of *A. arborescens* showed overall inhibitions of fungal growth on inoculations in vitro test while the use of OFI mucilage did not show a positive effect on concentration of pseudomonads and yeasts during storage of breba fig fruit [10]. OFI mucilage is present in cladodes, in the Golgi apparatus, and the mucilage synthesis probably takes place in it and in the vesicles derived from it [11]. These polysaccharides swell when dissolved in water, or in some cases, form colloidal and very viscous suspensions or jellied masses [12]. In particular, OFI mucilage has complex polymeric substances of a carbohydrate nature; a highly branched structure [13]; and is composed of arabinose, galactose, rhamnose, xylose and galacturonic acid [14]. The use of a natural edible coating could replace calcium-based treatments and compounds made from different fermentation or polymerization reactions (e.g., HPMC, Xanthan gum, etc.). OFI mucilage and *A. vera* coatings have been previously studied by Allegra [14] and Benitez [3], regarding the maintenance of quality in kiwifruit slices during storage. However, those studies showed that non-combined treatments have not improved the physical and microbiological qualities of fresh-cut kiwi during storage. Benitez showed that, with kiwi slices, coated and uncoated samples were given similar scores during storage, except in the bitterness attributes.

The objective of the present work was to evaluate the efficacy of *O. ficus-indica* coatings with *A. arborescens* (2%) in prolonging the shelf life of kiwifruit slices. We focused our study on the effects of OFI mucilage with *A. arborescens* on the visual quality score, firmness, and microbial load of minimally processed kiwifruit slices, during 9 days of storage.

2. Materials and Methods

2.1. Fruits and Preparation of the Edible Coating

'Hayward' kiwifruits (*Actinia deliciosa*) were purchased from a supermarket belonging to a large, very well-known organized distributor in Italy, which provides medium-high quality products (Conad, Palermo, Italy). Kiwifruits were selected based on their firmness (49 N $\pm$ 2.2) and soluble solid total (12.5 °Brix $\pm$ 0.2) stored in air at 5 $\pm$ 0.5 °C (RH = 85%) for 24 h. After storage, the fruits were dipped in chlorinated water (100 ppm of free chlorine) for 6 min. Damaged fruits (bruised or showing other physical decays) were removed, and a total of 400 fruits with an average firmness of 48.2 $\pm$ 3.5 N and weight of 100 $\pm$ 4.2 g were processed.

Cactus pear (*O. ficus-indica* (L.) Mill.) cladodes were collected from an experimental field near Roccapalumba (37°48′ N; 13°38′ E; 350 m asl), western Sicily, Italy. Promptly after harvest, cladodes were transported to the post-harvest laboratory of the University of Palermo and were cut and cubed (2 cm^3). Samples were homogenized (20% *w*/*v*) in distilled water with water ratio 1:1.5. The solution was maintained at 40 °C for 90 min and centrifuged (model CS6R) at 3000 rpm $\times$ 20 min. The supernatant was boiled to halve the initial volume, and ethanol at 99% was added in ratio 1:2, in order to reduce the amount of alcohol used in the precipitation [15]. Afterwards, the solution was stored at 4 $\pm$ 1 °C for 48 h to allow a better aggregation of the mucilage. The last phase involved the elimination of the supernatant and the soaking of the pure mucilage.

Matured leaves of the *A. arborescens* plant were harvested at the experimental field of the University of Palermo (Italy). *A. arborescens* gel was prepared from 1 kg of leaves taken from 10-year-old plants. The leaves were cleaned externally with a knife, removing the margin, and were then cut lengthwise. The parenchyma (from which the gel is obtained) was separated from the epidermis. The gelatinous parenchyma was homogenized with Ultra-Turrax (Ultra-Turax T25, Janke and Kunkle, IKa Labortechnik, Breisgau, Germany) for 5 min at 24,500 rpm, thus obtaining a mucilaginous gel, and was subsequently filtered to eliminate the fibrous portion [16].

2.2. Fruit Processing

Kiwifruits were peeled manually and cut into slices with a semiautomatic machine. Slices were 2.2 ± 0.3 cm thick and 5.4 ± 1.1 cm diameter, characterized by the Hunter's parameters L* = 66.6 ± 2.4, a* = 7.8 ± 1.3, and b* = 25.4 ± 2.1, with a total solid soluble content (SST) of 12.9 ± 2.2 °Brix and titratable acidity (TA) of 1.5 ± 0.4 (g L^{-1}). Fresh-cut slices were dipped in the coating solution for 60 s; the excess coating was drained, and the coated slices were dried in a forced-air dryer (20 °C) for 10 min.

The coating treatment consisted of: (a) 30 g of pure mucilage extract, 500 mL distilled water with 2% of *A. arborescens* and 50 mL glycerol as a plasticizer (MC+AL); (b) 30 g of mucilage extract, 500 mL distilled water, 50 mL glycerol (MC); (c) 500 mL distilled water with 2% of *A. arborescens* and 50 mL glycerol as a plasticizer. The control trial (CTR) was set up by sliced fruits dipped in distilled water. Approximately 100 g ± 0.8 of kiwifruit slices were packed in polyethylene terephthalate (PET) packages and sealed with a composite film (PP-PET, 64 μm, O_2 permeability = 5.30×10^{-8} μL m^{-2} s^{-1} Pa^{-1}). Packages were stored at 5 °C and 90% relative humidity (RH) for 9 d. Physico-chemical and microbiological quality parameters were analyzed on six slices used as single replicates (4 treatments × 5 time of storage × 6 replicates = 120 box), at the beginning of the experiment (after coating/dipping = day 0) and at 3, 5, 7, and 9 d after storage.

2.3. Firmness

Firmness was evaluated by a puncture test on kiwifruit slice flesh using a TA-XT Plus texture analyzer (Stable Micro Systems, Surrey, England). Firmness measurements were taken as the medium force value obtained during the test by a stainless steel probe with 4 mm diameter penetrating inside the fruit for 4 mm, at 1 mm/s. Average values were calculated from the results of at least six measurements in different slices for each sample. Measures were taken in the fruit outer pericarp (green flesh), where the fast rate of softening compromised fruit quality.

2.4. Weight Loss (%)

The weight of individual bags was recorded immediately after the treatment (day 0) and at the different sampling times (3, 5, 7, and 9 d during storage). Weight loss was expressed as the percentage reduction with respect to initial time, using the following equation:

$$\% \text{ Weight loss} = [(\text{Initial fruit bags weight} - \text{Final fruit bags weight}) \times 100]/\text{Initial fruit bags weight}$$

2.5. Total Soluble Solids Content and Titratable Acidity

Soluble solids total concentration was determined with a hand-held refractometer, and pH was determined by a pH meter. Titratable acidity (expressed as % citric acid) was determined by titration of 10 mL of juice with 0.1 M NaOH to an endpoint of pH 8.1 [3].

2.6. Visual Appearance Score

To measure the effect of cold storage on kiwifruit sensory traits at each storage time (0, 3, 5, 7, and 9 d), six slices, used as single replicates for treatment (MC+AL, MC, AL, and CTR), were scored by each of a six-judge trained panel, who generated a list of descriptors in a few preliminary meetings. All panelists participated in 10 d of training and 6 d of tasting, lasting for a total of 4 weeks. Visual appearance was focused on color, visible structural integrity, and visual appearance [17]. The different descriptors were quantified using a subjective 5 pt. scoring scale with 5 = very good, 4 = good, 3 = sufficient (limit of marketability), 2 = poor (limit of usability), and 1 = very poor (inedible).

2.7. Package O_2 and CO_2 Analysis

CO_2 and O_2 levels (Kpa) were measured on each package at the beginning of each experiment and after 3, 5, 7, and 9 d of storage, using a PBI Dansensor Checkpoint O_2 and CO_2 analyzer (Topac, Hingham, MS, USA) with zirconium and infrared detectors, respectively.

2.8. Color

The superficial color of the kiwifruit slices was determined using a Minolta colorimeter CR-400 model (Minolta Camera, Osaka, Japan) in the CIElab space. The L* (lightness), a* (greenness (−) to redness (+)), and b* (blueness (−) to yellowness (+)) values were recorded for each treatment at 3, 5, 7, and 9 d [5]. Total color difference (ΔE*) expressed the magnitude of the difference between the initial non-aged color pulp (zero time) and storage-aged samples. Total color difference (ΔE*) was calculated according to the following:

$$\Delta E^* = [(\Delta L^*)^2 + (\Delta a^*)^2 + (\Delta b^*)^2]^{1/2} \quad (1)$$

where $\Delta a^* = a^* - a_0^*$, $\Delta b^* = b^* - b_0^*$, $\Delta L^* = L^* - L_0^*$; a_0^*, b_0^*, and L_0^* are the corresponding blank values of control sample, and $a_0^* = -2.00$, $b_0^* = 4.00$, and $L_0^* = 80.00$, respectively [5]. All data were reported as mean ± standard error (SE, $n = 6$).

2.9. Microbiological Analysis

Fruit samples and mucilage were microbiologically investigated for total mesophilic microorganisms (TMM) and the undesired (spoilage and/or pathogenic) microbial groups. The fruits (25 g), *A. arborescens* (10 mL), and mucilage (10 mL) were suspended in Ringer's solution (Sigma-Aldrich, Milan, Italy) with a ratio 1:10 (fruit:diluent), homogenized for 2 min at the highest speed with a stomacher (BagMixer® 400, Interscience, Saint Nom, France), and serially diluted. The cell suspensions were inoculated as follows: TMM on plate count agar (PCA), incubated at 30 °C for 72 h; Pseudomonas on Pseudomonas agar base (PAB), supplemented with 10 mg/mL cetrimide fucidin, incubated at 20 °C for 48 h; yeasts on yeast potato dextrose (YPD) agar, incubated at 25 °C for 48 h. All materials and supplements were purchased from Oxoid (Milan, Italy). Count plates were carried out in duplicate for each trial.

2.10. Pectin Analysis

The pectic substances from fruits were extracted according to the methods described by Rouse and Atkins [18,19]. Frozen 'Hayward' *A. deliciosa* were thawed at 4 °C for 24 h; approximately 20 g were weighed into a plastic cup and homogenized at low speed with an Ultra-Turrax tissue mixer for 2 min. The tissue sample (5 g) was weighed into a 50 mL round-bottom plastic centrifuge tube. Hot 100% ethanol (30 mL) was added to the tubes. The samples were stirred thoroughly with a glass rod, heated in a boiling water bath for 10 min, cooled, and centrifuged at 10,000 rpm for 10 min. The alcoholic supernatant was decanted and discarded. The precipitate was extracted with 30 mL 100% ethanol, centrifuged, and the supernatant was discarded. The residue was transferred from the centrifuge tube to a 57 mm aluminum weighing dish (Fisher Scientific Inc. Italia, Milan, Italy). Samples were dried for 24 h in a conventional oven at 35 °C, weighed, and ground in a 65 mortar with a pestle. The precipitate from the alcohol solution was designated as alcohol-insoluble solids (AIS).

2.10.1. Water Soluble Pectin

Dried alcohol-insoluble solid (AIS) (80 mg) was weighed into a 50 mL centrifuge tube. Distilled water (20 mL) was added, and samples were stirred with a glass rod for 1 min. Samples were centrifuged at 10,000 rpm for 10 min and filtered through Whatman No. 1 filter paper to obtain water-soluble pectin (WSP). The extraction procedure was repeated once. The supernatants were collected and combined in a 100 mL volumetric flask. Distilled water was added to dilute the tube extract solution to volume.

2.10.2. Oxalate Soluble Pectin

The residue was dispersed in 20 mL of an aqueous solution containing 0.25% ammonium oxalate and 0.25% oxalic acid and stirred with a glass rod for 1 min. Samples were refluxed in a boiling water bath for 1 h, centrifuged at 10,000 rpm for 10 min, and filtered through Whatman No. 1 filter paper to obtain oxalate-soluble pectin (OSP). The oxalate extraction was repeated once. Supernatants were collected and diluted to 100 mL with distilled water in a volumetric flask.

2.10.3. Total Pectin

Total pectin (TP) was extracted following the method of Ahmed and Labavitch [20]. Dried AIS (5 mg) was weighed into a 30 mL beaker containing a magnetic stir bar. Concentrated sulfuric acid (2 mL) was added to the beaker, and the mixture was swirled gently. The beaker was placed on a stir plate and stirred gently, and 0.5 mL of distilled water were added dropwise. Stirring continued for 5 min; an additional aliquot of 0.5 mL distilled water was added dropwise. Stirring continued further for approximately 30 min until the dissolution of the AIS was complete. The dissolved sample was filtered through glass wool into a 25 mL volumetric flask. Each beaker was rinsed several times with distilled water, combined in a 25 mL flask, and diluted to volume. The solution was filtered through glass wool once more before use.

2.10.4. Non-Extractable Pectin

The difference between TP and the sum of WSP and OSP was used to determine the amount of non-extractable pectin (NXP), which was the protopectin fraction. TP, WSP, and OSP extractions were completed in duplicate for all samples. All pectin extracts were stored at 5 °C for 12 h before analysis.

2.10.5. Pectin Content Determination

Pectin extracts contents were analyzed by the m-hydroxydiphenyl method [21]. The extract from each sample (1 mL) was pipetted into a 16 × 150 mm test tube. Sulfuric acid tetraborate solution (0.0125 M sodium tetraborate in concentrated sulphuric acid) (6 mL) was added to each of the tubes in an ice water bath and mixed carefully using a Vortex mixer at moderate speed, with intermittent stopping to assure complete mixing. Duplicate samples were prepared for each pectin measurement with a corresponding blank. Tubes were heated in a boiling water bath for 5 min and immediately placed in ice water to cool. To duplicate tubes, 0.1 mL aliquot of 0.15% m-hydroxydiphenyl, was added to develop color. To the blank tube, 0.1 mL 0.5% sodium hydroxide was added. All samples and blanks were mixed using the Vortex mixer and allowed to stand for 15 min at room temperature. The absorbance of the samples following chromogen formation was measured at the wavelength of 520 nm using a Spectrophotometer. Galacturonic acid was used as a standard. A solution consisting of 1 mL distilled water, 6 mL sulfuric acid/tetraborate, and 0.1 mL 0.5% sodium hydroxide was used as reagent blank. The determination of the pectin was carried out according to Yu et al. [22].

2.11. Statistical Analysis

The experimental design consisted of three coating treatments and the untreated control, with observations made at 0, 3, 5, 7, and 9 d after coating. Analysis of Variance (ANOVA) to compare collected data was used (Systat 13.0 (Systat Software, Inc. San Jose, CA USA) was used as statistical software). Significant differences ($p \leq 0.05$) were evaluated using the Tukey's test.

3. Results and Discussion

3.1. Solid Soluble Total (SST), Titratable Acidity (TA), Color and Visual Score

Kiwifruit slice SST did not change significantly between treatments during storage time, but after 3 d, the sample slices treated with mucilage (MC) showed significant differences compared to the other treatments (CTR, AL, MC+AL). The control slices (CTR) and MC treatment showed an increase of mean values from 12.9 at 14.1 °Brix and 12.9 at 13.9 °Brix, respectively (Table 1). This is related to the sharp increase of ethylene and respiration rate [23]. Jordan et al. [24] observed that most of the carbohydrates in kiwifruit at harvest were starch, which was hydrolyzed into sugar during further ripening and may lead to the increase of TSS. Untreated kiwifruit slices showed a sharp, significant ($p \leq 0.05$) increase in SST, while no significant differences were observed in "MC+AL" and AL treatments during storage time (Table 1). Similar results were obtained by Sicari et al. [25] during conservation of strawberry fruits coated with *A. arborescens* gel. A reverse trend during storage time was observed (Table 2) in terms of values of TA; CTR and AL treatments both showed a significant decrease ($p \leq 0.05$) from cutting at day 9, while slices coated with MC+AL and MC showed no significant decrease during storage time. The reduction of acidity content in kiwifruit is related to softening and therefore may be associated with the consumption of organic acids under the mechanism of respiration and gluconeogenesis. Generally, the titratable acid content of most fruits increases or remain unchanged during the ripening process, and decreases only at a very ripe stage [26]. The treatment with MC+AL and MC influenced TA content by maintaining the mean values between 1.6 and 1.4%. In breba fig fruit, the malic acid, glutamic acid, citric acid, pipecolic acid, benzoic acid, malonic acid, pelargonic acid, stearic acid, and linoleic acid were higher in sample coated with OFI mucilage than in control samples analyzed at commercial harvest time [27].

Table 1. Solid soluble total of fruit slices of *Actinidia deliciosa* (Liang. Ferguson) coated with *O. ficus-indica* (OFI) mucilage (MC), or mucilage + *A. arborescens* (MC+AL) or *A. arborescens* (AL) not treated (CTR), just after being coated (0) and at 3, 5, 7, 9 d of storage at 5 °C. Different lowercase letters indicate significant differences between treatments at each sampling date. Different capital letters indicate significant differences between sampling date for the same treatment. Tukeys' significant test was applied at $p \leq 0.05$. Data are means ± S.E. (n = 3).

Time of Storage (Days)	Solid Soluble Total (°Brix)			
	CTR	MC	MC+AL	AL
0 days	12.9 ± 0.2 A nsA	12.9 ± 0.2 NS	12.9 ± 0.2 A	12.9 ± 0.2 NS
3 days	13.2 ± 0.2 nsA	13.1 ± 0.1	13.3 ± 0.2 A	13.0 ± 0.3
5 days	13.5 ± 0.2 aA	13.0 ± 0.1 b	13.5 ± 0.4 aA	13.2 ± 0.4 a
7 days	13.7 ± 0.4 nsA	13.3 ± 0.3	13.5 ± 0.2 A	13.3 ± 0.2
9 days	14.1 ± 0.1 nsAB	13.9 ± 0.4	13.6 ± 0.5 AB	13.5 ± 0.2

Table 2. Titratable acidity of fruit slices of *Actinidia deliciosa* (Liang. Ferguson) coated with *O. ficus-indica* (OFI) mucilage (MC), or mucilage + *A. arborescens* (MC+AL) or *A. arborescens* (AL) not treated (CTR), just after being coated (0) and at 3, 5, 7, 9 d of storage at 5 °C. Different lowercase letters indicate significant differences between treatments at each sampling date. Different capital letters indicate significant differences between sampling date for the same treatment. Tukeys' significant test was applied at $p \leq 0.05$. Data are means ± S.E. (n = 3).

Time of Storage (Days)	Titratable Acidity (g L^{-1})			
	CTR	MC	MC+AL	AL
0 days	1.8 ± 0.2 nsA	1.8 ± 0.2 NS	1.8 ± 0.2 NS	1.8 ± 0.2 A
3 days	1.5 ± 0.2 aA	1.7 ± 0.3 a	1.6 ± 0.2 b	1.7 ± 0.5 bA
5 days	1.3 ± 0.1 aAB	1.6 ± 0.2 b	1.6 ± 0.1 b	1.6 ± 0.3 bA
7 days	1.2 ± 0.3 aAB	1.5 ± 0.1 b	1.6 ± 0.4 b	1.5 ± 0.2 bA
9 days	0.9 ± 0.6 B	1.4 ± 0.5 b	1.4 ± 0.2 b	1.0 ± 0.3 aB

The brightness values (L*) showed a significant difference between treatments at all storage time points (Table 3). After cutting, MC+AL samples showed a slow decrease from the 3rd to the 7th day, while at the 9th day there was a significant decrease of 30% of brightness. The L* value of untreated kiwifruit slices showed an increase from the 5th to the 7th day: 58.4 and 70.4, respectively. The development of translucency, as a sign of tissue injury, could be triggered by an excessive in-package accumulation of CO_2 [28] or by water loss from the inside to the outside of cut fruit.

Table 3. Color L* of fruit slices of *Actinidia deliciosa* (Liang. Ferguson) coated with *O. ficus-indica* (OFI) mucilage (MC), or mucilage + *A. arborescens* (MC+AL) or *A. arborescens* (AL) not treated (CTR), just after being coated (0) and at 3, 5, 7, 9 d of storage at 5 °C. Different lowercase letters indicate significant differences between treatments at each sampling date. Different capital letters indicate significant differences between sampling date for the same treatment. Tukeys' significant test was applied at $p \leq 0.05$. Data are means ± S.E. (n = 3).

Time of Storage (Days)	**L* (CieLab)**			
	CTR	**MC**	**MC+AL**	**AL**
0 days	66.6 ± 2.4 nsA	66.6 ± 2.4 nsA	66.6 ± 2.4 nsA	66.6 ± 2.4 nsA
3 days	60.3 ± 1.9 cB	67.4 ± 1.9 abA	68.4 ± 5.3 aA	68.2 ± 2.1 aA
5 days	58.4 ± 1.9 dB	60.1 ± 2.9 cB	67.4 ± 3.2 bA	69.9 ± 2.0 aA
7 days	70.4 ± 2.8 aC	58.9 ± 3.4 cC	67.1 ± 1.9 bA	51.4 ± 5.8 dB
9 days	51.5 ± 1.9 bD	55.2 ± 5.3 aD	45.7 ± 3.7 cB	52.1 ± 4.7 bB

The color change measured with ΔE in slices coated with mucilage + *A. arborescens* (MC+AL) showed no relevant differences during the first 3 d of storage compared to other treatments, while after 3 d the mean values of ΔE showed significant differences between CTR, MC, and AL treatments (Table 4). During storage, no significant ($p \leq 0.05$) changes in slices coated with MC+AL were observed; in fact, only after 9 d was a change loss of 3.5% seen. A change of color occurred in untreated slices (CTR), showing significant difference between MC and AL.

Table 4. ΔE% of fruit slices of *Actinidia deliciosa* (Liang. Ferguson) coated with *O. ficus-indica* (OFI) mucilage (MC), or mucilage + *A. arborescens* (MC+AL) or *A. arborescens* (AL) not treated (CTR), just after being coated (0) and at 3, 5, 7, 9 days of storage at 5 °C. Different lowercase letters indicate significant differences between treatments at each sampling date. Different capital letters indicate significant differences between sampling date for the same treatment. Tukeys' significant test was applied at $p \leq 0.05$. Data are means ± S.E. (n = 3).

Time of Storage (Days)	**ΔE (%)**			
	CTR	**MC**	**MC+AL**	**AL**
3 days	1.2 ± 0.3 nsA	0.8 ± 0.2 A	0.7 ± 0.2 A	1.3 ± 0.5 A
5 days	3.1 ± 0.4 aB	1.2 ± 0.4 bA	0.9 ± 0.1 cA	1.8 ± 0.4 bA
7 days	4.7 ± 0.8 aB	2.2 ± 0.3 bAB	1.1 ± 0 cA	2.1 ± 0.4 bA
9 days	7.9 ± 1.9 aC	5.2 ± 1.1 bB	3.3 ± 0.9 cB	5.1 ± 0.7 bB

Kiwifruit slice control showed a continuous and significant ΔE increase throughout the storage period ($p \leq 0.05$) and, at the end of the storage period, showed higher color loss between the coated slices (Table 4). Because a ΔE value > 4 is considered perceptible to human eyes (CIE 2004), all differences, between the color at cutting time and the measured one during storage, were visible only after 7 d of storage.

MC+AL and MC treatments had the best visual appearance scores for all sampling times, while CTR and AL treatments had the fastest decay rates, becoming almost non-marketable, respectively, after 3 and 5 d of storage (Figure 1). In term of visual score, kiwifruit slices coated with MC+AL were still marketable after 7 d of storage, with a rapid decline of values similar to MC treatment.

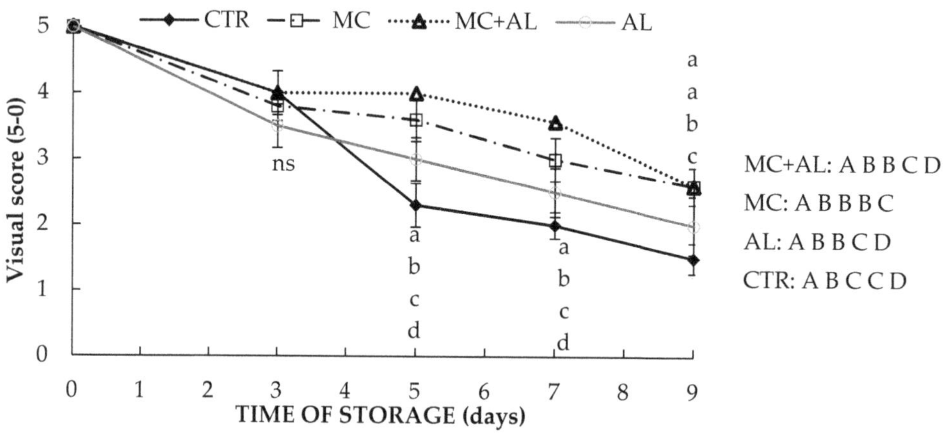

Figure 1. Visual score of fruit slices of *Actinidia deliciosa* (Liang, Ferguson) coated with *O. ficus-indica* (OFI) mucilage (MC), or mucilage + *A. arborescens* (MC+AL) or *A. arborescens* (AL) not treated (CTR), just after being coated (0) and at 3, 5, 7, 9 days of storage at 5 °C. Different lowercase letters indicate significant differences between treatments at each sampling date. Different capital letters indicate significant differences between sampling date for the same treatment. Tukeys' significant test was applied at $p \leq 0.05$. Data are means ± S.E. ($n = 3$).

Differences between treatments appeared 3 d after storage when untreated; CTR slices had the lowest value, while at the same time, no differences occurred between MC+AL and MC coated slices (Figure 1).

3.2. Firmness, Weight Loss and Pectin Content

Fruit softening is initiated and sustained by developmental and hormonal signals, and at the biochemical level it involves extensive remodeling and breakdown of the cell wall structure [29] with solubilization and depolymerisation of pectins [30]. The polysaccharide composition of *A. vera* gel [31] and OFI mucilage [14] has proved to be highly effective as a moisture barrier.

In our work, MC+AL kiwifruit slices had the highest firmness throughout the storage period (Figure 2). These results can be linked with the pectin total content detected in MC+AL samples. Indeed, fresh-cut processing causes the disruption of surface cells and injury stress of underlying tissues. The increases of pectin with OFI mucilage and *A. arborescens* coating could reduce the increase of membrane permeability and tissue disturbance. Moreover, calcium content into OFI mucilage [32] can maintain cell wall stability and integrity, as well as determining fruit quality [33]. Stabilization of membrane systems and formation of calcium pectates increased the rigidity of the middle lamella and cell walls and retarded polygalacturonase (PG) activity [34]. Composite coatings of polysaccharides could increase the lipid content, which, as a result, could further reduce the weight loss and firmness [35].

During the first 3 d of storage, there occurred a reduction of firmness from 48.2 N to 40–37 N in all treatments, and no significant differences were observed. After 5 d, kiwifruit slices MC+AL and MC treatments showed significant differences ($p \leq 0.05$) compared to untreated slices (CTR) and AL treatment. However, on the 7th day, the values of sample slices treated with MC+AL rapidly declined to values similar to CTR and AL treatment. At the last sampling time (Figure 2), significant differences occurred between MC and CTR, AL, and MC+AL treatments, showing, respectively, a firmness loss of 40.2, 54.8, 53.7, and 50.1% compared to firmness measured after cutting (48.2 N).

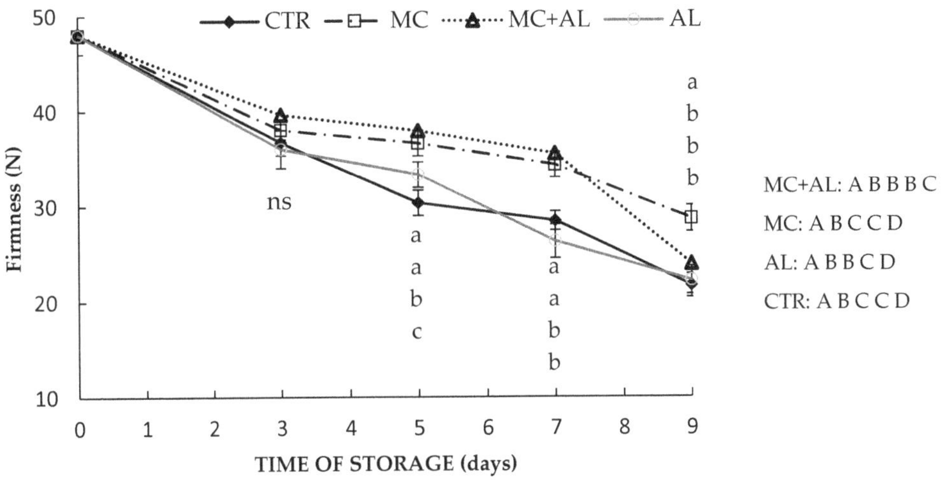

Figure 2. Firmness of fruit slices of *Actinidia deliciosa* (Liang, Ferguson) coated with *O. ficus-indica* (OFI) mucilage (MC), or mucilage + *A. arborescens* (MC+AL) or *A. arborescens* (AL) not treated (CTR), just after being coated (0) and at 3, 5, 7, 9 days of storage at 5 °C. Different lowercase letters indicate significant differences between treatments at each sampling date. Different capital letters indicate significant differences between sampling date for the same treatment. Tukeys' significant test was applied at $p \leq 0.05$. Data are means ± S.E. ($n = 3$).

After 3 d, weight loss was higher in untreated than in AL, MC, and MC+AL kiwifruit slices, while after the cutting of kiwifruit, no significant differences were observed between treatments (Figure 3).

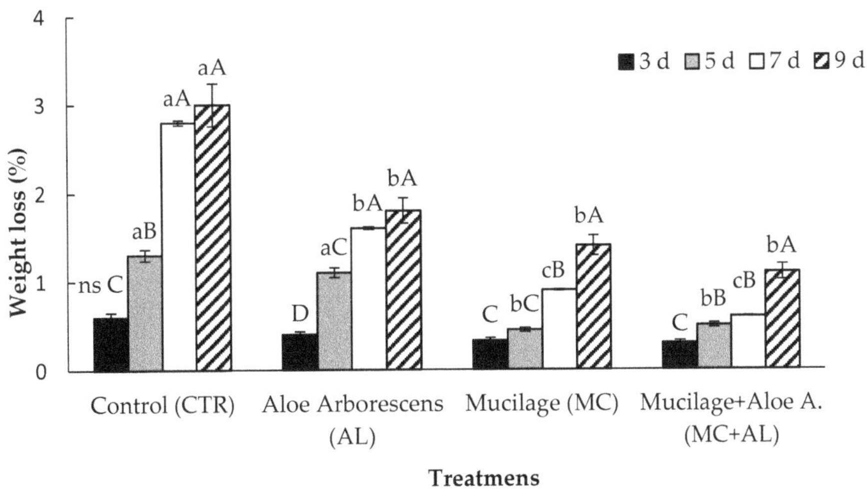

Figure 3. Weight loss (%) of fruit slices of *Actinidia deliciosa* (Liang, Ferguson) coated with *O. ficus-indica* (OFI) mucilage (MC), or OFI mucilage (MC), or mucilage + *A. arborescens* (MC+AL) or *A. arborescens* (AL) not treated (CTR), just after being coated (0) and at 3, 5, 7, 9 days of storage at 5 °C. Different lowercase letters indicate significant differences between treatments at each sampling date. Different capital letters indicate significant differences between sampling date for the same treatment. Tukeys' significant test was applied at $p \leq 0.05$. Data are means ± S.E. ($n = 3$).

The weight loss of CTR and AL samples slices significantly increased ($p \leq 0.05$) after 5 d of storage; at the last sampling time, no weight loss differences occurred between CTR and AL kiwifruit treatments (Figure 3).

In kiwifruit slices coated with MC or MC+AL, the weight loss (%) showed a significant difference ($p \leq 0.05$) compared to AL treatment and CTR slices. At the 7th and 9th day of storage, MC and MC+AL treatments showed a percentage increase of 0.90% and 0.60% and 1.40% and 1.20%, respectively. At the same time points, CTR and AL showed a percentage increase of 2.8% and 3.0% and 1.60% and 1.80%, respectively. In our work, we used the hydrophilic character of the OFI mucilage coating as a barrier to water transfer, retarding dehydration and, therefore, prolonging the firmness during fruit shelf life, while the use of aloe coating at 2% did not show effects on water loss fruit. There is also evidence for strong covalent interactions between pectin populations and cellulose [36,37]. The addition of *A. arborescens* (2%) to mucilage solution contributed to a significant effect on total pectin (Figure 4) on the 3th day of storage, but not during storage, compared to the MC treatment.

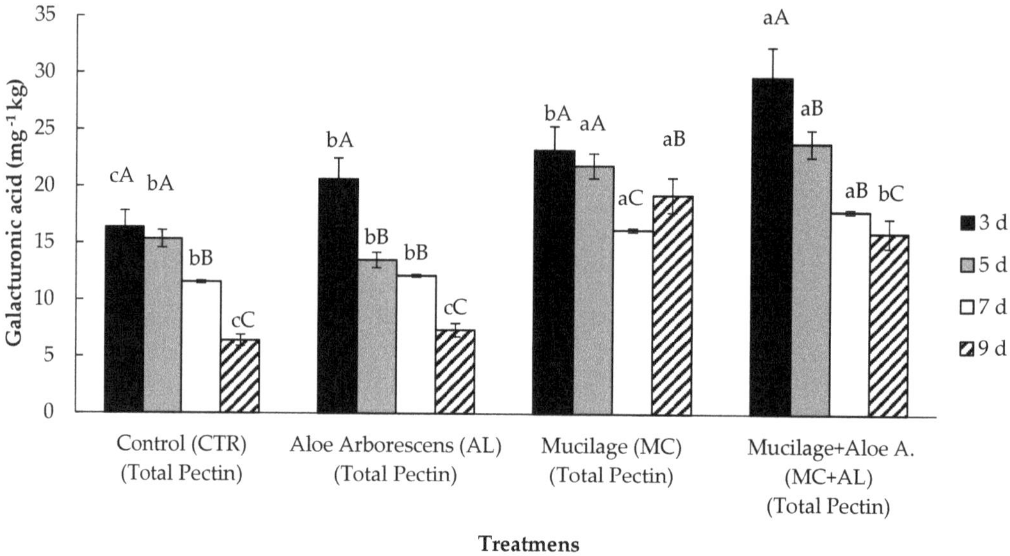

Figure 4. Total pectin (TP) of fruit slices of *Actinidia deliciosa* (Liang, Ferguson) coated with *O. ficus-indica* (OFI) mucilage (MC), or mucilage + *A. arborescens* (MC+AL), or *A. arborescens* (AL) not treated (CTR), just after being coated (0) and at 3, 5, 7, 9 days of storage at 5 °C. Different lowercase letters indicate significant differences between treatments at each sampling date. Different capital letters indicate significant differences between sampling date for the same treatment. Tukeys' significant test was applied at $p \leq 0.05$. Data are means ± S.E. ($n = 3$).

Pectin exists in the primary cell wall and the middle layer, which contribute to enhance the adhesion between cells and the mechanical strength of cells [38]. Kiwifruit slices coated with MC+AL and MC showed higher total pectin content than CTR and AL treatments for up to 9 d of storage (Figure 4). The pectin trends (Figures 4 and 5) influence the loss of firmness (Figure 2) of kiwifruit during storage, as reported by Allegra et al. [14]. Indeed, after cutting, the solubilization and depolymerization of the middle lamella of fruits might contribute to cell wall loosening and disaggregation [39]. During fruit storage, the middle lamella and primary cell wall structures were disassembled; meanwhile, pectin side chains depolymerized due to enzyme effects. These changes led to the loss of firmness, softening of fruit, and an increase of fruit postharvest decay, and decreased the quality of fresh fruit [40]. Therefore, the addition of OFI mucilage + *A. Arborescens* could significantly interfere with the assembly of structures by reducing the loss of cellular turgidity during storage. Indeed, lack of membrane integrity allows the leakage of cellular osmotic solutes into the apoplastic space, which then results in water movements and turgor loss.

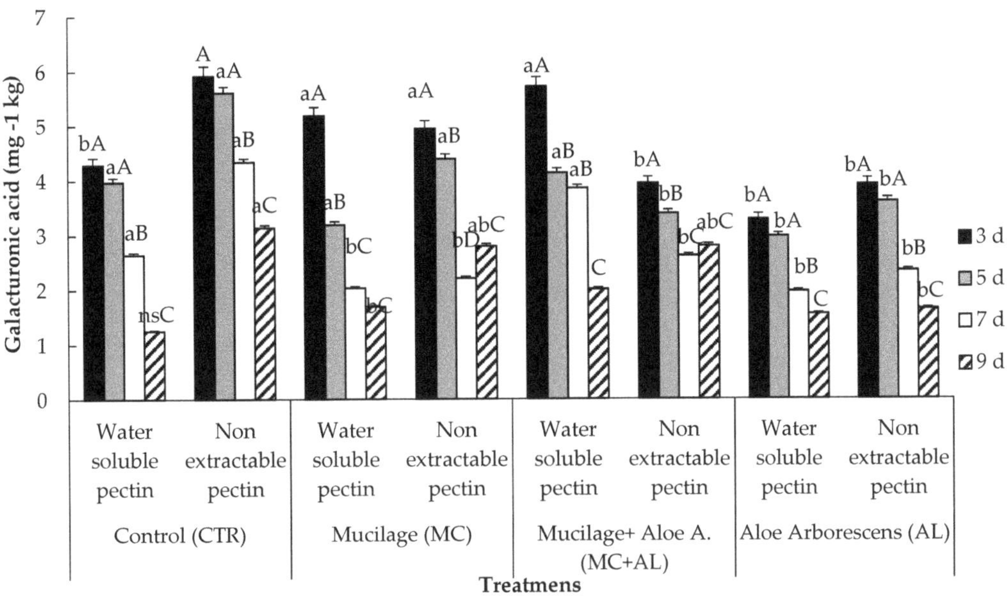

Figure 5. Water soluble pectin (WSP) and non-extractable pectin (NXP) of fruit slices of Actinidia deliciosa (Liang, Ferguson) coated with *O. ficus-indica* (OFI) mucilage + glycerol (MC), or OFI mucilage (MC), or mucilage + *A. arborescens* (MC+AL), or *A. arborescens* (AL) not treated (CTR), just after being coated (0) and at 3, 5, 7, 12 days of storage at 5 °C. Different lowercase letters indicate significant differences between treatments at each sampling date. Different capital letters indicate significant differences between sampling date for the same treatment. Tukeys' significant test was applied at $p \leq 0.05$. Data are means ± S.E. ($n = 3$).

Soluble pectin content in MC+AL and MC treatments showed a significant difference after cutting compared to CTR and AL treatment. At the 7th day of storage, kiwifruit coated with MC+AL showed a higher value than other treatments, but at the 9th day the values from the same treatment showed a reduction of 47.9% of soluble pectin content. Non-extractable pectin (Figure 5) significantly decreased in CTR and AL treatments during storage, while MC+AL and MC treatments after 7 d increased by 27.1 and 6.2%, respectively.

Similar results have been reported from Allegra e al. [14] on kiwifruit coated by mucilage extracted of *Opuntia ficus-indica*. Our results showed that the OFI effect influences the increase of pectin content during storage; on the contrary, when aloe coating has been used, a decrease of the pectin content is observed. In other works, *A. vera* and mucilage coating proved to be effective in reducing pectin depolymerization [3,26], while *A. arborescens* coating proved to reduce firmness loss in peach fruit during cold storage [16].

3.3. Head Spaces Gas Composition

Surface coatings can increase a fruit's skin resistance to gas diffusion, modify its internal atmosphere composition, and depress its respiration rate. In fresh fruit, the effect of covering the cuticle and blocking pores were important for resistance to water vapor diffusion [41]. However, the use of cover on skin could change internal atmosphere, respiration, and transpiration rates. This could be important in fresh-cut fruit because the use of coating substitute the removal of cuticle and pores that regulate changes of oxygen and carbon dioxide. In our work, after cutting and packaging in passive atmosphere, kiwifruit showed a progressive significant increase of CO_2 and a significant decrease in O_2. The O_2 (Kpa) content inside packages showed significant differences between samples coated with MC and MC+AL compared to CTR and AL treatment after 3 d of storage. Untreated slices showed lower values than the MC+AL, AL, and MC samples (Figure 6) after 5 and 7 days. An increase of CO_2 production was observed after cutting in all sample coated and uncoated, but MC treatment showed the lowest values between MC+AL, AL treatments and CTR. At the 5th day of storage no significant difference occurred between all kiwifruit samples, while at the 7th day kiwifruit coated with MC+AL and CTR showed a significant difference compared to MC and AL treatments. After 7 d all samples showed CO_2 production (<15.8 Kpa). Our results showed that kiwi slices treated with the MC+AL solution had a lower CO_2 concentration in bags until 7 d. A similar result was observed by Benitez et al. [3] on kiwi slices coated with *A. vera* combined with package under passive atmosphere and low temperature (2 ± 1 °C). The reduction of respiration was observed also in coated sweet cherry [42] and in table grapes [43].

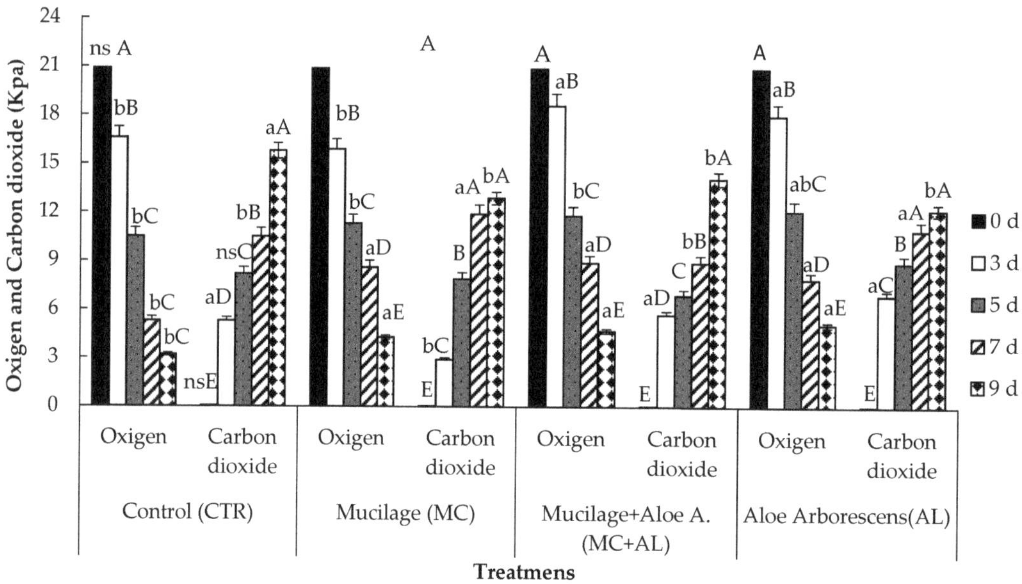

Figure 6. Oxygen and Carbon dioxide content (kPa) inside packages with fruit slices of Actinidia deliciosa (Liang, Ferguson) coated with *O. ficus-indica* (OFI) mucilage (MC), or mucilage + *A. arborescens* (MC+AL), or *A. arborescens* (AL) not treated (CTR), just after being coated (0) and at 3, 5, 7, 9 days of storage at 5 °C Different lowercase letters indicate significant differences between treatments at each sampling date. Different capital letters indicate significant differences between sampling date for the same treatment and parameter (Oxigen, carbon dioxide). Tukeys' significant test was applied at $p \leq 0.05$. Data are means ± S.E. (n = 3).

3.4. Microbiological Analysis

Aloe gel is made of a wide range of constituents that are mainly responsible for inhibiting the the growth of both Gram-positive and Gram-negative bacteria such as *Bacillus cereus, Salmonella typhimurium, Escherichia coli*, and *Klebsialla pneumonia* [44]. Allegra et al. [14] showed that the use of OFI mucilage on kiwifruit slices does not inhibit significantly the grown of yeast and bacteria; on the contrary, *A. arborescens* coating reduced the microbial spoilage of kiwifruit during 7 d of storage [3]. The loads of aerobic mesophilic bacteria, pseudomonas, and yeast detected on the different samples collected during the experimentations are reported in Table 3. The microbiological analyses of solutions of different coatings (MC, MC+AL, and AL) did not evidence the presence of any of the microbial groups observed. The groups of aerobic mesophilic bacteria were detected after cutting and at the 3rd day of storage in all treatments with low concentration (<2 Log CFU g^{-1}). No differences occurred between MC+AL coated slices and AL at the 5th and 7th d of storage. At the same times, the values of CTR and AL showed an increase in this population, with a load of 3.2 and 3.6, and 4.2 and 5.4 Log CFU g^{-1}, respectively. The presence of pseudomonads were detected after 7 d of storage (Table 5).

Table 5. Total mesophilic, *Pseudomonads*, and yeast content (Log CFU g^{-1}) in fruit slices of *Actinidia deliciosa* (Liang. Ferguson) coated with *O. ficus-indica* (OFI) mucilage (MC), or mucilage + *A. arborescens* MC+AL, or *A. arborescens* (AL) not treated (CTR), just after being coated (0) and at 3, 5, 7, 9 days of storage at 5 °C. Different lowercase letters indicate significant differences between treatments at each sampling date. Different capital letters indicate significant differences between sampling date for the same treatment. Tukeys' significant test was applied at $p \leq 0.05$. Data are means ± S.E. (n = 3).

Time of Storage (Days)	Treatments											
	Aerobic Mesophile Bacteria (Log CFU g^{-1})				Pseudomonads (Log CFU g^{-1})				Yeast Content (Log CFU g^{-1})			
	CTR	MC	MC+AL	AL	CTR	MC	MC+AL	AL	CTR	MC	MC+AL	AL
0 days	0 nsD	0 D	0 B	0 B	0 nsC	0 C	0 C	0 C	0 nsC	0 D	0 C	0 b
3 days	0.9 aC	1.1 aC	0 bB	0 bB	0 nsC	0 C	0 C	0 C	0 nsC	0 D	0 C	0 B
5 days	1 bC	2 aB	0 cB	0 cB	0 nsC	0 C	0 C	0 C	0 cC	2.9 aC	0.1 bC	0 cB
7 days	1.9 bB	4.3 aA	0 cB	0 cB	2 bB	3.5 aB	1 cB	0.5 cB	1 bB	3.3 aB	1.1 bB	0 cB
9 days	3.3 bA	5.4 aA	1 cA	1 cA	4.5 aA	5 aA	1.5 bA	1 bA	2 bA	4 aA	1.5 bA	1.3 cA

Differences among treatments appeared after 7 and 9 d of storage when CTR, MC, and AL treatments were characterized by different concentrations of pseudomonads. No differences occurred between MC coated slices and CTR after 5 d of storage. Pseudomonad's content (Table 5) did not show significant differences in all samples during 5 d of storage, while a significant increase in this population occurred in MC, showing a load of 1.77 Log CFU g^{-1} after 7 d. A similar trend was also observed for yeasts. In this case, the highest increase in concentration was registered on MC and MC+AL kiwifruit slices after 5 and 7 days. The addition of *A. arborescens* to the mucilage coating solution significantly reduced the aerobic mesophilic bacteria, pseudomonads, and yeast contents on kiwifruit slices during storage time. However, the cell densities of pseudomonads and yeasts were not able to determine a microbial decay of the sliced kiwifruits (Table 5).

4. Conclusions

This work reported the efficiency of *A. arborescens* + *Opuntia ficus-indica* coating on fresh-cut kiwi fruit. *A. arborescens* + *Opuntia ficus-indica* mucilage coating showed significant benefits in the postharvest quality of fresh-cut 'Hayward' kiwifruit, retarding fruit ripening and maintaining visual score and firmness. Moreover, it maintained the titratable acidity and total soluble solids and reduced the weight loss, in addition avoiding the microbial activity. *A. arborescens* + *Opuntia ficus-indica* coating mucilage allowed a gain of three days for commercial shelf life and 7 days for visual score parameter, as compared to the uncoated fresh-cut fruit.

Author Contributions: Conceptualization, G.S., A.A. and P.I.; methodology, G.S. and A.A.; validation, V.F., A.A. and G.S.; formal analysis, G.S., A.A. and R.P.; investigation, A.A., V.F. and G.S.; resources, A.A., V.F. and G.S.; software, A.A., R.P. and G.S.; data curation, V.F., A.A. and G.S.; writing—original draft preparation, A.A., P.I. and G.S.; writing—review and editing, G.S., A.A. and P.I.; visualization G.S. and A.A.; supervision, A.A., G.S. and P.I. All authors have read and agreed to the published version of the manuscript.

Funding: This research was funded by FFR2021 Giuseppe Sortino.

Institutional Review Board Statement: Not applicable.

Informed Consent Statement: Not applicable.

Data Availability Statement: All data generated or analyzed during this study are included in this published article.

Conflicts of Interest: The authors declare no conflict of interest.

References

1. Schroder, R.; Atkinson, R.G. Kiwifruit cell walls: Towards an understanding of softening? *N. Z. J. For. Sci.* **2006**, *36*, 112–129.
2. Karakurt, Y.; Huber, D.J. Activities of several membrane and cell-wall hydrolases, ethylene biosynthetic enzymes, and cell wall polyuronide degradation during low-temperature storage of intact and fresh-cut papaya (*Carica papaya*) fruit. *Postharvest Biol. Technol.* **2003**, *28*, 219–229. [CrossRef]
3. Benitez, S.; Achaerandio, I.; Sepulcre, F.; Pujola, M.A. Aloe *vera* based edible coatings improve the quality of minimally processed 'Hayward' kiwifruit. *Postharv. Biol. Technol.* **2013**, *81*, 29–36. [CrossRef]
4. Mastromatteo, M.; Mastromatteo, M.; Conte, A.; Del Nobile, M.A. Combined effect of active coating and MAP to prolong the shelf life of minimally processed kiwifruit (*Actinidia deliciosa* cv. *Hayward*). *Food Res. Int.* **2011**, *44*, 1224–1230. [CrossRef]
5. Passafiume, R.; Gaglio, R.; Sortino, G.; Farina, V. Effect of Three Different A. vera Gel-Based Edible coatings on the Quality of Fresh-Cut "Hayward" Kiwifruits. *Foods* **2020**, *9*, 939. [CrossRef]
6. Del Valle, V.; Hernández-Muñoz, P.; Guarda, A.; Galotto, M.J. Development of a cactus-mucilage edible coating (*Opuntia ficus indica*) and its application to extend strawberry (*Fragaria ananassa*) shelf life. *Food Chem.* **2005**, *91*, 751–756. [CrossRef]
7. Passafiume, R.; Perrone, A.; Sortino, G.; Gianguzzi, G.; Saletta, F.; Gentile, C.; Farina, V. Chemical-physical characteristics, polyphenolic content and total antioxidant activity of three Italian-grown pomegranate cultivars. *NFS J.* **2019**, *16*, 9–14. [CrossRef]
8. Lai, Q.; Wang, H.; Guo, X.; Abbasi, A.M.; Wang, T.; Li, T.; Liu, R.H. Comparison of phytochemical profiles, antioxidant and cellular antioxidant activities of seven cultivars of Aloe. *Int. J. Food Sci. Technol.* **2016**, *51*, 1489–1494. [CrossRef]
9. Zapata, P.J.; Navarro, F.D.; Guillén, S.; Castillo, D.; Martínez-Romero, D.; Valero, D.; Serrano, M. Characterisation of gels from different Aloe spp. as antifungal treatment: Potential crops for industrial applications. *Ind. Crops Prod.* **2013**, *42*, 223–230. [CrossRef]
10. Allegra, A.; Sortino, G.; Inglese, P.; Settanni, L.; Todaro, A.; Gallotta, A. The effectiveness of Opuntia ficus-indica mucilage edible coating on postharvest maintenance of 'Dottato' fig (*Ficus carica* L.) fruit. *Food Packag. Shelf Life* **2017**, *12*, 135–141. [CrossRef]
11. Trachtenberg, S.; Mayer, A.M. Composition and Properties of Opuntia Ficus-indica mucilage. *Phytochemistry* **1981**, *20*, 2665–2668. [CrossRef]
12. Sáenz, C.; Montoya, L.C. Nopalitos: Nueva hortaliza para Chile. *El Campesino* **1999**, *130*, 4–7.
13. Medina-Torres, L.; Brito-De La Fuente, E.; Torrestiana-Sanchez, B.; Katthain, R. Rheological properties of the mucilage gum (*Opuntia ficus indica*). *Food Hydrocoll.* **2000**, *14*, 417–424. [CrossRef]
14. Allegra, A.; Inglese, P.; Sortino, G.; Settanni, L.; Todaro, A.; Liguori, G. The influence of Opuntia ficus-indica mucilage edible coating on the quality of 'Hayward'kiwifruit slices. *Postharvest Biol. Technol.* **2016**, *120*, 45–51. [CrossRef]
15. Sáenz, C.; Vásquez, M.; Trumper, S.; Fluxá, C. Extractión y Composición Química del Mucílago de Tuna (*Opuntia ficus indica*). In Proceedings of the II Congreso Internacional de Tuna y Cochinilla, Santiago, Chile, 22–25 September 1992; pp. 93–96.
16. Sortino, G.; Saletta, F.; Puccio, S.; Scuderi, D.; Allegra, A.; Inglese, P.; Farina, V. Extending the shelf life of white peach fruit with 1-methylcyclopropene and A. arborescens edible coating. *Agriculture* **2019**, *10*, 151. [CrossRef]
17. Chironi, S.; Sortino, G.; Allegra, A.; Saletta, F.; Caviglia, V.; Ingrassia, M. Consumer assessment on sensory attributes of fresh table grapes cv 'Italia' and 'Red globe' after long cold storage treatment. *Chem. Eng. Trans.* **2017**, *58*, 421–426.
18. Rouse, A.H.; Atkins, C.D. Pectinesterase and pectin in commercial citrus juices as determined by methods used at the Citrus Experiment Station. *Calif. Agric. Exp. Stn. Bull.* **1955**, *570*, 1–9.
19. Rouse, A.H.; Atkins, C.D. Heat inactivation of pectinesterase in citrus juices. *Food Technol.* **1952**, *6*, 291–294.
20. Ahmed, A.E.R.; Labavitch, J.M. A simplified method for accurate determination of cell wall uronide content. *J. Food Biochem.* **1978**, *1*, 361–365. [CrossRef]
21. Kintner, P.K.; Buren, J.P. Carbohydrate Interference and Its Correction in Pectin Analysis Using the m-Hydroxydiphenyl Method. *J. Food Sci.* **1982**, *47*, 756–759. [CrossRef]
22. Yu, L.; Reitmeier, C.A.; Love, M.H. Strawberry texture and pectin content as affected by electron beam irradiation. *J. Food Sci.* **1996**, *61*, 844–846. [CrossRef]
23. Xu, F.; Liu, S.; Liu, Y.; Xu, J.; Liu, T.; Dong, S. Effectiveness of lysozyme coatings and 1-MCP treatments on storage and preservation of kiwifruit. *Food Chem.* **2019**, *288*, 201–207. [CrossRef] [PubMed]
24. Jordan, R.B.; Walton, E.F.; Klages, K.U.; Seelye, R.J. Postharvest fruit density as an indicator of dry matter and ripened soluble solids of kiwifruit. *Postharvest Biol. Technol.* **2000**, *20*, 163–173. [CrossRef]
25. Sicari, V.; Loizzo, M.R.; Pellicanò, T.M.; Giuffrè, A.M.; Poiana, M. Evaluation of A. arborescens gel as new coating to maintain the organoleptic and functional properties of strawberry (*Fragaria× ananassa cv. Cadonga*) fruits. *Int. J. Food Sci. Technol.* **2019**, *55*, 861–870. [CrossRef]

26. Famiani, F.; Battistelli, A.; Moscatello, S.; Cruz-Castillo, J.G.; Walker, R.P. The organic acids that are accumulated in the flesh of fruits: Occurrence, metabolism and factors affecting their contents—A review. *Rev. Chapingo Ser. Hortic.* **2015**, *21*, 97–128. [CrossRef]
27. Allegra, A.; Gallotta, A.; Carimi, F.; Mercati, F.; Inglese, P.; Martinelli, F. Metabolic profiling and post-harvest behavior of "Dottato" fig (Ficus carica L.) fruit covered with an edible coating from O. ficus-indica. *Front. Plant Sci.* **2018**, *9*, 1321. [CrossRef]
28. Soliva-Fortuny, R.; Martín-Belloso, O. Fresh-cut fruits: Apples and pears. In *Controlled and Modified Atmospheres for Fresh and Fresh-Cut Produce*; Gil, M.I., Beaudry, R., Eds.; Elsevier Inc.: London, UK, 2020; pp. 487–494.
29. Wang, D.; Yeats, T.H.; Uluisik, S.; Rose, J.K.; Seymour, G.B. Fruit softening: Revisiting the role of pectin. *Trends Plant Sci.* **2018**, *23*, 302–310. [CrossRef]
30. Brummell, D.A.; Cin, V.D.; Crisosto, C.H.; Labavitch, J.M. Cell wall metabolism during maturation, ripening and senescence of peach fruit. *J. Exp. Bot.* **2008**, *55*, 2029–2039. [CrossRef]
31. Hamman, J.H. Composition and applications of A. vera leaf gel. *Molecules* **2008**, *13*, 1599–1616. [CrossRef]
32. Espino-Díaz, M.; De Jesús Ornelas-Paz, J.; Martínez-Téllez, M.A.; Santillán, C.; Barbosa-Cánovas, G.V.; Zamudio-Flores, P.B.; Olivas, G.I. Development and characterization of edible films based on mucilage of Opuntia ficus-Indica (l.). *J. Food Sci.* **2010**, *75*, E347–E352. [CrossRef]
33. Dave, R.K.; Ramana Rao, T.V.; Nandane, A.S. Improvement of post-harvest quality of pear fruit with optimized composite edible coating formulations. *J. Food Sci. Technol.* **2017**, *54*, 3917–3927. [CrossRef] [PubMed]
34. Gol, N.B.; Patel, P.R.; Rao, T.R. Improvement of quality and shelf-life of strawberries with edible coatings enriched with chitosan. *Postharvest Biol. Technol.* **2013**, *85*, 185–195. [CrossRef]
35. Pérez-Gago, M.B.; Rojas, C.; Del Río, M.A. Effect of lipid type and amount of edible hydroxypropyl methylcellulose-lipid composite coatings used to protect postharvest quality of Mandarins cv. Fortune. *J. Food Sci.* **2002**, *67*, 2903–2910. [CrossRef]
36. Park, Y.B.; Cosgrove, D.J. Changes in cell wall biomechanical properties in the xyloglucan-deficient xxt1/xxt2 mutant of Arabidopsis. *Plant Physiol.* **2012**, *158*, 465–475. [CrossRef] [PubMed]
37. Broxterman, S.E.; Schols, H.A. Interactions between pectin and cellulose in primary plant cell walls Carbohydrate. *Polymers* **2018**, *192*, 263–272.
38. Cybulska, J.; Zdunek, A.; Kozioł, A. The self-assembled network and physiological degradation of pectins in carrot cell walls. *Food Hydrocoll.* **2015**, *43*, 41–50. [CrossRef]
39. Billy, L.; Mehinagic, E.; Royer, G.; Renard, C.M.; Arvisenet, G.; Prost, C. Relationship between texture and pectin composition of two apple cultivars during storage. *Postharvest Biol. Technol.* **2008**, *47*, 315–324. [CrossRef]
40. Rosli, H.G.; Civello, P.M.; Martínez, G.A. Changes in cell wall composition of three *Fragaria x ananassa* cultivars with different softening rate during ripening. *Plant Physiol. Biochem.* **2004**, *42*, 823–831. [CrossRef]
41. Banks, N.H.; Dadzie, B.K.; Cleland, D.J. Reducing gas exchange of fruits with surface coatings. *Postharv. Biol. Technol.* **1993**, *3*, 269–284. [CrossRef]
42. Martinez-Romero, D.; Alburquerque, N.; Valverde, J.; Guillen, F.; Castillo, S.; Valero, D.; Serrano, M. Postharvest sweet cherry quality and safety maintenance by A. vera treatment: A new edible coating. *Postharv. Biol. Technol.* **2006**, *39*, 93–100. [CrossRef]
43. Tripathi, P.; Dubey, N. Exploitation of natural products as an alternative strategy to control postharvest fungal rotting of fruit and vegetables. *Postharv. Biol. Technol.* **2004**, *32*, 235–245. [CrossRef]
44. Misir, J.; Brishti, F.H.; Hoque, M.M.A. Aloe vera gel as a novel edible coating for fresh fruits: A review. *Am. J. Food Sci. Technol.* **2014**, *2*, 93–97. [CrossRef]

horticulturae

MDPI

Article

Influence of Ripening Stage on Quality Parameters of Five Traditional Tomato Varieties Grown under Organic Conditions

María Concepción Ayuso-Yuste [1,2,*], Francisco González-Cebrino [3], Mercedes Lozano-Ruiz [3], Ana María Fernández-León [3] and María Josefa Bernalte-García [1,2]

1 Agriculture Engineering School, University of Extremadura, Avda. Adolfo Suárez s/n, 06007 Badajoz, Spain; bernalte@unex.es
2 University Research Institute for Agricultural Resources (INURA), University of Extremadura, 06007 Badajoz, Spain
3 Technological Institute of Food and Agriculture (CICYTEX), Avda. Adolfo Suárez s/n, 06007 Badajoz, Spain; francisco.gonzalezce@juntaex.es (F.G.-C.); mercedes.lozano@juntaex.es (M.L.-R.); anaferleon@hotmail.com (A.M.F.-L.)
* Correspondence: cayuso@unex.es; Tel.: +34-924-286-200; Fax: +34-924-286-201

Abstract: Consumers demand traditional, tasty tomatoes in contrast to new hybrid cultivars that have poor sensory characteristics. Some physicochemical parameters (total soluble solids, pH, titratable acidity, color and firmness), functional properties (vitamin C, lycopene, β-carotene and total antioxidant activity) and sensory characteristics of five traditional tomato varieties (T1-T5) in three ripening stages, grown under organic conditions, were evaluated. These were compared to the commercial hybrid 'Baghera', grown in the same conditions. Firmness of all varieties declined and the color parameters L* and H* decreased, whereas a* and a*/b* increased progressively with ripening. Lycopene also increased with ripening in all varieties, with the highest content being 132.64 mg kg^{-1} fw for T4. All traditional tomato varieties were richer in lycopene than commercial ones in the two last ripening stages. Vitamin C content ranged widely between 27.33 and 267.27 mg kg^{-1} fw. Among the traditional varieties, T2 (BGV003524) stood out due to its highest total soluble solids and vitamin C contents, total antioxidant activity, and H* and a*/b*; this variety was also the most appreciated by panelists. These traditional varieties could be an alternative to commercial cultivars, as they have good quality characteristics and can satisfy consumer demand for organic produce.

Keywords: antioxidant capacity; ascorbic acid; lycopene; organic production; *Solanum lycopersicum*

Citation: Ayuso-Yuste, M.C.; González-Cebrino, F.; Lozano-Ruiz, M.; Fernández-León, A.M.; Bernalte-García, M.J. Influence of Ripening Stage on Quality Parameters of Five Traditional Tomato Varieties Grown under Organic Conditions. *Horticulturae* **2022**, *8*, 313. https://doi.org/10.3390/horticulturae8040313

Academic Editors: Yang Bi, Yongcai Li, Di Gong and Maria Dulce Antunes

Received: 25 February 2022
Accepted: 6 April 2022
Published: 8 April 2022

1. Introduction

Tomato (*Solanum lycopersicum* L.) is one of the most widely cultivated vegetables both for fresh and processed produce. In 2020, tomato production for fresh consumption was 186.8 million tons in the world, of which 4.3 million tons were produced in Spain, which is the second largest producer in Europe after Italy [1]. Commercial hybrids are usually employed in this crop; one of these is the cultivar 'Baghera', that is widely cultivated in Spain due to its early production, medium-size, uniformly colored fruits and resistance to the main diseases of this crop.

The widespread consumption of tomatoes makes them a major source of health-related compounds [2]. Tomato antioxidants include vitamin C and carotenoids, such as lycopene and β-carotene. These antioxidants play an important role in protecting cells and cellular components against oxidative damage [3,4]. Lycopene is the carotenoid responsible for the red color in tomatoes [5], and accounts for more than 80% of the total carotenoids in ripe tomatoes [6]. The reported benefits of lycopene include prevention of degenerative diseases, such as certain types of cancer [7,8] and cardiovascular diseases [9,10]. Tomato antioxidant concentrations are related to genotype, but also to factors such as ripening stage, cultivation practices and climatic environment [11,12].

The most important external criterion to assess ripeness in tomato is color, which is directly related to the perception of quality. Tomatoes are usually consumed at their maximum organoleptic quality, i.e., when red color appears but before excessive softening occurs [13].

There is a growing interest in promoting the survival of traditional tomato varieties, as consumers find them more tasty and healthy [14,15]. Organic horticulture represents a sustainable and alternative agricultural model with the potential to provide both environmental improvements and high quality output [16]. In many regions, traditional landraces of tomato are still cultivated on a small scale for local consumption [17], and they are sold at higher prices than modern hybrids, as they are greatly appreciated by consumers [15,18,19]. It is necessary to characterize the quality and nutritional attributes of traditional tomato varieties, because only some phenotypic descriptors are available [20,21].

The aim of this work was to assess variations in the physicochemical parameters and bioactive compounds of five traditional tomato varieties, grown under organic conditions, in three ripening stages, and to compare these with the commercial hybrid 'Baghera' produced in the same conditions.

2. Materials and Methods

2.1. Plant Material

Tomato fruits were grown in open air under organic conditions in the south west of Spain, according to Spanish regulations, in one season. The study was carried out in an experimental field at the Scientific and Technological Research Center of Extremadura (CICYTEX) at an altitude of 217 m above sea level (38°51′ N; 6°40′ W). The climate is semi-temperate, with around 550 mm of rainfall and mild winters. The soil texture is sandy loam, pH 6.8, and low organic matter content. Plants of all cultivars were nursery produced and transplanted outdoors on the same date, in a randomized complete block design with three replications. The experimental unit was a plot of 15 m^2. Trellising was used to grow tomato plants that were pruned to a single stem.

The traditional tomato varieties studied were T1, T2, obtained from the Centre for the Conservation and Improvement of Agrodiversity (COMAV, Valencia, Spain), T3, T4, and T5 from the Institute of Agricultural and Food Research and Development (IMIDA, Murcia, Spain). The commercial hybrid used in this study was 'Baghera' from (Clause, Portes lès Valence, France) (Table 1, Figure 1). Tomato samples were picked in a single harvest and visually classified at three different ripening stages: turning (***turning***), light-red (***light-red***) and red (***red***), according to the color ripeness chart established by the United States Standards for Grades of Fresh Tomatoes [22]. Ten fruits per variety and ripening stage were randomly selected and analyzed in the laboratory for total soluble solids content (TSS), titratable acidity (TA), pH, firmness and color. Firstly, color and firmness measurements were performed on whole tomatoes, and then the fruits were homogenized for analysis of TSS, TA and pH. In addition, 10 fruits per variety and ripening stage were packed and stored at −80 °C for carotenoids, vitamin C, and total antioxidant activity (TAA) analyses. Finally, a sensory evaluation was carried out with tomato samples in the ***light-red*** stage.

Table 1. Sources, origins and fruit type of the studied varieties.

Variety	Code	Source	Origin	Local Name [1]	Fruit Type [2]
BGV001000	T1	COMAV	Granada	Tomate del país	2 */3 **
BGV003524	T2	COMAV	León	–	5/1
MUL15	T3	IMIDA	Murcia	Tomate de pera	4/1
MUL24	T4	IMIDA	Murcia	Tomate murciano	2/1
MUL36	T5	IMIDA	Murcia	Tomate murciano	3/1
Baghera F1	Baghera	Commercial hybrid	Clause (France)	–	2/1

[1] –: Without local name. [2] Descriptors for tomato: * "Predominant fruit shape" (7.2.2.5) / ** "Fruit cross-sectional shape" (7.2.2.29), according to IPGRI [23].

Figure 1. Traditional varieties and commercial hybrid: BGV001000 (T1), BGV003524 (T2), MUL15 (T3), MUL24 (T4), MUL36 (T5) and 'Baghera' F1 (Baghera).

2.2. Total Soluble Solids, Titratable Acidity and pH

Tomato fruits were divided into four groups and homogenized using a Thermomix 31–1 (Vorwerk España M.S.L., S.C Madrid, Spain). Total soluble solids (TSS) were determined with a digital refractometer (Mettler-Toledo S.A.E., Barcelona, Spain) and the results were expressed as °Brix. Tritatable acidity (TA) was assayed by automatic titration with 0.1 N sodium hydroxide up to pH 8.1 with a DL50 Graphix automatic titrator (Mettler-Toledo S.A.E., Barcelona, Spain) and expressed as g citric acid 100 g^{-1}. In addition, for each variety, Acceptability or Maturity Index (AI = TSS/TA) and Flavor Index (FI = (TSS/(20 × TA)) + TA) were calculated [24].

2.2.1. Firmness

For each variety and ripening stage, firmness was evaluated on ten fruits by a compression test using a TA.XT2i Texture Analyzer (Aname, S.L., Madrid, Spain). Measurements were taken at two opposite points on the equator of each fruit with a plane compression plate (100 mm diameter), a speed of 2 mm s^{-1} and a fruit deformation of 2%. Maximum force (N) and slope of the force/deformation curve (N mm^{-1}) were registered.

2.2.2. Color

Color measurements were performed on the surface of ten tomatoes per variety and ripening stage with a Konica Minolta CM-3500d Spectrophotometer (Aquateknica, S.A., Valencia, Spain). Reflectance measurements were made with the primary illuminant D65 and a circular measurement area of 30 mm diameter. Color readings were taken on two different positions of the equatorial region of each fruit. The CIEL*a*b* parameters were directly obtained and then Hue angle ($H^* = \tan^{-1} (b^*/a^*)$), Chroma ($C^* = (a^{*2} + b^{*2})^{1/2}$) and a*/b* ratio were calculated.

2.2.3. Carotenoids

Lycopene and β-carotene were determined by HPLC, according to the method described by Sabio et al. [25] in an Agilent 1100 HPLC chromatograph (Agilent Technologies, Inc., Palo Alto, CA, USA) with an Agilent LiChrosorb RP-18 4.6 × 200 mm–10 µm column maintained at 30 °C and after extraction with acetone: ethanol: hexane. The flow rate was 1 mL min^{-1}, peaks were detected by DAD at 460 nm and quantification was carried out using lycopene and β-carotene standards solutions. Extractions were carried out in triplicate and the results were expressed as mg kg^{-1} fw.

2.2.4. Vitamin C

Vitamin C was assessed as ascorbic acid (AA) by HPLC with UV detection according to González-Cebrino et al. [16]. Fruits were sliced and homogenized using a domestic blender, and then mixed and extracted with EDTA/metaphosphoric solution (85%) on ice for 1 min (Omni Mixer Homogenizer, Omni International, Kennesaw, GA, USA). The extracted sample was filtered (Millipore 0.45 μm) and injected into a 1050 HPLC chromatograph (Agilent Technologies, Inc., Palo Alto, CA, USA) with an Agilent Zorbax SB-C8 4.6 × 250 mm–5μm column maintained at 30 °C. A buffer solution of 50 mM acetic acid/acetate (pH 4) was used as the mobile phase. The flow rate was 0.5 mL min^{-1} and detection was at 260 nm. Quantification was carried out using AA standards solutions. Analyses were carried out in triplicate and the results were expressed as mg of vitamin C kg^{-1} fw.

2.3. Total Antioxidant Activity

Total antioxidant activity (TAA) was estimated in triplicate, according to the method established by Cano et al. [26] with some modifications. Extractions were carried out using 1 g of tomato homogenate diluted in 50 mM Na phosphate buffer (pH 7.5). Then, 20 μL of diluted homogenate was placed in a spectrometric cuvette and 1 mL of the radical cation ABTS (2,2′-azino-bis(3-ethylbenzothiazoline-6-sulfonic acid)) diammonium salt (Sigma-Aldrich, Madrid, Spain) was added. The reaction was monitored in a UV-2401 PC Shimadzu spectrophotometer (Shimadzu Scientific Instruments, Inc., Columbia, MD, USA) at 730 nm, and the decrease in absorbance was determined after 20 min. The results were expressed as mg of Trolox 100 g^{-1} fw.

2.4. Sensory Evaluation

The sensory panel comprised eight trained judges from our institution, all of whom were regular consumers of fresh tomatoes. Each panelist was asked to rate the following sensory attributes: color, sourness, sweetness, juiciness, and consistency. The evaluation was carried out with samples in the ***light-red*** stage. Firstly, tomatoes were cut into slices of approximately 1 cm thickness and labeled with a random number. The sensory panel evaluated the intensity of the attributes on a 10-point scale, where 0 represented "no intensity" and 10 represented "strong intensity". No information about tomato varieties was provided to the panelists.

2.5. Statistical Analysis

The results were analyzed using SPSS 21.0 statistical software (SPSS Inc, Chicago, IL, USA) by one-way analysis of variance (ANOVA) by ripening stage and variety, and when differences were detected, mean values were compared with Tukey's test ($p < 0.05$). Correlations were performed using Pearson's correlation coefficient (r, $p < 0.01$) to identify significant interactions among studied variables.

3. Results and Discussion

3.1. Total Soluble Solids, Titratable Acidity and pH

The total soluble solid content, pH and tritatable acidity results are presented in Table 2, and acceptability and flavor indexes in Figure 2. TSS content increased with ripening for all varieties, showing the highest values at the last stage (***red***), except in T5, which tended to decrease with maturity. At ***turning*** and ***light-red***, T2 had the highest TSS values, whereas in the ***red*** stage 'Baghera' showed the highest value together with T2, although the latter was without significant differences with T1 and T4 (Table 2). Ruiz et al. [19] also recorded a higher mean value for a commercial variety in the ***red*** stage. Although pH values showed differences among varieties, there was a narrow range of variation: T4 had the lowest value at ***turning***, whereas the maximum was recorded for both T2 and T5 in the ***light-red*** stage. T2 and T3 showed a clear decrease in tritatable acidity with ripening, and together with T5, had the lowest TA values in the ***red*** stage.

Table 2. Tomato quality parameters: TSS, pH, tritatable acidity and firmness.

	Ripening Stage	T1	T2	T3	T4	T5	Baghera
TSS (°Brix)	*turning*	4.55 ± 0.10 c	5.10 ± 0.00 d	4.11 ± 0.21 a	4.45 ± 0.05 bc	4.27 ± 0.15 ab	4.67 ± 0.08 c
	light-red	4.08 ± 0.05 a	5.13 ± 0.10 e	4.33 ± 0.14 ab	4.40 ± 0.00 bc	4.63 ± 0.05 cd	4.80 ± 0.27 d
	red	5.06 ± 0.49 bc	5.13 ± 0.15 cd	4.67 ± 0.10 b	4.93 ± 0.12 bc	4.03 ± 0.06 a	5.55 ± 0.10 d
pH	*turning*	4.41 ± 0.02 bc	4.35 ± 0.01 abc	4.29 ± 0.13 ab	4.24 ± 0.05 a	4.46 ± 0.00 c	4.31 ± 0.05 abc
	light-red	4.40 ± 0.05 a	4.52 ± 0.10 b	4.40 ± 0.02 a	4.32 ± 0.01 a	4.52 ± 0.03 b	4.35 ± 0.03 a
	red	4.38 ± 0.13 a	4.50 ± 0.01 b	4.51 ± 0.02 b	4.33 ± 0.03 a	4.35 ± 0.01 a	4.35 ± 0.01 a
TA (g citric acid 100 g^{-1})	*turning*	0.30 ± 0.01 ab	0.34 ± 0.01 ab	0.37 ± 0.07 b	0.37 ± 0.03 b	0.26 ± 0.01 a	0.34 ± 0.01 b
	light-red	0.32 ± 0.04 b	0.26 ± 0.01 a	0.31 ± 0.01 b	0.34 ± 0.02 b	0.31 ± 0.03 b	0.35 ± 0.02 b
	red	0.33 ± 0.05 bc	0.28 ± 0.01 ab	0.26 ± 0.02 a	0.35 ± 0.02 c	0.29 ± 0.00 ab	0.36 ± 0.02 c
Firmness (force, N)	*turning*	7.54 ± 1.28 b	11.98 ± 1.30 d	6.22 ± 1.06 ab	9.55 ± 1.69 c	5.66 ± 1.38 a	11.32 ± 1.12 d
	light-red	6.70 ± 0.93 bc	6.53 ± 1.53 bc	4.55 ± 0.77 a	6.33 ± 0.91 bc	5.50 ± 1.27 ab	6.79 ± 1.88 c
	red	4.15 ± 1.11 b	5.79 ± 1.79 c	2.75 ± 0.49 a	2.62 ± 0.83 a	3.43 ± 0.84 ab	2.91 ± 1.18 a

Data are expressed as mean values ± standard deviation. In a given row, for a ripening stage, values followed by different letters are significantly different (Tukey's test, $p < 0.05$).

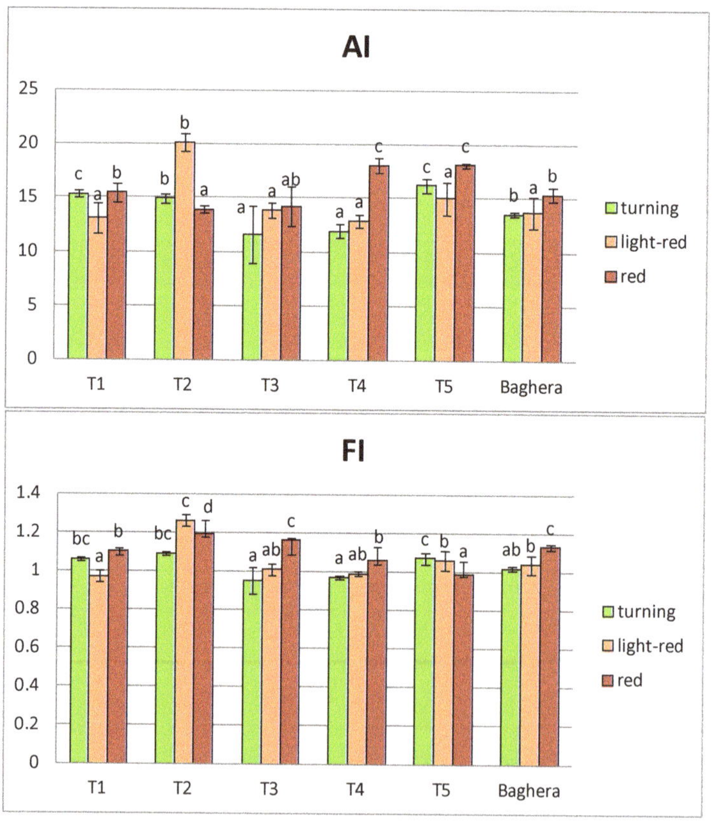

Figure 2. Tomato Aceptability (**AI**) and Flavor (**FI**) Indexes. For each ripening stage, different letters represent significant differences according to Tukey's test ($p < 0.05$). Mean values and standard deviation.

In most varieties, the Acceptability and Flavor Indexes were higher in the ***red*** stage. T5 and T1 had the highest values of the Acceptability Index at ***turning***, and T2 at ***light-red***. The latter also had the highest Flavor Index at all ripening stages (Figure 2). For all varieties and ripening stages, Flavor Index values were higher than 0.85, which was described by Hernández et al. [24] as the minimum required value for an acceptable flavor in tomato.

3.1.1. Firmness

It is known that firmness is a tomato quality parameter closely associated with ripening stage [27]. It can be seen in Table 2 and Figure 3 that the firmness of each tomato variety was strongly affected by the ripening stage, and force (N) and slope (Nmm^{-1}) values exhibited a clear decrease as ripening progressed. Therefore, firmness is a useful criterion for assessing the ripening stage of tomato fruits. Although tomatoes harvested in the red stage could be considered the best-tasting tomatoes, these fruits are more fragile and can become easily damaged; thus it is better to harvest them at an earlier stage in order to extend their shelf-life and prevent damage from occurring during handling.

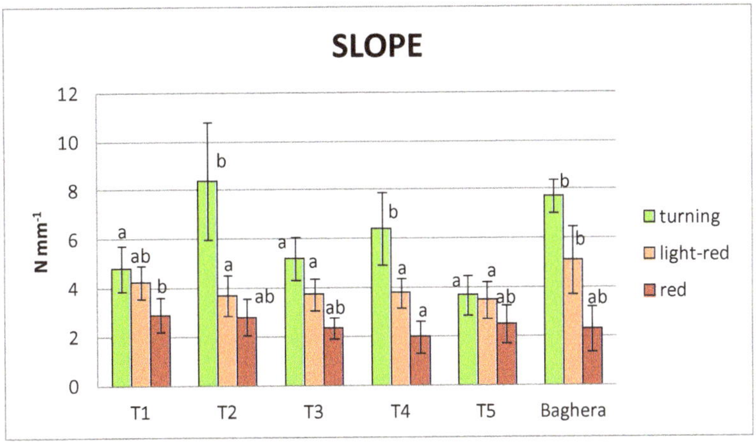

Figure 3. Tomato firmness. For each ripening stage, different letters represent significant differences according to Tukey's test ($p < 0.05$). Mean values and standard deviation.

Differences in maximum force (N) and slope (Nmm^{-1}) were observed among the tomato varieties studied (Table 2, Figure 3). T2 and 'Baghera' recorded the highest firmness values at ***turning***. In the ***red*** stage, all the traditional tomato varieties had similar or higher firmness values than the commercial hybrid 'Baghera'.

3.1.2. Color

Most of the external color parameters decreased with ripening except a* and a*/b*, which increased (Table 3 and Figure 4). The decrease of L* with ripening reflected the darkening of the fruits with carotenoid synthesis, the degradation of chlorophylls and, consequently, the disappearance of greenness [28,29]. As in other studies [13,28], a* values increased from ***turning*** to ***red*** as a consequence of lycopene synthesis. C* is not a good indicator of tomato ripening because it is essentially an expression of the saturation of a single color, and as a result, different colors may have the same chroma values [13], as was the case for T4 and 'Baghera'.

Among the studied varieties, there were differences in mean values for all color parameters. At ***turning***, T2 revealed the lowest significant values for a*, C* and a*/b* ratio, and the highest significant values for L* and H*. The a $*$ value was slightly negative (−2.87), corresponding to an under-ripe stage and associated with a major accumulation of chlorophylls. In contrast, at ***light-red***, T2 showed the lowest values for L*, b*, C*, and H*, and the highest values for a* and a*/b* ratio. Finally at ***red***, T2 presented the significantly highest value of a*/b* (1.74), indicating a more intense red color (Figure 4).

Table 3. Tomato external color parameters.

	Ripening Stage	T1	T2	T3	T4	T5	Baghera
L*	*turning*	49.09 ± 4.30 a	59.89 ± 1.96 c	53.02 ± 3.42 ab	56.34 ± 1.85 bc	52.36 ± 6.02 ab	56.21 ± 4.91 bc
	light-red	48.03 ± 3.42 b	43.30 ± 2.98 a	45.33 ± 4.16 ab	45.70 ± 4.96 ab	46.46 ± 6.41 ab	47.10 ± 2.10 ab
	red	39.73 ± 2.52 b	36.86 ± 1.63 a	36.52 ± 2.94 a	38.12 ± 2.49 ab	36.73 ± 1.79 a	38.25 ± 1.23 ab
a*	*turning*	11.86 ± 4.60 c	−2.87 ± 4.69 a	7.52 ± 2.79 b	7.01 ± 5.14 b	4.13 ± 3.21 b	4.51 ± 1.69 b
	light-red	18.12 ± 4.18 a	22.35 ± 3.08 b	17.54 ± 3.41 a	18.34 ± 4.06 a	19.02 ± 5.55 ab	17.94 ± 3.82 a
	red	27.51 ± 2.33 b	24.21 ± 2.13 a	28.18 ± 1.96 b	24.74 ± 2.95 a	24.59 ± 2.87 a	29.87 ± 2.80 b
b*	*turning*	34.62 ± 4.99 cd	22.97 ± 5.69 a	26.91 ± 5.65 ab	30.77 ± 5.83 bc	43.76 ± 7.61 e	38.95 ± 9.98 de
	light-red	38.07 ± 5.38 d	17.34 ± 1.95 a	31.66 ± 3.38 bc	27.92 ± 4.55 b	35.84 ± 10.88 cd	34.69 ± 2.71 cd
	red	27.33 ± 3.15 cd	13.97 ± 1.36 a	26.26 ± 3.83 bc	25.52 ± 4.42 bc	23.49 ± 2.01 b	29.51 ± 2.65 d
C*	*turning*	36.95 ± 4.14 cd	23.62 ± 5.53 a	28.10 ± 5.47 ab	31.99 ± 5.58 bc	44.08 ± 7.43e	39.28 ± 9.80de
	light-red	42.49 ± 4.10 d	28.44 ± 2.11 a	36.31 ± 3.78 bc	33.66 ± 4.36 b	41.32 ± 9.15 d	39.26 ± 2.01 cd
	red	38.84 ± 3.28 c	27.97 ± 2.30 a	38.59 ± 3.57 cd	35.62 ± 4.74 bc	34.05 ± 3.05 b	42.01 ± 3.64 d

Data are expressed as mean values ± standard deviation. In a given row, for a ripening stage, values followed by different letters are significantly different (Tukey's test, $p < 0.05$).

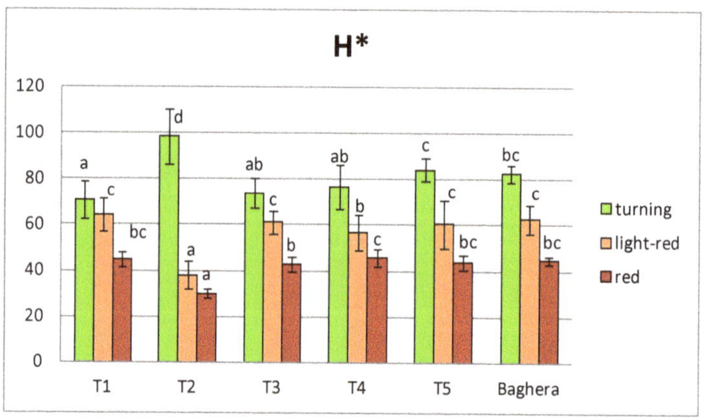

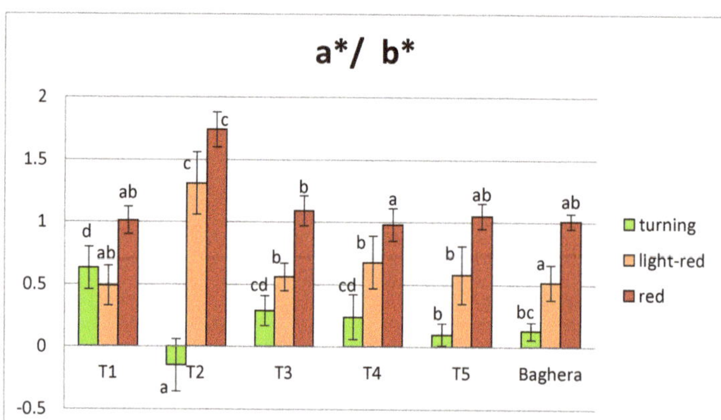

Figure 4. Tomato **Hue angle** and **a*/b*** ratio. For each ripening stage, different letters represent significant differences according to Tukey's test ($p < 0.05$). Mean values and standard deviation.

The ratio a*/b* and H* are often used as indicators of tomato color changes [13,30]. For all varieties, the a*/b* ratio and H* values showed differences among the three ripening stages. There was a progressive decrease in H*, whereas a*/b* ratio increased, as observed in other studies [28,29].

3.1.3. Carotenoids

For lycopene and β-carotene content, differences were found among the studied varieties (Figure 5). Lycopene ranged from 6.57 mg kg^{-1} fw for T2 (***turning***) to 132.64 mg kg^{-1} fw for T4 (***red***). The highest significant lycopene content at ***turning*** and ***light-red*** was found for T5, whereas T4 was the variety with the highest concentration at ***red***. All the traditional varieties showed higher lycopene content than the commercial hybrid 'Baghera' in the two last stages, although other authors found no differences between commercial and traditional varieties in the red stage [14]. Our tomato varieties showed higher carotenoid content than those reported by Pinela et al. [31] in local farmers' varieties from Portugal. Lycopene content varied significantly among tomato varieties [32] and can be also affected by environmental factors and agricultural techniques [11,33]. Roselló et al. [34] reported values from 2.72 to 73.77 mg kg^{-1} fw for five traditional varieties. In organic production, the average lycopene content has been reported at 56 mg kg^{-1} fw for 13 commercial varieties harvested at the fully ripe stage [30], and Murariu et al. [33] found that the lycopene content of red tomatoes ranged from 9.53 to 3.60 mg 100 g^{-1} fw depending on the genotype and the production system. The values obtained in this work were considerably higher than those reported by these authors, mainly in the ***red*** stage.

Lycopene content is an internal quality attribute associated with the color and ripening process of tomatoes [28,35–37]. In the studied samples, the lycopene content increased with ripening, on average more than four-fold from the ***turning*** to ***red*** stages, reaching the maximum concentration in the ***red*** stage in all varieties (Figure 5). Lycopene showed a strong Pearson's correlations with different color and firmness parameters. There were positive correlations with a* (0.832) and a*/b* ratio (0.772), and negative with L* (−0.799), H* (−0.801), force (−0.775) and slope (−0.811).

Although lycopene is the most abundant carotenoid in red-ripe tomatoes, β-carotene content was also evaluated. This carotenoid is important because of its pro-vitamin A activity, and it accumulates during tomato fruit ripening [38]. For the studied samples, β-carotene content varied from 5.66 to 16.75 mg kg^{-1} fw. T5 showed the highest content in all stages, but no significant differences were found with T1 at ***turning***. Roselló et al. [34] reported β-carotene values from 1.18 to 8.76 mg kg^{-1} fw for five traditional varieties, and Pinela et al. [31] values from 0.30 to 0.51 mg 100 g^{-1} for Portuguese local varieties. The highest concentrations of β-carotene were found in all varieties at ***light-red***, except for T2, which recorded the maximum level in the ***red*** stage, as Gautier et al. [39] reported for other varieties.

3.1.4. Vitamin C

Vitamin C content ranged between 27.33 for T2 (***turning***) and 267.27 mg kg^{-1} fw for T1 (***red***), with differences depending on variety and ripening stage (Table 4).

Table 4. Vitamin C concentration and total antioxidant activity in tomato.

	Ripening Stage	T1	T2	T3	T4	T5	Baghera
Vitamin C (mg kg^{-1} fw)	***turning***	175.05 ± 1.95 d	27.33 ± 1.25 a	96.17 ± 6.23 b	164.28 ± 6.36 c	176.22 ± 9.22 d	155.77 ± 2.61 c
	light-red	137.18 ± 6.92 a	183.73 ± 5.03 c	159.14 ± 5.66 b	159.12 ± 5.08 b	168.54 ± 2.16 b	187.43 ± 2.35 c
	red	267.27 ± 2.90 f	173.85 ± 2.74 b	132.91 ± 3.48 a	249.36 ± 6.27 e	226.45 ± 3.50 d	186.00 ± 1.47 c
TAA (mg Trolox 100 g^{-1} fw)	***turning***	62.32 ± 0.78 b	28.05 ± 0.51 a	60.01 ± 0.91 b	69.83 ± 2.28 c	64.01 ± 1.56 b	63.59 ± 1.68 b
	light-red	63.75 ± 1.07 a	85.41 ± 1.80 c	63.12 ± 1.93 a	72.01 ± 3.19 b	73.81 ± 5.10 b	62.56 ± 1.70 a
	red	63.12 ± 1.47 ab	68.88 ± 1.29 c	61.37 ± 1.77 ab	69.47 ± 1.15 c	65.24 ± 3.70 bc	58.73 ± 1.80 a

Data are expressed as mean values ± standard deviation. In a given row, for a ripening stage, values followed by different letters are significantly different (Tukey's test, $p < 0.05$).

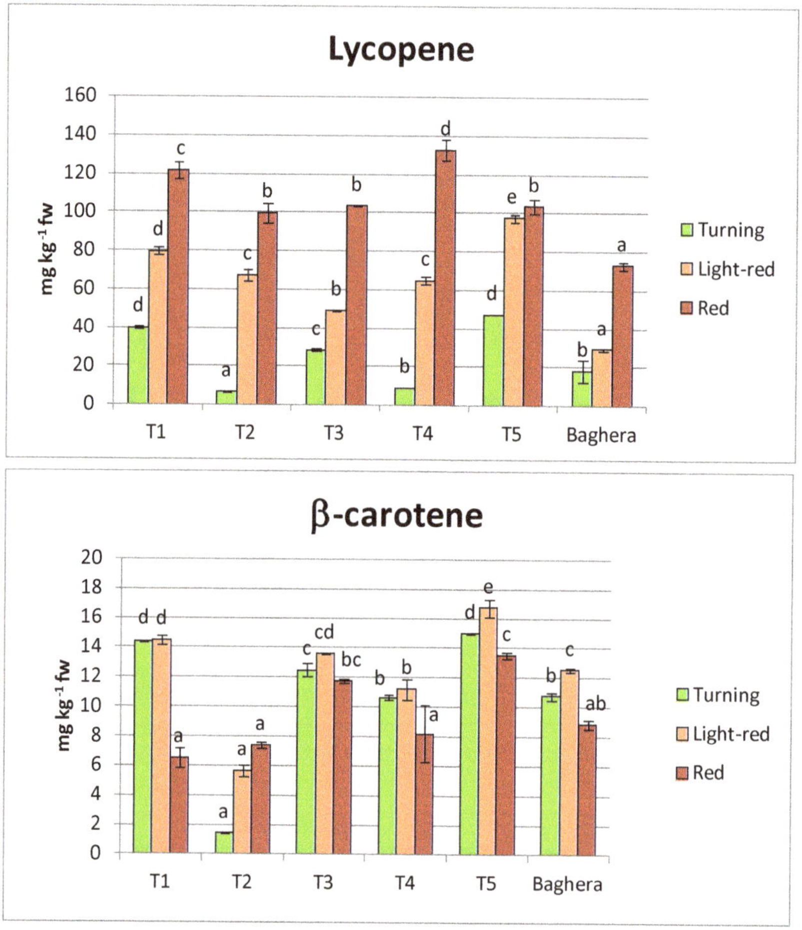

Figure 5. Tomato carotenoids (**lycopene** and **β-carotene**) content. For each ripening stage, different letters represent significant differences according to Tukey's test ($p < 0.05$). Mean values and standard deviation.

At ***turning*** both T5 and T1 revealed the highest significant concentrations (176.22 and 175.05 mg kg^{-1} fw, respectively), at ***light-red*** 'Baghera' and T2 (187.43 and 183.73 mg kg^{-1} fw, respectively), and at ***red*** T1 (267.27 mg kg^{-1} fw). These values were higher than those found by Adalid et al. [2], who reported concentrations of between 51 and 183 mg kg^{-1} for 15 traditional varieties, mainly at a pink stage. For organic tomatoes grown in open air, the vitamin C content observed by Murariu et al. [33] ranged from 11.61 to 29.21 mg 100 g^{-1} fw in the red stage.

As ripening progressed, changes in vitamin C content were observed in the studied varieties (Table 4), in contrast to the results of Raffo et al. [40], who indicated some ascorbic acid variation for cherry tomatoes during ripening. In T1, T4 and T5, vitamin C content decreased slightly from ***turning*** to ***light-red*** stage and then increased at ***red***. On the other hand, in T2, T3 and 'Baghera' vitamin C accumulated during the first stages and then decreased at ***red***, according to other authors [41]. A positive Pearson's correlation between vitamin C and lycopene (0.642) and a negative with H* (−0.608) were found in the studied tomato samples.

3.2. Total Antioxidant Activity (TAA)

TAA values showed differences depending on variety and ripening stage (Table 4), ranging from 28.05 to 85.41 mg Trolox 100 g^{-1} fw, corresponding to T2 at ***turning*** and ***light-red***, respectively. The highest TAA values were found in T4 at ***turning*** (69.83 mg Trolox 100 g^{-1} fw), in T2 at ***light-red*** (85.41 mg Trolox 100 g^{-1} fw) and in both T4 and T2 at ***red***. On the other hand, the commercial hybrid 'Baghera' showed the lowest values in the ***light-red*** and ***red*** stages, although without significant differences with T1 and T3.

Ascorbic acid, together with phenolic compounds, represents the main water-soluble antioxidants in tomato [40], but it must be taken into account that the antioxidant properties of tomatoes largely depend on their lycopene content [37]. Significant correlations between lycopene content and antioxidant activity have been previously reported [38,42]. However, in our study, no such significant correlation was observed.

Most of the tomato varieties did not show differences in TAA values as ripening advanced, as reported by Cano et al. [42]. T2 and T5 exhibited an increase from the ***turning*** to the ***red*** stage, with the highest value at ***light-red***. In general, higher TAA levels were observed in tomatoes harvested at ***light-red***, as found by Raffo et al. [40].

There was also a negative Pearson's correlation (−0.617) between H* and TAA, as previously described for lycopen and vitamin C, so H* could be used as a good indicator of functional quality in tomato.

3.3. Sensory Evaluation

As far as we know, very few sensory analyses of traditional tomato varieties have been carried out. The results of the sensory evaluations carried out with the tomato varieties in the ***light-red*** stage are depicted in Figure 6. There were significant differences for color, sweetness and juiciness. However, no significant differences were found for sourness and consistency. T4 showed the best value for sourness (7.22) and T2 for consistency (8.16). With respect to color, T3 was the best considered variety (8.16), while no significant differences were found with T1, T2 and 'Baghera'. The highest average score in juiciness corresponded to T2 (8.12), without significant differences with T1, T3 and T4.

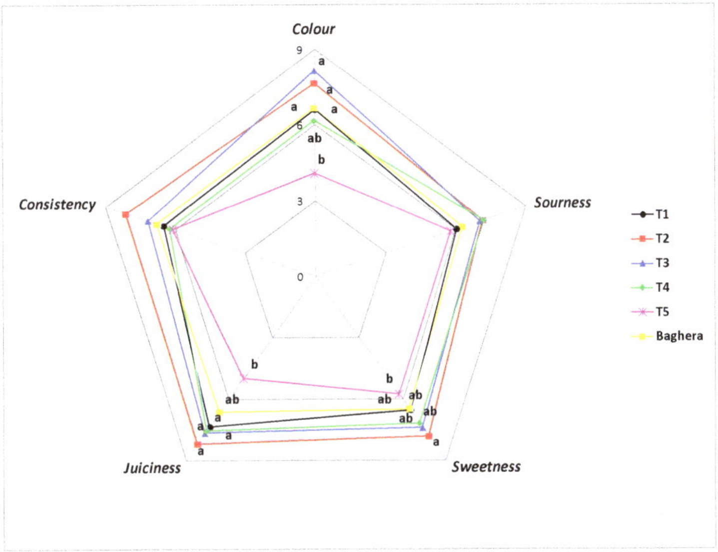

Figure 6. Sensory evaluation results of tomato in the ***light-red*** stage. For each quality parameter, samples with different letters differed at a significance level of 0.05 according to the Tukey's test (n = 8).

T5 was the variety that was least appreciated by panelists. In contrast, T2 achieved the highest scores for sourness and sweetness, which is in accordance with its high values of AI and FI, since these indexes are related to sweetness, sourness and flavor perception.

4. Conclusions

Variety and ripening stage have a strong influence on the physicochemical variables and bioactive compound contents of tomatoes. In general, the traditional varieties showed similar physicochemical characteristics and Acceptability and Flavor Indices to those of 'Baghera', although in the ***red*** stage, they presented concentrations of lycopen, vitamin C and TAA higher than those 'Baghera'. As expected, the firmness of all the varieties declined and the color parameters L* and H* decreased with ripening, whereas a* and a*/b* increased progressively.

In the ***turning*** stage, T2 together with 'Baghera' showed the highest values of TSS and firmness, while T1 and T5 stood out in terms of their lycopene and vitamin C contents. In the ***light-red*** stage, T2 was notable for its highest TSS and vitamin C content, as well as TAA, whereas T5 presented the greatest carotenoid content. In the ***red*** stage, T4 exhibited remarkable levels of lycopene, vitamin C and TAA. The traditional tomato T2 (at ***light-red***) was the variety that was most appreciated by panelists for sweetness and sourness, two major descriptors related to the final taste of tomatoes.

In summary, most of the traditional studied varieties had good quality characteristics and could meet the increasing consumer demand for tasty and healthy food, since they also presented higher concentrations of bioactive compounds at full ripeness.

Author Contributions: Conceptualization, M.C.A.-Y., F.G.-C. and M.L.-R.; methodology, M.C.A.-Y., F.G.-C., M.L.-R. and A.M.F.-L.; formal analysis, M.C.A.-Y., F.G.-C., A.M.F.-L. and M.J.B.-G.; investigation, M.C.A.-Y., F.G.-C., A.M.F.-L. and M.J.B.-G.; resources, M.L.-R.; data curation, M.C.A.-Y., F.G.-C. and A.M.F.-L.; writing-original draft preparation, F.G.-C.; writing—review & editing, M.C.A.-Y. and M.J.B.-G.; visualization, M.C.A.-Y. and M.J.B.-G.; supervision, M.J.B.-G.; project administration, M.L.-R. All authors have read and agreed to the published version of the manuscript.

Funding: This work was supported by project INIA RTA200600144C02-02 and by the Government of Extremadura (Dirección General de Ciencia y Tecnología) and FEDER's funds.

Institutional Review Board Statement: Not applicable.

Informed Consent Statement: Not applicable.

Data Availability Statement: Not applicable.

Acknowledgments: Authors thank to the Centre for the Conservation and Improvement of Agrodiversity (COMAV, Valencia, Spain) and the Institute of Agricultural and Food Research and Development (IMIDA, Murcia, Spain) for providing the landrace seeds and to J. Gragera-Facundo, from the CICYTEX, who conducted the field production of the tomatoes.

Conflicts of Interest: The authors declare no conflict of interest.

References

1. FAOSTAT. Crops and Livestock Products Data. Available online: http://www.fao.org/faostat/es/ (accessed on 2 March 2020).
2. Adalid, A.M.; Roselló, S.; Nuez, F. Evaluation and selection of tomato accessions (Solanum section Lycopersicon) for content of lycopene, β-carotene and ascorbic acid. *J. Food Compos. Anal.* **2010**, *23*, 613–618. [CrossRef]
3. Borguini, R.G.; Ferraz da Silva Torres, E.A. Tomatoes and Tomato Products as Dietary Sources of Antioxidants. *Food Rev. Int.* **2009**, *25*, 313–325. [CrossRef]
4. Dillingham, B.L.; Rao, A.V. Biologically active lycopene in human health. *Int. J. Naturop. Med.* **2009**, *4*, 23–27.
5. Brandt, S.; Pék, Z.; Barna, E.; Lugasi, A.; Helyes, L. Lycopene content and colour of ripening tomatoes as affected by environmental conditions. *J. Sci. Food Agric.* **2005**, *86*, 568–572. [CrossRef]
6. Shi, J.; Le Maguer, M. Lycopene in Tomatoes: Chemical and Physical Properties Affected by Food Processing. *Crit. Rev. Biotechnol.* **2000**, *20*, 293–334. [CrossRef]
7. Giovannucci, E. Tomato Products, Lycopene, and Prostate Cancer: A Review of the Epidemiological Literature. *J. Nutr.* **2005**, *135*, 2030S–2031S. [CrossRef]

8. Limpens, J.; Schröder, F.H.; De Ridder, C.M.A.; Bolder, C.A.; Wildhagen, M.F.; Obermüller-Jevic, U.C.; Krämer, K.; Van Weerden, W.M. Combined Lycopene and Vitamin E Treatment Suppresses the Growth of PC-346C Human Prostate Cancer Cells in Nude Mice. *J. Nutr.* **2006**, *136*, 1287–1293. [CrossRef]
9. Riccioni, G.; Manzini, B.; di Ilio, E.; Bucciarelli, T.; D'Orazio, N. Protective effect of lycopene in cardiovascular disease. *Eur. Rev. Med. Pharmacol. Sci.* **2008**, *12*, 183–190.
10. Willcox, J.K.; Catignani, G.L.; Lazarus, S. Tomatoes and cardiovascular health. *Crit. Rev. Food Sci. Nutr.* **2003**, *43*, 1–18. [CrossRef]
11. Dumas, Y.; Dadomo, M.; Di Lucca, G.; Grolier, P. Effects of environmental factors and agricultural techniques on anti-oxidant content of tomatoes. *J. Sci. Food Agric.* **2003**, *83*, 369–382. [CrossRef]
12. Aldrich, H.T.; Salandanan, K.; Kendall, P.; Bunning, M.; Stonaker, F.; Külen, O.; Stushnoff, C. Cultivar choice provides options for local production of organic and conventionally produced tomatoes with higher quality and antioxidant content. *J. Sci. Food Agric.* **2010**, *90*, 2548–2555. [CrossRef] [PubMed]
13. López, A.F.; Gómez, P.A. Comparison of color indexes for tomato ripening. *Hortic. Bras.* **2004**, *22*, 534–537. [CrossRef]
14. Berni, R.; Romi, M.; Parrotta, L.; Cai, G.; Cantini, C. Ancient Tomato (*Solanum lycopersicum* L.) Varieties of Tuscany Have High Contents of Bioactive Compounds. *Horticulturae* **2018**, *4*, 51. [CrossRef]
15. Parisi, M.; Scalzo, R.L.; Migliori, C.A. Postharvest Quality Evolution in Long Shelf-Life "Vesuviano" Tomato Landrace. *Sustainability* **2021**, *13*, 11885. [CrossRef]
16. Gonzalez-Cebrino, F.; Lozano, M.; Ayuso, M.C.; Bernalte, M.J.; Vidal-Aragon, M.C.; González-Gómez, D. Characterization of traditional tomato varieties grown in organic conditions. *Span. J. Agric. Res.* **2011**, *9*, 444. [CrossRef]
17. Passam, H.C.; Karapanos, I.C.; Bebeli, P.J.; Savvas, D. A review of recent research on tomato nutrition, breeding and post-harvest technology with reference to fruit quality. *Eur. J. Plant Sci. Biotechnol.* **2007**, *1*, 1–21.
18. Alonso, A.; García-Aliaga, R.; García-Martínez, S.; Ruiz, J.; Carbonell-Barrachina, A.A. Characterization of Spanish Tomatoes using Aroma Composition and Discriminant Analysis. *Food Sci. Technol. Int.* **2009**, *15*, 47–55. [CrossRef]
19. Ruiz, J.J.; Martínez, N.; García-Martínez, S.; Serrano, M.; Valero, M.; Moral, R. Micronutrient Composition and Quality Characteristics of Traditional Tomato Cultivars in Southeast Spain. *Commun. Soil Sci. Plant Anal.* **2005**, *36*, 649–660. [CrossRef]
20. Ercolano, M.R.; Carli, P.; Soria, A.; Cascone, A.; Fogliano, V.; Frusciante, L.; Barone, A. Biochemical, sensorial and genomic profiling of traditional Italian tomato cultivars. *Euphytica* **2008**, *164*, 571–582. [CrossRef]
21. Athinodorou, F.; Foukas, P.; Tsaniklidis, G.; Kotsiras, A.; Chrysargyris, A.; Delis, C.; Kyratzis, A.C.; Tzortzakis, N.; Nikoloudakis, N. Morphological Diversity, Genetic Characterization, and Phytochemical Assessment of the Cypriot Tomato Germplasm. *Plants* **2021**, *10*, 1698. [CrossRef]
22. U.S. Department of Agriculture. United States Standards for Grade of Fresh Tomatoes. 1991. Available online: https://www.ams.usda.gov/sites/default/files/media/Tomato_Standard%5B1%5D.pdf (accessed on 20 November 2021).
23. International Plant Genetic Resources Institute. Descriptors for Tomato (*Lycopersicon* spp.). 1996. Available online: http://pdf.usaid.gov/pdf_docs/PNACH866.pdf (accessed on 7 November 2021).
24. Hernández, M.; Rodríguez, E.M.; Díaz, C. Chemical composition of tomato (*Lycopersicon esculentum*) from Tenerife, the Canary Islands. *Food Chem.* **2008**, *106*, 1046–1056. [CrossRef]
25. Sabio, E.; Lozano, M.; de Espinosa, V.M.; Mendes, R.L.; Pereira, A.P.; Palavra, A.F.; Coelho, J.A. Lycopene and β-Carotene Extraction from Tomato Processing Waste Using Supercritical CO_2. *Ind. Eng. Chem. Res.* **2003**, *42*, 6641–6646. [CrossRef]
26. Cano, A.; Ruiz, J.H.; Garcia-Canovas, F.; Acosta, M.; Arnao, M.B. An end-point method for estimation of the total antioxidant activity in plant material. *Phytochem. Anal.* **1998**, *9*, 196–202. [CrossRef]
27. Subedi, P.; Walsh, K. Non-invasive techniques for measurement of fresh fruit firmness. *Postharvest Biol. Technol.* **2009**, *51*, 297–304. [CrossRef]
28. Arias, R.; Lee, T.-C.; Logendra, L.; Janes, H. Correlation of Lycopene Measured by HPLC with the L*, a*, b* Color Readings of a Hydroponic Tomato and the Relationship of Maturity with Color and Lycopene Content. *J. Agric. Food Chem.* **2000**, *48*, 1697–1702. [CrossRef]
29. Radzevičius, A.; Viskelis, P.; Bobinas, C. Quality and physiological parameters of tomato (*Lycopersicon esculentum* Mill.) fruits of Lithuanian selection. *Biologija* **2008**, *54*, 108–111. [CrossRef]
30. Perkins-Veazie, P.; Roberts, W.; Collins, J.K. Lycopene Content among Organically Produced Tomatoes. *J. Veg. Sci.* **2007**, *12*, 93–106. [CrossRef]
31. Pinela, J.; Barros, L.; Carvalho, A.M.; Ferreira, I. Nutritional composition and antioxidant activity of four tomato (*Lycopersicon esculentum* L.) farmer' varieties in Northeastern Portugal homegardens. *Food Chem. Toxicol.* **2012**, *50*, 829–834. [CrossRef]
32. Kuti, J.O.; Konuru, H.B. Effects of genotype and cultivation environment on lycopene content in red-ripe tomatoes. *J. Sci. Food Agric.* **2005**, *85*, 2021–2026. [CrossRef]
33. Murariu, O.C.; Brezeanu, C.; Jităreanu, C.D.; Robu, T.; Irimia, L.M.; Trofin, A.E.; Popa, L.-D.; Stoleru, V.; Murariu, F.; Brezeanu, P.M. Functional Quality of Improved Tomato Genotypes Grown in Open Field and in Plastic Tunnel under Organic Farming. *Agriculture* **2021**, *11*, 609. [CrossRef]
34. Roselló, S.; Adalid, A.M.; Cebolla-Cornejo, J.; Nuez, F. Fast indirect quantification of lycopene and β-carotene content in traditional tomato varieties. (Cuantificación indirecta rápida del contenido en licopeno y β-caroteno en variedades tradicionales de tomate) Original text in Spanish. *Actas Port. Hortic.* **2005**, *4*, 429–435.

35. Helyes, L.; Pék, Z.; Lugasi, A. Tomato Fruit Quality and Content Depend on Stage of Maturity. *HortScience* **2006**, *41*, 1400–1401. [CrossRef]
36. Kusumiyati, T.; Akinaga, M.T.; Kawasaki, S. On-tree and after-harvesting evaluation of firmness, color and lycopene content of tomato fruit using portable NIR spectroscopy. *J. Food Agric. Environ.* **2008**, *6*, 327–332.
37. Guil-Guerrero, J.; Rebolloso-Fuentes, M. Nutrient composition and antioxidant activity of eight tomato (*Lycopersicon esculentum*) varieties. *J. Food Compos. Anal.* **2009**, *22*, 123–129. [CrossRef]
38. Ilahy, R.; Hdider, C.; Lenucci, M.S.; Tlili, I.; Dalessandro, G. Phytochemical composition and antioxidant activity of high-lycopene tomato (*Solanum lycopersicum* L.) cultivars grown in Southern Italy. *Sci. Hortic.* **2011**, *127*, 255–261. [CrossRef]
39. Gautier, H.; Diakou-Verdin, V.; Bénard, C.; Reich, M.; Buret, M.; Bourgaud, F.; Poëssel, J.L.; Caris-Veyrat, C.; Génard, M. How does tomato quality (sugar, acid, and nutritional quality) vary with ripening stage, temperature, and irradiance? *J. Agric. Food Chem.* **2008**, *56*, 1241–1250. [CrossRef]
40. Raffo, A.; Leonardi, C.; Fogliano, V.; Ambrosino, P.; Salucci, M.; Gennaro, L.; Bugianesi, R.; Giuffrida, F.; Quaglia, G. Nutritional Value of Cherry Tomatoes (*Lycopersicon esculentum* Cv. Naomi F1) Harvested at Different Ripening Stages. *J. Agric. Food Chem.* **2002**, *50*, 6550–6556. [CrossRef]
41. Giovanelli, G.; Lavelli, V.; Peri, C.; Nobili, S. Variation in antioxidant components of tomato during vine and post-harvest ripening. *J. Sci. Food Agric.* **1999**, *79*, 1583–1588. [CrossRef]
42. Cano, A.; Acosta, M.; Arnao, M.B. Hydrophilic and lipophilic antioxidant activity changes during on-vine ripening of tomatoes (*Lycopersicon esculentum Mill.*). *Postharvest Biol. Technol.* **2003**, *28*, 59–65. [CrossRef]

horticulturae

MDPI

Article

Physicochemical Characteristics and Nutritional Composition during Fruit Ripening of *Akebia trifoliata* (Lardizabalaceae)

Shuaiyu Zou [1], Puxin Gao [1], Tianjiao Jia [1] and Hongwen Huang [1,2,*]

1 Lushan Botanical Garden, Chinese Academy of Sciences, Jiujiang 332900, China; zousy@lsbg.cn (S.Z.); gaopx@lsbg.cn (P.G.); jiatj@lsbg.cn (T.J.)
2 South China Botanical Garden, Chinese Academy of Sciences, Guangzhou 510650, China
* Correspondence: huanghw@scbg.ac.cn

Abstract: *Akebia trifoliata* is a high-value medicinal and edible fruit crop in China, and it has begun to be widely cultivated as a new fruit crop in many areas of China. Its fruits crack longitudinally when fully ripe and should be harvested before fruit cracking. Physicochemical characteristics and nutritional composition of the ripening process are prerequisites to establishing proper harvest maturity windows. In the current study, we have investigated the fruit quality characteristics of two *A. trifoliata* clonal lines ('Luqing' and 'Luyu') that were harvested at four time points (S1: 120 days after full bloom (DAFB), S2: 134 DAFB, S3: 148 DAFB, S4: 155 DAFB). An increase in fruit size (fruit weight, fruit length, and fruit diameter) was associated with delayed harvest maturity. The firmness of *A. trifoliata* fruit exhibited a decreasing trend with delaying the harvest stage. In particular, the firmness decreased sharply from S2 to S3 stage. The TSS, fructose, and glucose content in *A. trifoliata* fruit continuously increased from the S1 to S4 stage and accumulated sharply from S2 to S3 stage. However, the sucrose and starch content showed an increasing trend from the S1 to S2 stage but declined sharply in the S3 or S4 stage. Ascorbic acid progressively increased with the advancement of *A. trifoliata* maturity stages, while total phenolics and total flavonoids levels declined with fruit ripening. Considering the results of all quality parameters mentioned above, the *A. trifoliata* fruit harvested at the S3 maturity stage was the ideal harvest maturity for long-distance transportation and higher consumer acceptability before fruit cracking. Our research reveals the dynamic changes in physicochemical characteristics and nutritional composition during fruit ripening of *A. trifoliata*. Results in this study reflect the importance of maturity stages for fruit quality and provide basic information for optimal harvest management of *A. trifoliata*.

Keywords: *Akebia trifoliata*; maturity stage; physicochemical; nutritional; ripening

Citation: Zou, S.; Gao, P.; Jia, T.; Huang, H. Physicochemical Characteristics and Nutritional Composition during Fruit Ripening of *Akebia trifoliata* (Lardizabalaceae). *Horticulturae* **2022**, *8*, 326. https://doi.org/10.3390/horticulturae8040326

Academic Editor: Daniela Farinelli

Received: 16 March 2022
Accepted: 10 April 2022
Published: 13 April 2022

1. Introduction

Akebia trifoliata is a perennial woody vine that belongs to the family Lardizabalaceae, has large and edible fruits, and is mainly distributed in East Asia [1]. Its fruits are known as "Bayuezha" in China; ripe fruit could crack longitudinally in the Chinese lunar August. The fully ripe fruit of *A. trifoliata* has a soft pulp texture and sweet flavor, tasting like a mixture of banana and litchi. The fruits of *A. trifoliata* are rich in sugars, crude proteins, amino acids, vitamins, minerals, ash, and fiber [2,3]. As an edible fruit plant, the fruit of *A. trifoliata* can also be processed into commodities such as desserts, wine, juices, vinegar, tea, edible oil, etc. [4,5]. The value of *A. trifoliata* fruit lies not only in its nutrition or commercialization but also in its bioactive ingredients with anti-inflammatory, diuretic, antimicrobial, anticancer, anti-obesity, and antioxidative properties [6–11]. In recent years, *A. trifoliata*, as a high-value medicinal and edible fruit crop, has been widely cultivated in China [12,13]. Moreover, the genomic data of *A. trifoliata* subsp. *australis* were published last year, which will greatly accelerate the extensive and intensive research on this new fruit crop [14].

The fruit of *A. trifoliata* cracks open longitudinally when completely ripe on the vine. However, pericarp cracking causes a series of problems in that pulp is easily contaminated by impurities, infected by pathogens, or eaten by birds, resulting in short shelf life, low visual quality, and decreasing acceptance by consumers and the commercial value. Namely, the fruit quality decreases rapidly when it is harvested after full ripening and cracking. The maturity stage at harvest is a fundamental preharvest factor that could determine the final fruit quality and storage potential. For example, in order to prolong the shelf life, winter jujube was harvested at the white maturity stage and stored at a low temperature for a period of time and then moved to market [15]; strawberries were frequently harvested at the turning stage (75% red), or even at the green stage (50% red) for long-distance transport or export markets [16]; mulberries harvested at the fully ripe stage are suitable for fresh consumption while those harvested at semi ripe stage are appropriate for the processing industry [17]. Therefore, it is of great practical significance for the fruit of *A. trifoliata* to explore a proper maturity stage at harvest which can meet the demand for good fruit quality, shelf life, and market value simultaneously.

The fruit maturity stage is usually judged by external attributes such as size, appearance, texture, and color, or internal attributes such as total soluble solids, acidity, and starch. The colors of *A. trifoliata* fruit are varied in different cultivars at maturity, from green to cyan, from yellow to brown, from purple to pink, etc. Thus, the maturity stage of *A. trifoliata* fruit is hardly assessed just by appearance or color in some cultivars before fruit cracking. Thus, we should investigate the changes in physicochemical characteristics and nutritional composition during *A. trifoliata* fruit development to better understand the fruit ripening process of *A. trifoliata*.

Contents of nutrients, phytochemical compounds, and volatiles in fruits are usually influenced by numerous preharvest or postharvest factors, but the maturity stage at harvest is probably the most important factor in determining the final fruit quality [18–20]. Studies showed that the physical properties, biochemical properties, and proximate composition of tomatoes were more affected by maturity stages than genotype [21]. For winter jujube, the physicochemical properties during fruit ripening were influenced by the maturity stage at harvest [22,23]. However, there is little information on the changes in physicochemical characteristics and nutritional composition of *A. trifoliata* at different fruit maturity stages. Thus, the objective of this study was to investigate the changes in physicochemical characteristics, nutritional composition, and antioxidant content of *A. trifoliata* fruit at different stages of on-the-vine ripening. The results of this study can help us to understand the basic dynamic change patterns of fruit quality and to identify the optimum harvest maturity stage of *A. trifoliata* with better quality and longer marketability before fruit cracking.

2. Materials and Methods

2.1. Plant Material

Fruits of two *A. trifoliata* clonal lines ('Luqing' and 'Luyu' selected from our genetic improvement programs of *Akebia* [12,13]) free from pests, insects, and diseases were randomly harvested at four time points (S1: 120 days after full bloom (DAFB), S2: 134 DAFB, S3: 148 DAFB, S4: 155 DAFB) until the fruits begin cracking in the experimental farm located at Zhangjiajie county (29°41′ N, 111°09′ E), China (Figure 1). At the S3 time point, the fruits of two clonal lines begin to soften to some extent with no fruit cracking. At the S4 time point, the fruits begin to crack (about 50%), and the S4 time point was set as the fruit cracking time (Figure 1D). The *A. trifoliata* trees were five years old, and these plants were freely pollinated, and the sprinkling irrigation system was used for field water balance; the vines were pruned in the December or January and applied 4–5 kg organic fertilizer per plant in the winter. Because the fruit from one tree of *A. trifoliata* did not provide enough samples for four stages, fruits were harvested from different stages of ten trees in each clonal line at the same time and considered as a sample. Three fruits pulp were mixed into one biological replicate for sugars, starch, ascorbic acid, total phenolics, total flavonoids, and proximate analysis, and the fruit pulp was immediately treated with liquid nitrogen and stored at −80 °C until further analysis. Sampled fruits from each stage were immediately evaluated for fruit weight, length, diameter, and moisture on 5 fruits. In total, three biological replicates were collected for each maturity stage.

Figure 1. The fruits of *A. trifoliata* at different maturity stages. (**A–D**) the fruit of Luqing at S1, S2, S3, and S4 stages, respectively. The pericarp is pale green when it cracks. (**E–H**) the fruit of Luyu at S1, S2, S3, and S4 stages, respectively. The pericarp is light pink when it cracks. S1: 120 days after full bloom (DAFB), S2: 134 DAFB, S3: 148 DAFB, S4: 155 DAFB.

2.2. Physical Parameters

The average fruit weight of each sample (5 fruits) was recorded by using an electronic balance and expressed in grams. Fruit length (FL) and fruit diameter (FD) were determined using a digital caliper with 0.01 mm accuracy, and their values were expressed in millimeters. The firmness of *A. trifoliata* was determined independently for each of 5 fruit per replicate using a digital fruit firmness tester (GY-4, Zhejiang, China) outfitted with a 3.5-mm cylinder probe. A knife was to peel off the cuticle of the fruit peel, then the cylinder probe was aligned vertically to the test position of the fruit peel, and pressure was slowly applied to make the probe insert the fruit peel. The results of the firmness values were expressed as the kg/cm^2.

2.3. Biochemical Parameters

Total soluble solids content (TSS) of *A. trifoliata* fruit pulp was determined by a digital refractometer (PAL-1, Atago, Tokyo, Japan), and the results were expressed in °Brix. Titratable acidity (TA) was determined by titrating with 0.1 N NaOH until reaching pH 8.1, and the values were expressed as % of citric acid. The pulp pH was calculated using a pH meter (Jenway, Staffordshire, UK).

2.4. Determination of Sugars

The levels of individual sugars (glucose, fructose, and sucrose) were simultaneously determined by high-performance liquid chromatography (HPLC). Firstly, the fruit pulp sample (accurate to 0.001 g) was weighed and placed in a 100 mL triangle flask. Then 10 mL of extraction solution (0.02 mol/L NaOH) was added accurately. The extraction solution was fully oscillated at 180 r/min for 40 min, and the qualitative filter paper was used for filtration. Finally, 2 mL filtrate was filtered through a 0.45 μm aqueous phase filter membrane and then used for HPLC analysis with an evaporative light scattering detector. The detection conditions were as follows: the carrier gas was nitrogen, the gain was 500, the spray mode was cooling, the drift tube temperature was 50 °C, the gas pressure was 40 psi, and the analysis time was 30 min. Shodex's AsahiPak NH2P-50 4E column (150 mm × 4.6 mm) was used for analysis. The mobile phase consisted of water and acetonitrile at the flow rate of 0.8 mL/min. The column temperature was 35 °C, and the injection volume was 10 μL.

2.5. Determination of Starch

About 1 g of sample of fruit pulp was extracted with 20 mL of 80% ethanol using a high-speed dispersion machine for 30 min. Then the sample was transferred to a water bath at 80 °C for 30 min, stirred for 30 min until cool and filtered. The above step was repeated once, and then the sample was extracted with 80% ethanol and a filter. The filter residue was washed into a centrifuge tube with 20 mL of hot distilled water and placed in a boiling water bath for gelatinization for 15 min. Then, 2 mL of cold 9.2 mol/L perchloric acid was added into the centrifuge tube and stirred continuously, extracted for 15 min, and filtered. The filtrate was poured into a 100 mL volumetric flask and 10 mL distilled water was added to the filter residue. Then, 1 mL 9.2 mol/L perchloric acid was added, and the filtrate residue was placed in a boiling water bath and heated for 15 min, and filtered. The filter residue was washed twice with distilled water, then poured into the voluminous flask. Absorbance was recorded at 490 nm. The calibration curve was derived by the phonel-sulfate method.

2.6. Determination of Ascorbic Acid, Total Phenolics, and Total Flavonoids

The ascorbic acid (AsA) content of *A. trifoliata* fruit pulp was measured by the indophenol's titration method [24]. Briefly, 2 g of fruit pulp was mixed with distilled water (10 mL), and a solution of 0.4% oxalic acid (90 mL) was taken in a 100 mL volumetric flask. Then the mixture was filtered with filter paper. After that, 5 mL filtrated aliquot was taken and titrated against 2,6-dichlorophenolindophenol to a light pink color endpoint. The amount of AsA was expressed in milligrams per 100 g of fresh weight (FW).

Total phenolics content in the *A. trifoliata* fruit pulp was determined with Folin-Ciocalteu reagent by the method of Razzaq et al. [25]. A total of 1.0 g of fruit pulp was extracted with 97% (v/v) methanol. Then, 1.0 mL of extract solution was mixed with 1.0 mL Folin–Ciocalteu solvent, let stand for 5 min, and 8.0 mL of Na_2CO_3 (75 mg/mL) was added to the mixture. After being kept in the dark for 30 min at room temperature, the absorbance was measured at the wavelength of 765 nm. A standard curve of gallic acid was used for calculating the total phenolics in *A. trifoliata* fruit pulp and the results were expressed as mg gallic acid equivalents (GAE) per 100 g of FW.

Total flavonoids content was measured as described by Zhao et al. [26]. Briefly, fruit pulp (2.0 g) was extracted with 80% (v/v) ethanol and centrifuged at 10,000 g for 30 min at 4 °C. Then, 0.5 mL of supernatant was mixed with 0.1 mL of 5% $NaNO_2$. After 5 min, 0.1 mL of 10% $AlCl_3$ solution was added, followed by incubating for 6 min, then 1 mL 1 mM NaOH was added. The reaction solution was kept for 30 min at room temperature, and the absorbance was assayed at 510 nm. The total flavonoids content in *A. trifoliata* fruit pulp was calculated using a standard curve of rutin and was expressed as mg rutin equivalents per 100 g of FW.

2.7. Proximate Analysis

Dry matter content and moisture content in *A. trifoliata* fruit were measured by oven drying at 70 °C till its weight became constant. The crude fiber was measured by extraction of fruit sample in the presence of sulphuric acid and potassium hydroxide reagents, based on the method outlined by Abdullahi et al. [27]. Fat content was measured by the Soxhlet method based on the solubility of the free lipid content in organic solvents such as ethyl ether and petroleum ether. The Kjeldahl method was used for the determination of the protein content in *A. trifoliata* fruit pulp and protein content were calculated as N × 6.25 based on the method of Imran et al. [28].

2.8. Statistical Analysis

Statistical analyses were performed using one-way analysis of variance (ANOVA), and the significant differences between means were determined by the least significant difference (LSD) test at $p < 0.05$ by using IBM SPSS Statistics 20.0 software. The results were expressed as means ± standard errors.

3. Results

3.1. Changes in the Physical Parameters of *A. trifoliata* Fruit during Maturity Stages

The changes in the physical parameters of *A. trifoliata* (clonal lines 'Luqing' and 'Luyu') at four different maturity stages (S1–S4) are shown in Table 1. In both clonal lines of 'Luqing' and 'Luyu', fruit weight, fruit length, and fruit diameter increased significantly with increasing maturity. For the 'Luqing', the values of fruit weight increased from 203.19 ± 11.59 g at the S1 stage to 536.33 ± 18.93 g at the S4 stage. The fruit length increased significantly from 128.17 ± 2.62 mm at the S1 stage to 172.35 ± 4.29 mm at the S4 stage. The fruit diameter increased significantly from 52.44 ± 1.23 mm at the S1 stage to 72.04 ± 2.02 mm at the S4 stage. For the 'Luyu', the values of fruit weight, fruit length, and fruit diameter all increased at various extents from the S1 stage to the S4 stage. For example, fruit weight increased from 152.93 ± 5.81 g to 302.82 ± 8.22 g; the fruit length increased significantly from 117.86 ± 1.67 mm to 139.90 ± 2.37 mm. Meanwhile, the fruit diameter increased significantly from 48.19 ± 0.95 mm to 62.12 ± 1.34 mm.

Table 1. Effect of harvest maturity on physical parameters of two *Akebia trifoliata* clonal lines 'Luqing' and 'Luyu' fruit.

Clonal Lines	Maturity Stages	Fruit Weight/g	Fruit Length/mm	Fruit Diameter/mm
Luqing	S1	203.19 ± 11.59 d	128.17 ± 2.62 c	52.44 ± 1.23 c
	S2	289.54 ± 13.38 c	148.10 ± 3.03 b	59.45 ± 1.43 b
	S3	463.10 ± 14.66 a	168.27 ± 3.32 a	69.69 ± 1.56 a
	S4	536.33 ± 18.93 a	172.35 ± 4.29 a	72.04 ± 2.02 a
Luyu	S1	152.93 ± 5.81 d	117.86 ± 1.67 d	48.19 ± 0.95 c
	S2	191.37 ± 6.22 c	123.77 ± 1.79 c	51.76 ± 1.01 b
	S3	259.50 ± 6.71 b	129.41 ± 1.93 b	61.49 ± 1.09 a
	S4	302.82 ± 8.22 a	139.90 ± 2.37 a	62.12 ± 1.34 a

Values are the means ± standard errors. Means with different letters within the same column indicate statistical differences at the $p < 0.05$ level. S1: 120 days after full bloom (DAFB), S2: 134 DAFB, S3: 148 DAFB, S4: 155 DAFB.

3.2. Changes in Fruit Firmness, Total Soluble Solids, and Titratable Acidity during Maturity Stages

The fruit firmness of *A. trifoliata* was significantly ($p < 0.05$) affected by the harvest maturity stages during fruit development. The firmness of *A. trifoliata* fruit at four maturity stages decreased continuously with the delaying of the harvest stage (Figure 2A). The firmness of 'Luqing' fruit in S1, S2, S3 and S4 stages were 29.32 kg/cm^2, 26.29 kg/cm^2, 8.58 kg/cm^2 and 5.00 kg/cm^2, respectively. The firmness decreased significantly from S2 to S3 stage, while there were no significant differences between the S1 and S2 stage or S3 and S4 stage. The firmness of 'Luyu' fruit at S1, S2, S3, and S4 stages was 33.13 kg/cm^2, 28.02 kg/cm^2, 4.74 kg/cm^2, and 3.41 kg/cm^2, respectively, and the firmness decreased significantly from S1 to S3 stage, while there was no significant difference between S3 and S4 stage. On the whole, the firmness of 'Luqing' fruit at S1 and S2 stages was lower than 'Luyu', while higher than 'Luyu' at S3 and S4 stages, which meant that the firmness of 'Luqing' fell faster than 'Luyu' in the early mature stage, whereas in late maturity it declined more slowly than 'Luyu'.

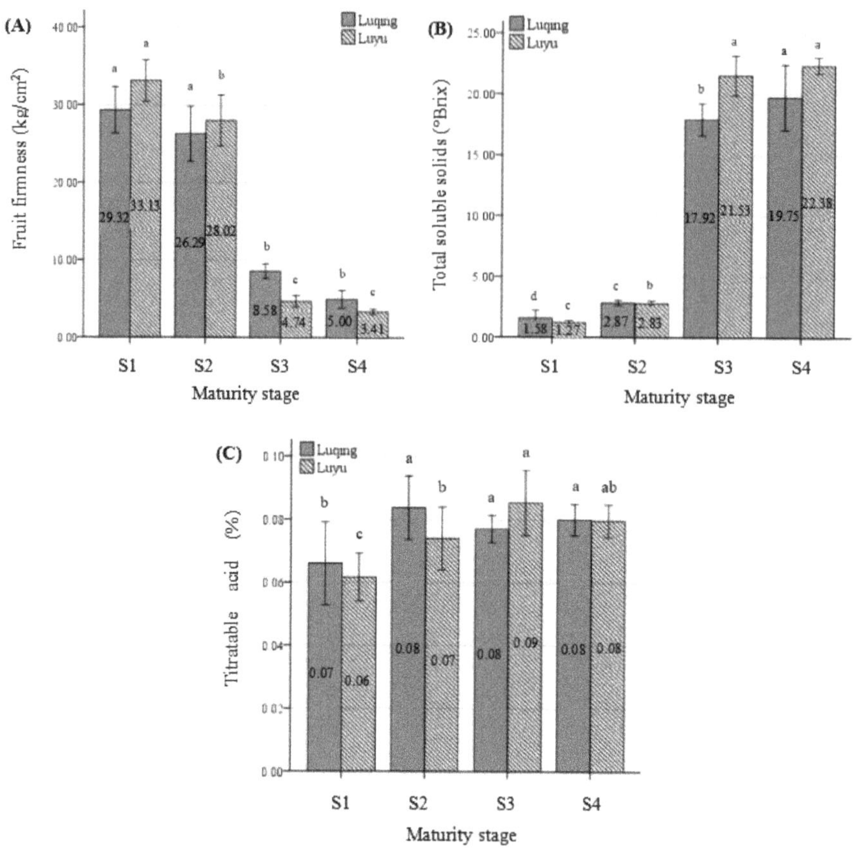

Figure 2. Fruit firmness (**A**), total soluble solids (**B**), and titratable acidity (**C**) in *Akebia trifoliata* fruit at four stages of ripening. Error bars indicate standard error from 10 replicates. Means with different letters within each clonal line indicate statistical differences at the $p < 0.05$ level using the LSD test. S1: 120 days after full bloom (DAFB), S2: 134 DAFB, S3: 148 DAFB, S4: 155 DAFB.

During the fruit maturation and ripening period, the changes in TSS content of 'Luqing' and 'Luyu' in four mature stages displayed a similar pattern, namely continuously increased and reached maximum values at the fourth harvest stage (Figure 2B). The TSS content of 'Luqing' fruit at S1, S2, S3, and S4 stages was 1.58 °Brix, 2.87 °Brix, 17.92 °Brix, and 19.75 °Brix, respectively. For 'Luyu' fruit, the TSS content at S1, S2, S3, and S4 stages was 1.27 °Brix, 2.83 °Brix, 21.52 °Brix, and 22.38 °Brix, respectively. For both clonal lines, TSS initially increased slightly as the fruit developed from the S1 to S2 stage; thereafter, TSS rose sharply from S2 to S3 stage, and after that, it increased slightly from S3 to S4 stage.

The TA content of *A. trifoliata* fruit was kept at low levels, and the values were in the range of 0.06–0.08% from the S1 to S4 maturity stage (Figure 2C).

3.3. Changes in Carbohydrates Contents during Maturity Stages

Carbohydrate contents in *A. trifoliata* fruit displayed significant changes during fruit growth and development (Figure 3). The fructose contents of 'Luqing' fruit at S1, S2, S3, and S4 stages were 0.31 g/100 g FW, 0.68 g/100 g FW, 7.39 g/100 g FW, and 7.80 g/100 g FW, respectively. From the graph of the fructose content of 'Luqing', it was obvious that fructose accumulated rapidly, mainly from S2 to S3 stage (Figure 3A). For 'Luyu' fruit, fructose content at S1, S2, S3, and S4 stages were 0.16 g/100 g FW, 0.21 g/100 g FW, 5.33 g/100 g FW, and 8.83 g/100 g FW, respectively, which showed 'Luyu' fruit accumulated a large amount of fructose from S2 to S4 stage. Similarly, glucose indicated the same accumulation pattern as the fructose (Figure 3B). That is, the glucose of 'Luqing' mainly accumulated from the S2 to S3 stage, and 'Luyu' from the S2 to S4 stage.

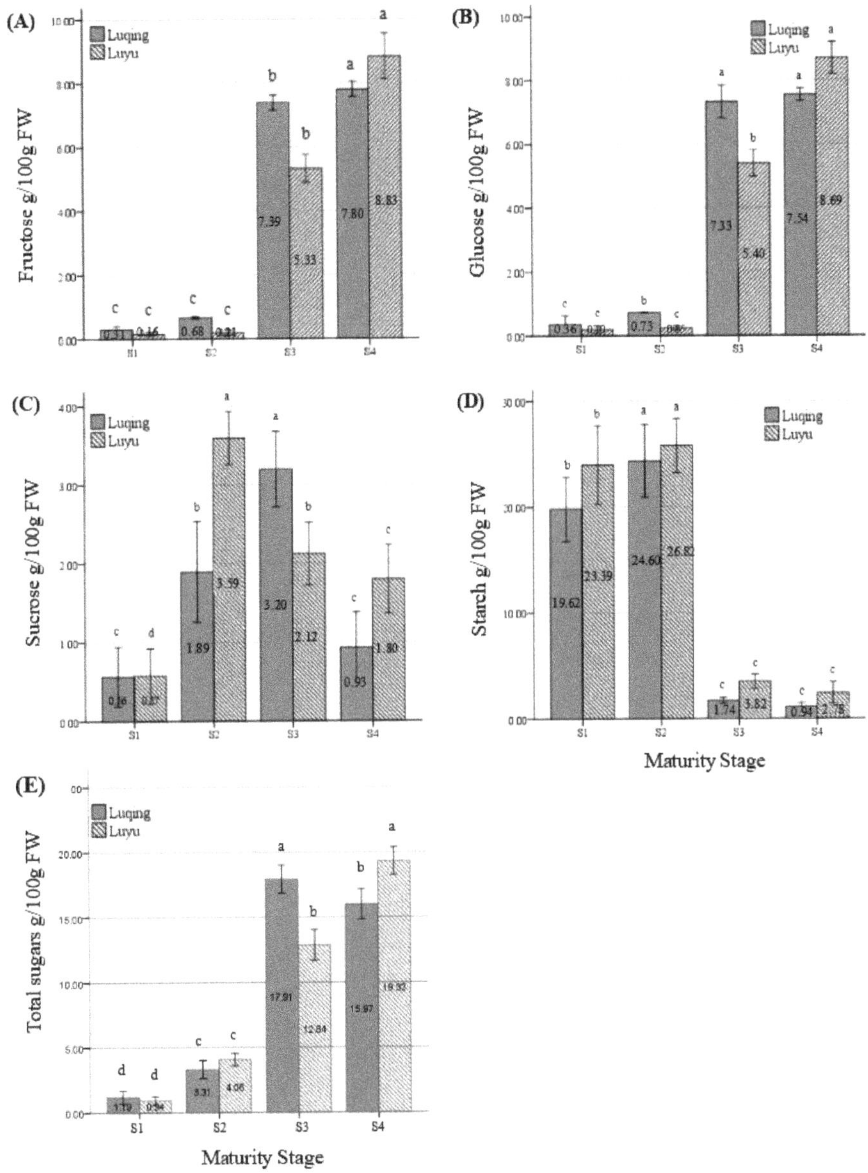

Figure 3. Fructose (**A**), glucose (**B**), sucrose (**C**), starch (**D**), and total sugars (**E**) (the sum of fructose, glucose, and sucrose) concentrations in *Akebia trifoliata* fruit at four stages of ripening. Error bars indicate standard error from 5 replicates. Means with different letters within each clonal line indicate statistical differences at the $p < 0.05$ level using the LSD test. S1: 120 days after full bloom (DAFB), S2: 134 DAFB, S3: 148 DAFB, S4: 155 DAFB.

The sucrose content of 'Luqing' fruit continuously accumulated from the S1 to S3 stage (0.56–3.20 g/100 g FW) and reached maximum values at the S3 stage but had a sharp, significant decline from S3 to S4 stage (3.20–0.93 g/100 g FW) (Figure 3C). While sucrose content of 'Luyu' fruit increased from S1 to S2 stage (0.57–3.59 g/100 g FW), thereafter sucrose declined from S2 to S4 stage (3.59–1.80 g/100 g FW). Starch concentrations of 'Luqing' and 'Luyu' fruit both increased significantly from the S1 to S2 stage (19.62–24.60 g/100 g FW and 23.30–26.82 g/100 g FW, respectively) (Figure 3D), while starch declined sharply from S2 to S3 stage (24.60–1.74 g/100 g FW and 26.82–3.82 g/100 g FW, respectively) and reached minimum values at the S4 stage (0.94 g/100 g FW and 2.78 g/100 g FW, respectively). The total sugars of 'Luqing' increased significantly from the S1 to S3 stage (1.19–17.92 g/100 g FW) while showing some decline at the S4 stage (15.97 g/100 g FW) (Figure 3E). For the 'Luyu', the total sugars continuously increased from the S1 to S4 stage and reached maximum values at the S4 stage (0.94–19.32 g/100 g FW).

3.4. Changes in AsA, Total Phenolics, and Total Flavonoids during Maturity Stages

As displayed in Figure 4A, on the whole, the AsA content of both clonal lines fruit continuously increased during the whole ripening period, except for a slight decline at the S2 stage. The AsA content of 'Luqing' fruit at S1, S2, S3, and S4 stages were 0.87 mg/100 g FW, 0.83 mg/100 g FW, 1.10 mg/100 g FW, and 1.26 mg/100 g FW, respectively. The AsA contents of 'Luyu' fruit at S1, S2, S3, and S4 stages were 0.87 mg/100 g FW, 0.86 mg/100 g FW, 1.07 mg/100 g FW, and 1.16 mg/100 g FW, respectively.

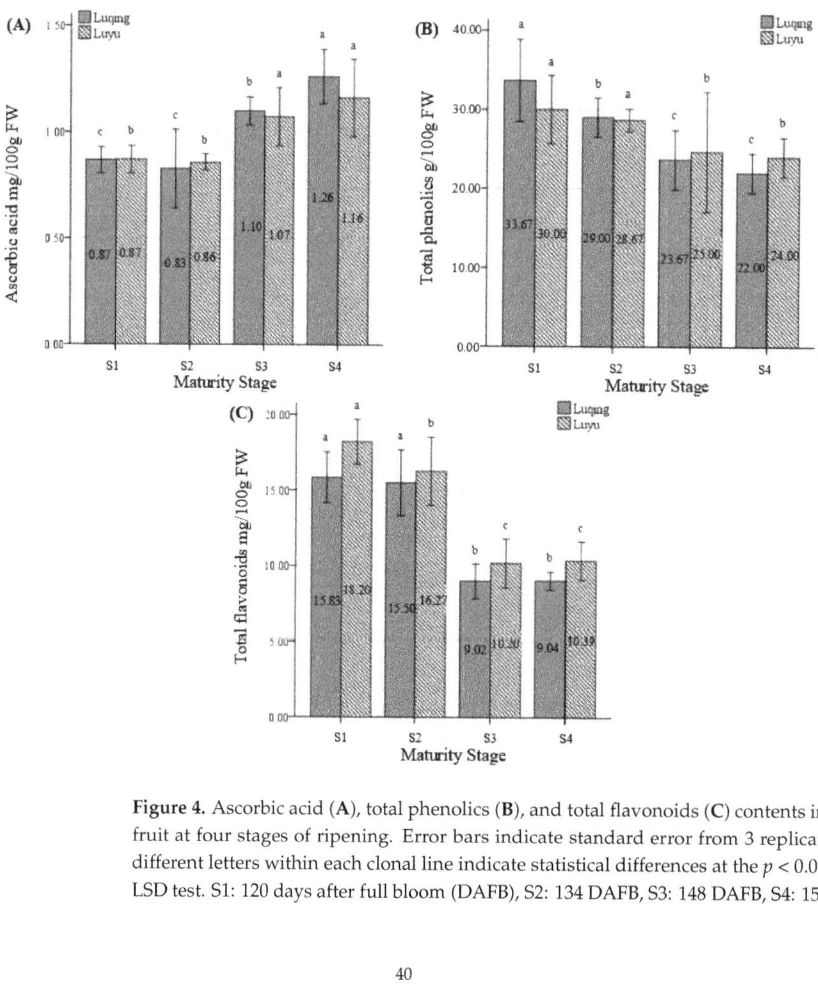

Figure 4. Ascorbic acid (**A**), total phenolics (**B**), and total flavonoids (**C**) contents in *Akebia trifoliata* fruit at four stages of ripening. Error bars indicate standard error from 3 replicates. Means with different letters within each clonal line indicate statistical differences at the $p < 0.05$ level using the LSD test. S1: 120 days after full bloom (DAFB), S2: 134 DAFB, S3: 148 DAFB, S4: 155 DAFB.

The contents of the total phenolics of *A. trifoliata* fruit at four maturity stages are presented in Figure 4B. Overall, the total phenolics declined from the S1 to S4 stage. In comparison with 'Luyu', the content of total phenolics in 'Luqing' was slightly higher than 'Luyu' at S1 (33.67 mg/100 g FW and 30.00 mg/100 g FW, respectively) and S2 (29.00 mg/100 g FW and 28.67 mg/100 g FW, respectively) stages, but slightly lower than 'Luyu' ($p < 0.05$) at the S3 and S4 stages.

As shown in Figure 4C, the maturity stage had a significant influence on the total flavonoids in *A. trifoliata* fruit. During the ripening process, 'Luqing' and 'Luyu' fruit at four maturity stages exhibited similar total flavonoids change patterns. Specifically, 'Luqing' and 'Luyu' showed a slight decrease from the S1 to S2 stage (15.83 mg/100 g FW to 15.50 mg/100 g FW, 18.20 mg/100 g FW to 16.27 mg/100 g FW, respectively), whereas a sharp decrease was found from S2 to S3 stage (15.50–9.02 mg/100 g FW, 16.27–10.20 mg/100 g FW, respectively), thereafter, there was no significant difference between S3 and S4 stage.

3.5. Proximate Composition

The proximate composition of two *Akebia trifoliata* clonal lines is presented in Table 2. The dry matter content of 'Luqing' continuously decreased from the S1 to S4 stage (25.78 ± 1.21% to 17.19 ± 1.10%). For 'Luyu', the dry matter content of 'Luqing' continuously decreased from the S1 to S4 stage (29.68 ± 0.63% to 18.71 ± 1.28%). The moisture content of 'Luqing' continuously increased from the S1 to S4 stage (74.22 ± 1.21% to 82.81 ± 1.10%). Similarly, the change pattern of moisture content in 'Luyu' fruit, like 'Luqing's', increased from the S1 to S4 stage (70.32 ± 0.63% to 81.29 ± 1.28%). The crude fiber content of 'Luqing' fruit at S1, S2, S3, and S4 stages were 0.67 ± 0.03%, 0.60 ± 0.06%, 0.20 ± 0.00%, and 0.17 ± 0.03%, respectively, indicating a continuously decreasing pattern. However, the decrease of crude fiber content in 'Luyu' mainly occurred during the S2 and S3 stages. 'Luqing' and 'Luyu' fruits had a low fat content, and the fat content decreased as the fruits matured. The protein content of 'Luqing' and 'Luyu' fruits fell in half from the S1 stage to the S4 stage. At each stage of maturity, the protein content of 'Luqing' and 'Luyu' was basically the same. In summary, moisture content showed an increasing trend as ripening progressed, while crude fiber, fat, and protein content showed a decreasing pattern along with the maturity of the fruit.

Table 2. Proximate composition of *Akebia trifoliata* fruit at four maturity stages (mean ± standard error).

Clonal Lines	Maturity Stage	Dry Matter Content (%)	Moisture (%)	Crude Fiber(%)	Fat Content(%)	Protein Content(%)
Luqing	S1	25.78 ± 1.21 b	74.22 ± 1.21 b	0.67 ± 0.03 a	0.50 ± 0.00 a	1.14 ± 0.05 a
	S2	23.46 ± 1.23 b	76.54 ± 1.23 b	0.60 ± 0.06 a	0.40 ± 0.06 a	0.99 ± 0.01 b
	S3	19.63 ± 0.80 a	80.37 ± 0.80 a	0.20 ± 0.00 b	0.20 ± 0.00 b	0.56 ± 0.02 c
	S4	17.19 ± 1.10 a	82.81 ± 1.10 a	0.17 ± 0.03 b	0.17 ± 0.03 b	0.54 ± 0.01 c
Luyu	S1	29.68 ± 0.63 b	70.32 ± 0.63 b	0.70 ± 0.06 a	0.47 ± 0.03 a	1.05 ± 0.01 a
	S2	26.98 ± 1.89 b	72.90 ± 1.89 b	0.70 ± 0.06 a	0.43 ± 0.03 a	0.98 ± 0.03 b
	S3	21.26 ± 0.84 a	78.74 ± 0.84 a	0.20 ± 0.00 b	0.23 ± 0.03 b	0.52 ± 0.02 c
	S4	18.71 ± 1.28 a	81.29 ± 1.28 a	0.20 ± 0.00 b	0.20 ± 0.00 b	0.52 ± 0.01 c

Means in the same column for the same clonal line with different letters are significantly different at $p < 0.05$. S1: 120 days after full bloom (DAFB), S2: 134 DAFB, S3: 148 DAFB, S4: 155 DAFB.

4. Discussion

This study demonstrated that the maturity stage had significant effects on the physicochemical and nutritional properties of *A. trifoliata* fruit. An increase in fruit size (fruit weight, fruit length, and fruit diameter) was associated with delaying harvest maturity. The fruit weight of 'Luqing' at S2, S3, and S4 stages increased by 42.50%, 59.94%, and 15.81%, respectively. While in the 'Luyu' clonal line, the fruit weight increased by 25.14%, 35.60%, and 16.69%, respectively, and the fruit weight in both clonal lines increased rapidly from S2 to S3 stage. The fruit weight increased in 'Luqing' more quickly than 'Luyu' at all stages, except for the S4 stage, but the fruit weight of 'Luqing' was still higher than 'Luyu' at each stage. The highest growth rate of fruit length in 'Luqing' occurred at the S2 stage (15.55%), then the growth rate gradually declined at S3 and S4 stages (13.62% and 2.42%, respectively). While the increasing trends of fruit length in 'Luyu' were different from 'Luqing,' whose highest growth rate was found at the S4 stage (8.11%), the lowest growth rate arose at the S3 stage (4.56%) and increased by 5.01% at S2 stage. For the fruit diameter, the highest growth rate in both clonal lines occurred at the S3 stage (17.22% for 'Luqing' and 18.80% for 'Luyu'), and the lowest growth rates were found at the S4 stage (3.37% for 'Luqing' and 1.02% for 'Luyu'). In summary, the fruit weight and fruit diameter increased rapidly from the S2 to S3 stage, while the fruit length had different growth patterns between 'Luqing' and 'Luyu'. Fruit growth showed an S-shaped curve in both clonal lines, consistent with previous reports in *A. trifoliata* fruit [3]. In particular, the values of fruit weight, fruit length, and fruit diameter at the S3 and S4 stages were close to fully ripened fruit [12]. Previous studies have shown that the quality of ripe fruit and storage life are greatly influenced by the maturity stage at harvest [22,29,30]. Fruit firmness, as one of the foremost quality parameters, has a direct impact on the fruit shelf life and consumer acceptance [31]. The fruit of *A. trifoliata* starts softening and cracking with fruit ripening, associated with easy rotting mediated by bacterial infection and mechanical damage. Therefore, although the fruit of *A. trifoliata* could be picked after cracking, the fruit quality decreases rapidly within 3–5 days at room temperature after harvest. Therefore, harvesting fruits before softening and cracking may be an effective method to prolong the shelf life of *A. trifoliata*. In the current study, the firmness of both clonal lines of *A. trifoliata* exhibited a decreasing trend with the delaying of the harvest stage. The fruit firmness of 'Luqing' and 'Luyu' from S1 to S4 declined by 82.95% and 89.71%, respectively. In particular, the firmness decreased sharply from the S2 to S3 stage (67.36% for 'Luqing' and 83.08% for 'Luyu'), which indicated that fruit textural property has tremendous changes during the period. Niu et al. [32] found that the cell wall became thinner, looser, and showed a substantial breakdown in the pericarp of cracking fruit compared with that in noncracking fruit in *A. trifoliata*. In this study, the firmness of *A. trifoliata* decreased sharply before cracking, consistent with the pericarp structure observation results described by Niu et al. [32]. Fruit ripening and cracking are complex physiological, genetically programed processes that are accompanied by the transcription of many genes and the synthesis of large amounts of proteins [33]. Studies have indicated that pectin metabolism-related genes showed a strong link with cracking phenotypes in sweet cherry fruit [34]. Studies on atemoya pericarp cracking have suggested that starch decomposition into soluble sugars and cell wall polysaccharides metabolism are closely related to the ripening and cracking of African Pride atemoya [35]. Those results in this study may provide useful physiological data for advanced research about the fruit cracking of *Akebia*.

TSS and TA are not only the main compounds responsible for the fruit flavor but are also considered crucial indicators of fruit maturity and postharvest fruit quality evaluation during storage [36]. In this study, the TSS content in both clonal lines of *A. trifoliata* fruit showed an upward trend from the S1 stage to the S4 stage. Exactly, the TSS content of 'Luqing' at S2, S3, and S4 stages increased by 102.78%, 904.11%, and 2.86%, respectively. Furthermore, the TSS content of 'Luyu' at S2, S3, and S4 stages increased by 25.00%, 2068.00%, and 60.33%, respectively. Obviously, the TSS content accumulated sharply from the S2 stage to the S3 stage mainly because of the accumulation of sugars and the breakdown of starch and reached physiological maturity at the S3 stage. Niu et al. [37] found that the activity of beta-amylase (BAM) increased 10-fold when fruit begins to crack longitudinally along with its ventral suture, and BAM may play a significant role in starch degradation during fruit ripening. Meanwhile, the TA content of *A. trifoliata* fruit remained at a very low level during the fruit ripening process, which could be almost ignored and is consistent with previous reports in *A. trifoliata* fruit [38,39]. The continuous accumulation of TSS and low level of TA resulted in the very sweet taste of *A. trifoliata* fruit. Therefore,

the balance of sugars and acids in the fruit could strongly affect the taste of the fruit. For this reason, TSS and TA, along with firmness, are always considered the main parameters for determining fruit quality. At different maturity stages, the changes in TSS, TA, and TSS/TA have also been documented by many studies in different fruits [40–42]. For *A. trifoliata*, the firmness and TSS changed sharply between S2 and S3 stages and then tended to level off; we can consider firmness and TSS as the maturity indicators of *A. trifoliata* fruit.

Carbohydrates play important roles not only in the human diet but also in plants' stress responses, metabolic processes, and biological processes [43]. Sugars are important signaling molecules during plants development and are involved in modulating gene expression in plants [43,44]. Results in this study indicated dynamics of carbohydrates with fruit ripening in *A. trifoliata*. Fructose and glucose contents in 'Luqing' and 'Luyu' showed the same accumulation pattern, which initially increased slightly as the fruit developed from S1 to S2 stage, thereafter rose sharply from S2 to S3 stage (the fructose and glucose contents in 'Luqing' and 'Luyu' increased by 10-fold, 9-fold and 27-fold, 20-fold, respectively), after that increased slightly again from S3 to S4 stage. On the other hand, rapid accumulation of fructose and glucose at the S3 stage could be a signal response to reach physiological maturity in *A. trifoliata*. Cao et al. [45] reported that *A. trifoliata* fruit is a kind of climacteric fruit and is easy to rot and deteriorate after harvest. Furthermore, the increasing content of fructose and glucose at advanced stages of fruit ripening has been found in many climacteric fruit crops [46–49]. The sucrose content in 'Luqing' showed an increasing trend from the S1 to S3 stage but declined sharply at the S4 stage, while the turning point of descent in 'Luyu' occurred at the S3 stage, and the strain differences are mainly responsible for this difference. The contents of fructose, glucose, and sucrose at the S4 stage in 'Luyu' are higher than in 'Luqing'; thus, the TSS at the S4 stage was higher in 'Luyu' and tasted sweeter. The starch content in both clonal lines showed an increasing trend from the S1 to S2 stage but declined sharply at the S3 stage, indicating that the starch degraded rapidly as the fruit approached ripening. The increasing trend in fructose and glucose levels was due to the breakdown of carbohydrates (sucrose and starch) with the advancement of fruit maturity [40,50]. The fructose and glucose contents in *A. trifoliata* fruit increased sharply from the S3 stage, while starch content stayed at a low level from the S3 stage, which resulted in the sweet, soft, and glutinous flavor of *A. trifoliata* fruit pulp. The total sugars content of 'Luqing' increased significantly from the S1 to S3 stage, especially at the S3 stage the content of total sugars increased by 6-fold. It is worth noting that the total sugars content of 'Luqing' had a significant decline after the S3 stage, which may be due to hydrolysis or translocation of sugars. While the total sugars content of 'Luyu' continuously increased from the S1 to S4 stage, which showed a different accumulate dynamics of total sugars. In terms of total sugar content, 'Luqing' is more suitable for harvesting at the S3 stage.

Ascorbic acid is not only an essential nutrient for humans but is also considered a powerful antioxidant component [19]. In the present study, the AsA content of *A. trifoliata* fruit had no difference between the S1 and S2 stages while it significantly increased from S2 to S3 stage, indicating that AsA is rapidly synthesized before fruit ripening. In this study, the AsA content in both clonal lines is higher than that reported in wild *A. trifoliata* subsp. *australis* and *A. trifoliata* [38,39,51,52]. Phenolics and flavonoid compounds are important secondary metabolites of plants, which can protect cells from oxidative damage, increase plant resistance to pathogens, and provide antioxidative properties for the human diet as well [53,54]. Our work showed that total phenolics and total flavonoids levels declined with the advancement of *A. trifoliata* maturity stages. Such variations in phenolics and flavonoids have been found in many fruit crops [55,56], while structural genes, light intensity, and temperature have been reported to influence phenolics and flavonoid biosynthesis [57,58].

The dry matter content is a significant attribute that could indicate the carbon incorporation at different ripening stages of fruits. The decreasing trends of dry matter percentage as maturity progressed in the two clonal lines of *Akebia* were mainly due to the indirect relation with moisture content that increased with the advancement of maturity in fruits.

The moisture content of fruits is an essential parameter that determines their flavor character, storage ability, and suitability for consumption. It is worth noting that moisture content in both clonal lines increased significantly at the S3 stage and had maximum values at the S4 stage. The increase in moisture content with maturity could be due to the hydrolysis of starch to sugars as the fruit ripeness. A similar increasing trend of moisture content with maturity has been observed in mulberry fruit and cherry fruit [17,59]. Conversely, the crude fiber, fat, and protein content in fresh fruit decreased significantly at the S3 stage, which may be due to a significant increase in the moisture content of fruit at the S3 stage.

5. Conclusions

A. trifoliata is a delicate and perishable fruit having a short market and shelf life. For the two clonal lines of *A. trifoliata*, 'Luqing' has bigger fruit than 'Luyu', while the fruits' biochemical and nutritional attributes have no significant difference between them. This study suggested that the stage of maturity had a significant effect on physiochemical parameters and nutritional properties during *A. trifoliata* fruit on-vine ripening, with dramatic changes occurring at the S3 stage of ripening. Obviously, the values of firmness, TSS, fructose, glucose, sucrose, starch, ascorbic acid, total phenolics, and total flavonoids showed remarkable changes during the transition to physiological maturity. The earlier the harvest before fruit cracking, the harder the fruit is, more suitable for long-distance transportation and the longer shelf life. However, considering the nutrients content, sugars content, pulp texture, and other physiological parameters, the S3 maturity stage was considered a proper harvesting stage for long-distance transport of *A. trifoliata* fruit to have desirable fruit quality and consumer acceptability.

Author Contributions: Conceptualization, S.Z. and H.H.; methodology, P.G.; investigation, T.J.; writing—original draft preparation, S.Z.; writing—review and editing, H.H. All authors have read and agreed to the published version of the manuscript.

Funding: This research was funded by the Plant Germplasm Innovation Program, Biological Resources Programme, The Chinese Academy of Sciences (KFJ-BRP-007-001), and Youth Foundation of Lushan Botanical Garden, The Chinese Academy of Sciences (2021 ZWZX07).

Institutional Review Board Statement: Not applicable.

Informed Consent Statement: Not applicable.

Data Availability Statement: Not applicable.

Conflicts of Interest: The authors declare no conflict of interest.

References

1. Li, L.; Yao, X.H.; Zhong, C.H.; Chen, X.Z.; Huang, H.W. *Akebia*: A Potential New Fruit Crop in China. *Hortscience* **2010**, *45*, 4–10. [CrossRef]
2. Wang, D.; Li, F.; Yan, J.; Zhong, H. Study and application of nutritional components of wild plant *Akebia trifoliata* Var *australis* (Diels) Rehd. *Amino Acids Biot. Resour.* **2004**, *26*, 16–17.
3. Zhong, C.H.; Bu, F.W.; Wang, Z.Y.; Peng, D.F. Fruit development and biological characteristics of seedling progenies of *Akebia trifoliata*. *Hunan Agric. Sci.* **2006**, *1*, 27–29.
4. Deng, A.; Li, H.; Wang, Y.; Xie, P.; Cheng, L.; Peng, C. Study on fabrication of *Akebia trifoliata* fruit tea. *Guangzhou Chem. Ind.* **2018**, *46*, 120–123.
5. Luo, M.; Zhou, D.D.; Shang, A.; Gan, R.Y.; Li, H.B. Influences of Microwave-Assisted Extraction Parameters on Antioxidant Activity of the Extract from *Akebia trifoliata* Peels. *Foods* **2021**, *10*, 1432. [CrossRef]
6. Lu, W.L.; Yang, T.; Song, Q.J.; Fang, Z.Q.; Pan, Z.Q.; Liang, C.; Jia, D.W.; Peng, P.K. *Akebia trifoliata* (Thunb.) Koidz Seed Extract inhibits human hepatocellular carcinoma cell migration and invasion in vitro. *J. Ethnopharmacol.* **2019**, *234*, 204–215. [CrossRef]
7. Ouyang, J.K.; Dong, L.M.; Xu, Q.L.; Wang, J.; Liu, S.B.; Qian, T.; Yuan, Y.F.; Tan, J.W. Triterpenoids with α-glucosidase inhibitory activity and cytotoxic activity from the leaves of *Akebia trifoliata*. *RSC Adv.* **2018**, *8*, 40483–40489. [CrossRef]
8. Shin, S.; Son, D.; Kim, M.; Lee, S.; Roh, K.B.; Ryu, D.; Lee, J.; Jung, E.; Park, D. Ameliorating Effect of *Akebia quinata* Fruit Extracts on Skin Aging Induced by Advanced Glycation End Products. *Nutrients* **2015**, *7*, 9337–9352. [CrossRef]

9. Song, D.H.; Kim, G.J.; Chung, K.H.; Lee, K.J.; An, J.H. Ormosanine from *Akebia quinata* suppresses ethanol-induced inflammation and apoptosis and activates antioxidants via the mitogen activated protein kinase signaling pathway. *J. Funct. Foods* **2018**, *48*, 357–366. [CrossRef]
10. Sung, Y.Y.; Kim, D.S.; Kim, H.K. *Akebia quinata* extract exerts anti-obesity and hypolipidemic effects in high-fat diet-fed mice and 3 T3-L1 adipocytes. *J. Ethnopharmacol.* **2015**, *168*, 17–24. [CrossRef]
11. Wang, X.Y.; Yu, N.X.; Peng, H.L.; Hu, Z.Y.; Sun, Y.; Zhu, X.M.; Jiang, L.; Xiong, H. The profiling of bioactives in *Akebia trifoliata* pericarp and metabolites, bioavailability and in vivo anti-inflammatory activities in DSS-induced colitis mice. *Food Funct.* **2019**, *10*, 3977–3991. [CrossRef] [PubMed]
12. Zou, S.Y.; Yao, X.H.; Zhong, C.H.; Zhao, T.T.; Huang, H.W. Genetic analysis of fruit traits and selection of superior clonal lines in *Akebia trifoliate* (Lardizabalaceae). *Euphytica* **2018**, *214*, 111. [CrossRef]
13. Zou, S.Y.; Yao, X.H.; Zhong, C.H.; Zhao, T.T.; Huang, H.W. Effectiveness of recurrent selection in *Akebia trifoliata* (Lardizabalaceae) breeding. *Sci. Hortic.* **2019**, *246*, 79–85. [CrossRef]
14. Huang, H.; Liang, J.; Tan, Q.; Ou, L.F.; Li, X.L.; Zhong, C.H.; Huang, H.L.; Moller, I.M.; Wu, X.J.; Song, S.Q. Insights into triterpene synthesis and unsaturated fatty-acid accumulation provided by chromosomal-level genome analysis of *Akebia trifoliata* subsp. *australis*. *Hortic. Res.* **2021**, *8*, 33. [CrossRef]
15. Zozio, S.; Servent, A.; Cazal, G.; Mbeguie-A-Mbeguie, D.; Ravion, S.; Pallet, D.; Abel, H. Changes in antioxidant activity during the ripening of jujube (*Ziziphus mauritiana* Lamk). *Food Chem.* **2014**, *150*, 448–456. [CrossRef]
16. Aubert, C.; Bruaut, M.; Chalot, G.; Cottet, V. Impact of maturity stage at harvest on the main physicochemical characteristics, the levels of vitamin C, polyphenols and volatiles and the sensory quality of Gariguette strawberry. *Eur. Food Res. Technol.* **2021**, *247*, 37–49. [CrossRef]
17. Nayab, S.; Razzaq, K.; Ullah, S.; Rajwana, I.A.; Amin, M.; Faried, H.N.; Akhtar, G.; Khan, A.S.; Asghar, Z.; Hassan, H.; et al. Genotypes and harvest maturity influence the nutritional fruit quality of mulberry. *Sci. Hortic.* **2020**, *266*, 109311. [CrossRef]
18. Fawole, O.A.; Opara, U.L.; Theron, K.I. Chemical and Phytochemical Properties and Antioxidant Activities of Three Pomegranate Cultivars Grown in South Africa. *Food Bioprocess Technol.* **2012**, *5*, 2934–2940. [CrossRef]
19. Lee, S.K.; Kader, A.A. Preharvest and postharvest factors influencing vitamin C content of horticultural crops. *Postharvest Biol. Technol.* **2000**, *20*, 207–220. [CrossRef]
20. Opara, U.L.; Al-Ani, M.R. Antioxidant contents of pre-packed fresh-cut versus whole fruit and vegetables. *Br. Food J.* **2010**, *112*, 797–810. [CrossRef]
21. Leonardi, C.; Ambrosino, P.; Esposito, F.; Fogliano, V. Antioxidative activity and carotenoid and tomatine contents in different typologies of fresh consumption tomatoes. *J. Agric. Food Chem.* **2000**, *48*, 4723–4727. [CrossRef]
22. Song, J.X.; Bi, J.F.; Chen, Q.Q.; Wu, X.Y.; Lyu, Y.; Meng, X.J. Assessment of sugar content, fatty acids, free amino acids, and volatile profiles in jujube fruits at different ripening stages. *Food Chem.* **2019**, *270*, 344–352. [CrossRef] [PubMed]
23. Wang, C.T.; Cheng, D.; Cao, J.K.; Jiang, W.B. Antioxidant capacity and chemical constituents of Chinese jujube (*Ziziphus jujuba* Mill.) at different ripening stages. *Food Sci. Biotechnol.* **2013**, *22*, 639–644. [CrossRef]
24. Hussain, S.B.; Anjum, M.A.; Hussain, S.; Ejaz, S.; Ahmed, M. Physico-Chemical Profiling of Promising Sweet Orange Cultivars Grown under Different Agro-Climatic Conditions of Pakistan. *Erwerbs-Obstbau* **2017**, *59*, 315–324. [CrossRef]
25. Razzaq, K.; Khan, A.S.; Malik, A.U.; Shahid, M. Ripening period influences fruit softening and antioxidative system of 'Samar Bahisht Chaunsa' mango. *Sci. Hortic.* **2013**, *160*, 108–114. [CrossRef]
26. Zhao, Y.T.; Zhu, X.; Hou, Y.Y.; Pan, Y.F.; Shi, L.; Li, X.H. Effects of harvest maturity stage on postharvest quality of winter jujube (*Zizyphus jujuba* Mill. cv. Dongzao) fruit during cold storage. *Sci. Hortic.* **2021**, *277*, 109778. [CrossRef]
27. Abdullahi, I.I.; Abdullahi, N.; Muhammad Abdu, A.; Ibrahim, A.S. Proximate, Mineral and Vitamin Analysis of Fresh and Canned Tomato. *Biosci. Biotechnol. Res. Asia* **2016**, *13*, 1163–1169. [CrossRef]
28. Imran, M.; Khan, H.; Shah, M.; Khan, R.; Khan, F. Chemical composition and antioxidant activity of certain Morus species. *J. Zhejiang Univ. Sci. B* **2010**, *11*, 973–980. [CrossRef]
29. Zhao, Y.T.; Zhu, X.; Hou, Y.Y.; Wang, X.Y.; Li, X.H. Postharvest nitric oxide treatment delays the senescence of winter jujube (*Zizyphus jujuba* Mill. cv. Dongzao) fruit during cold storage by regulating reactive oxygen species metabolism. *Sci. Hortic.* **2020**, *261*, 109009. [CrossRef]
30. Mahmood, T.; Anwar, F.; Afzal, N.; Kausar, R.; Ilyas, S.; Shoaib, M. Influence of ripening stages and drying methods on polyphenolic content and antioxidant activities of mulberry fruits. *J. Food Meas. Charact.* **2017**, *11*, 2171–2179. [CrossRef]
31. Khodabakhshian, R.; Emadi, B.; Khojastehpour, M.; Golzarian, M.R. Determining quality and maturity of pomegranates using multispectral imaging. *J. Saudi. Soc. For. Agric. Sci.* **2015**, *16*, 322–331. [CrossRef]
32. Niu, J.; Shi, Y.L.; Huang, K.Y.; Zhong, Y.C.; Chen, J.; Sun, Z.M.; Luan, M.B.; Chen, J.H. Integrative transcriptome and proteome analyses provide new insights into different stages of *Akebia trifoliata* fruit cracking during ripening. *Biotechnol. Biofuels* **2020**, *13*, 149. [CrossRef]
33. Wu, J.; Xu, Z.; Zhang, Y.; Chai, L.; Yi, H.; Deng, X. An integrative analysis of the transcriptome and proteome of the pulp of a spontaneous late-ripening sweet orange mutant and its wild type improves our understanding of fruit ripening in citrus. *J. Exp. Bot.* **2014**, *65*, 1651–1671. [CrossRef] [PubMed]

34. Michailidis, M.; Karagiannis, E.; Bazakos, C.; Tanou, G.; Ganopoulos, I.; Molassiotis, A. Genotype- and tissue-specific metabolic networks and hub genes involved in water-induced distinct sweet cherry fruit cracking phenotypes. *Comput. Struct. Biotechnol. J.* **2021**, *19*, 5406–5420. [CrossRef] [PubMed]
35. Chen, J.; Duan, Y.; Hu, Y.; Li, W.; Sun, D.; Hu, H.; Xie, J. Transcriptome analysis of atemoya pericarp elucidates the role of polysaccharide metabolism in fruit ripening and cracking after harvest. *BMC Plant Biol.* **2019**, *19*, 219. [CrossRef] [PubMed]
36. Saquet, A.A. Storage of pears. *Sci. Hortic.* **2019**, *246*, 1009–1016. [CrossRef]
37. Niu, J.; Sun, Z.M.; Shi, Y.L.; Huang, K.Y.; Zhong, Y.C.; Chen, J.; Chen, J.H.; Luan, M.B. Comparative Analysis of *Akebia trifoliata* Fruit Softening at Different Flesh Ripening Stages Using Tandem Mass Tag Technology. *Front. Nutr.* **2021**, *8*, 684271. [CrossRef]
38. Wang, Y.J.; Ao, W.C.; He, X.S.; Duan, W.H.; Xin, W.N.; Gong, C. Comparative study of physical and chemical components of *Akebia trifoliata* from 9 places of production in China. *J. West China For. Sci.* **2016**, *45*, 43–48. [CrossRef]
39. He, X.S.; Gong, C.; Huang, J.J.; Ao, W.C.; Zuo, J.L.; Wang, Y.J. Comparison of differences between fruit economic characteristics in *Akebia trifoliata* (Thunb.) Koidz. Subsp. *australis* (Diels) T. Shimizu from different geographical provenances. *Non-Wood For. Res.* **2017**, *35*, 36–42.
40. Mahmood, T.; Anwar, F.; Abbas, M.; Boyce, M.C.; Saari, N. Compositional Variation in Sugars and Organic Acids at Different Maturity Stages in Selected Small Fruits from Pakistan. *Int. J. Mol. Sci.* **2012**, *13*, 1380–1392. [CrossRef]
41. Riaz, M.; Zamir, T.; Rashid, N.; Jamil, N.; Rizwan, S.; Masood, Z.; Mushtaq, A.; Tareen, H.; Khan, M.; Ali, M. Comparative study of nutritional quality of orange (*Citrus sinensis*) at different maturity stages in relation to significance for human health. *Am. Eur. J. Toxicol. Sci.* **2015**, *7*, 209–213.
42. Sanchez, E.M.; Calin-Sanchez, A.; Carbonell-Barrachina, A.A.; Melgarejo, P.; Hernandez, F.; Martinez-Nicolas, J.J. Physicochemical characterisation of eight Spanish mulberry clones: Processing and fresh market aptitudes. *Int. J. Food Sci. Technol.* **2014**, *49*, 477–483. [CrossRef]
43. Rolland, F.; Baena-Gonzalez, E.; Sheen, J. Sugar sensing and signaling in plants: Conserved and novel mechanisms. *Annu. Rev. Plant Biol.* **2006**, *57*, 675–709. [CrossRef]
44. Araya, T.; Noguchi, K.; Terashima, I. Effects of carbohydrate accumulation on photosynthesis differ between sink and source leaves of *Phaseolus vulgaris* L. *Plant Cell Physiol.* **2006**, *47*, 644–652. [CrossRef] [PubMed]
45. Cao, Y.; Xiong, D.S.; Zhu, J.T.; Yu, Z.H.; Li, G.Z. Study on the respiration physiology of Akebia trifoliate fruit and the suitable storage conditions. *J. Fruit Sci.* **2003**, *20*, 512–514.
46. Basson, C.E.; Groenewald, J.H.; Kossmann, J.; Cronje, C.; Bauer, R. Sugar and acid-related quality attributes and enzyme activities in strawberry fruits: Invertase is the main sucrose hydrolysing enzyme. *Food Chem.* **2010**, *121*, 1156–1162. [CrossRef]
47. Glew, R.H.; Ayaz, F.A.; Sanz, C.; Vanderjagt, D.J.; Huang, H.S.; Chuang, L.T.; Strnad, M. Changes in sugars, organic acids and amino acids in medlar (*Mespilus germanica* L.) during fruit development and maturation. *Food Chem.* **2003**, *83*, 363–369. [CrossRef]
48. Wu, B.H.; Liu, H.F.; Guan, L.; Fan, P.G.; Li, S.H. Carbohydrate metabolism in grape cultivars that differ in sucrose accumulation. *Vitis* **2011**, *50*, 51–57.
49. Zhang, Y.Z.; Li, P.M.; Cheng, L.L. Developmental changes of carbohydrates, organic acids, amino acids, and phenolic compounds in 'Honeycrisp' apple flesh. *Food Chem.* **2010**, *123*, 1013–1018. [CrossRef]
50. Sturm, K.; Koron, D.; Stampar, F. The composition of fruit of different strawberry varieties depending on maturity stage. *Food Chem.* **2003**, *83*, 417–422. [CrossRef]
51. Zhong, W.M.; Ma, Y.H. Fruit nutrition quality of a good *Akebia trifoliate* plant. *Guizhou Agric. Sci.* **2015**, *43*, 140–142.
52. Wan, M.C.; Liu, X.W.; Ban, X.C.; Luo, K.M.; Shi, L.J.; Zhang, C.J.; Yang, Y.Y.; Li, Z.J.; He, Y.C. The fruit character and nutrition composition of *Akebia trifoliata* (Thunb.) Koidz under the cultivation condition. *Guizhou Agric. Sci.* **2008**, *36*, 121–122.
53. Jin, P.; Wu, X.; Xu, F.; Wang, X.L.; Wang, J.; Zheng, Y.H. Enhancing Antioxidant Capacity and Reducing Decay of Chinese Bayberries by Essential Oils. *J. Agric. Food Chem.* **2012**, *60*, 3769–3775. [CrossRef] [PubMed]
54. Kirigia, D.; Winkelmann, T.; Kasili, R.; Mibus, H. Nutritional composition in African nightshade (*Solanum scabrum*) influenced by harvesting methods, age and storage conditions. *Postharvest Biol. Technol.* **2019**, *153*, 142–151. [CrossRef]
55. Fawole, O.A.; Opara, U.L. Changes in physical properties, chemical and elemental composition and antioxidant capacity of pomegranate (cv. Ruby) fruit at five maturity stages. *Sci. Hortic.* **2013**, *150*, 37–46. [CrossRef]
56. Fawole, O.A.; Opara, U.L. Effects of maturity status on biochemical content, polyphenol composition and antioxidant capacity of pomegranate fruit arils (cv. 'Bhagwa'). *S. Afr. J. Bot.* **2013**, *85*, 23–31. [CrossRef]
57. Zhang, Q.; Wang, L.L.; Liu, Z.G.; Zhao, Z.H.; Zhao, J.; Wang, Z.T.; Zhou, G.F.; Liu, P.; Liu, M.J. Transcriptome and metabolome profiling unveil the mechanisms of *Ziziphus jujuba* Mill. peel coloration. *Food Chem.* **2020**, *312*, 125903. [CrossRef] [PubMed]
58. Zoratti, L.; Karppinen, K.; Escobar, A.L.; Haggman, H.; Jaakola, L. Light-controlled flavonoid biosynthesis in fruits. *Front. Plant Sci.* **2014**, *5*, 534. [CrossRef]
59. Mahmood, T.; Anwar, F.; Bhatti, I.A.; Iqbal, T. Effect of maturity on proximate composition, phenolics and antioxidant attributes of cherry fruit. *Pak. J. Bot.* **2013**, *45*, 909–914.

horticulturae

Article

Plant Cover Stimulates Quicker Dry Matter Accumulation in "Early" Potato Cultivars without Affecting Nutritional or Sensory Quality

Nikola Major [1,*], Smiljana Goreta Ban [1,*], Josipa Perković [1], Dragan Žnidarčič [2], Anita Silvana Ilak Peršurić [3], Milan Oplanić [3], Gvozden Dumičić [4], Branimir Urlić [4] and Dean Ban [1]

1 Department of Agriculture and Nutrition, Institute of Agriculture and Tourism, 52440 Poreč, Croatia; josipa@iptpo.hr (J.P.); dean@iptpo.hr (D.B.)
2 Biotechnical Centre Naklo, 4202 Naklo, Slovenia; dragan.znidarcic@bc-naklo.si
3 Department of Economics and Agricultural Development, Institute of Agriculture and Tourism, 52440 Poreč, Croatia; anita@iptpo.hr (A.S.I.P.); milan@iptpo.hr (M.O.)
4 Department of Plant Sciences, Institute for Adriatic Crops and Karst Reclamation, 21000 Split, Croatia; gdumicic@krs.hr (G.D.); branimir.urlic@krs.hr (B.U.)
* Correspondence: nikola@iptpo.hr (N.M.); smilja@iptpo.hr (S.G.B.)

Abstract: "Early" potato crops are grown in the Mediterranean basin and are marketed from March to June, well before main-crop potato in the spring–summer growth period. Different growing technologies have been implemented to enhance potato "earliness" to achieve a better market price, but at the same time, the applied technologies may influence yield and quality of the 'early potato'. The main goal of this study was to investigate differences in "early" potato nutritional and sensory characteristics after oil-frying influenced by location and plant covering in five potato cultivars. The present investigation was carried out at two planting locations during two seasons. The application of a plant cover significantly increased the potato tubers' dry matter, starch, and sugar content in the second season. Sensory analysis of the oil-fried "early" potatoes revealed no differences between potatoes grown with or without the plant cover. We also observed significantly higher dry matter content in potatoes grown at the Split location in the second year, while no differences in the sensory scores between oil-fried potatoes grown at the investigated locations were observed. By employing a plant cover or by choosing a warmer planting location the desired potato maturity level could be reached in less time, and one could more effectively exploit the "early" potato market. By employing such techniques there should be no loss in yield, nutritional or sensory quality of potato tubers.

Keywords: potato; location; plant cover; mineral content; dry matter; *Solanum tuberosum* L.

Citation: Major, N.; Goreta Ban, S.; Perković, J.; Žnidarčič, D.; Peršurić, A.S.I.; Oplanić, M.; Dumičić, G.; Urlić, B.; Ban, D. Plant Cover Stimulates Quicker Dry Matter Accumulation in "Early" Potato Cultivars without Affecting Nutritional or Sensory Quality. *Horticulturae* **2022**, *8*, 364. https://doi.org/10.3390/horticulturae8050364

Academic Editor: Massimiliano Renna

Received: 17 March 2022
Accepted: 19 April 2022
Published: 21 April 2022

1. Introduction

Global potato production has a weight of 359 million tones, followed by other vegetables such as tomatoes, watermelon, and sweet potato [1]. Wide acceptance of potato as an edible crop and its adaptability to different climates led to its high consumption worldwide [2]. The cultivated potato belongs largely to one botanical species, with thousands of cultivars differing in size, shape, color, texture, cooking characteristics, and taste [3].

Early potatoes are defined as "potatoes harvested before they are completely mature, marketed immediately after harvesting and of which skin can be easily removed without peeling" [4]. "Early" potato crops are widely grown in the Mediterranean basin and are marketed from March to June, well before main-crop potatoes grown in the spring–summer growth period [5].

A raw potato tuber is rich in nutrients, vitamins and minerals that are essential health properties [6,7]. Potatoes also contain important amounts of other phytonutrients, including protein, fiber, complex carbohydrates, carotenoids, and are one of the primary suppliers of polyphenols in the human diet [8]. A medium-sized raw potato contains high levels of potassium and nearly half the daily adult requirement of vitamin C [9].

Modern potato cultivars are bred to fulfill consumer demands or for further processing in the food industry. They offer a wide range of cooking characteristics suitable for various types of thermal processing and are adjusted to highlight other ingredients or be consumed as a complete meal [3]. Potato is considered one of the most popular vegetables worldwide and can be consumed in many ways, including being boiled, roasted, baked, fried, or steamed [10]. "Early" tubers contain higher amounts of phytonutrients compared to mature tubers and are considered a delicacy when steamed or boiled [11]. In Croatia, "early" potatoes are regularly consumed oil-fried and are considered to be a "special" dish. In Italy, "early" potatoes are consumed in 51.7% of households, which spend 9.1% of their income on potatoes [12].

Technology for growing potatoes is adjusted to eliminate stress that may be caused by water and nutrient supply, but the environmental conditions during crop production have been reported to have significant impact on growth, yield, and quality of harvested potato [13].

In some parts of the Mediterranean basin, early potato production could be challenged by environmental temperatures in the first vegetative stages [14]. Enhancement of potato earliness could be achieved either by choosing early cultivars [15] or by using row covers [16]. Row covers are especially useful as protection from possible early spring frosts, since both covering and mulching with artificial materials enhances average air temperatures on the soil surface as well as 10 cm into the soil [17]. The application of different agricultural practices could affect the quality of potato [18].

Potato heat processing is responsible for the creation of characteristic sensory profiles of potato products [19]. During potato oil-frying, high temperatures cause partial evaporation of water and oil adsorption, replacing some of the lost water [20]. Therefore, potatoes with higher dry matter content (20–22%) and lower reducing sugar content (up to 3 mg/g) are preferred for oil-frying because they develop better texture, higher yields, and lower oil absorption during the process [21].

The objective of this study was to investigate the impact of plant covering and different planting locations on the early accumulation of dry matter in five potato cultivars, as well as the effects on the "early" potato nutritional properties and sensory characteristics after oil-frying.

2. Materials and Methods

2.1. Treatments and Experimental Design

A two-year field trial was conducted in the Mediterranean region of Croatia at two locations—Pula (44°51′ N, 13°51′ E) and Split (43°30′ N, 16°30′ E). At the first location, in Pula, planting was conducted on February 13th and February 23rd, and in Split it was conducted on February 26th and February 25th in the first and second years, respectively. A two-factorial experiment was designed as split-plot in three replications. The first factor being the main plot with different potato cultivars and the second factor being the sub-plot with or without direct covering. The main plot was 3.5 m in width and 10.5 m in length (36.75 m^2) and consisted of four rows.

Potato cultivars used in the first trial year were: Adora, Berber, Jerla, and Red Scarlet. In the second trial year, Adora was replaced by Vivaldi. Adora is classified as a very early cultivar, Berber and Jaerla as early, and Read Scarlet and Vivaldi are medium early cultivars (HZPC, Holland B.V., Joure, The Netherlands). Adora, Berber, Jerla and Vivaldi have a yellow skin color, whereas Red Scarlet has red skin color. The second factor in the experiment was usage of non-woven covers and comparisons with uncovered crops. Material used for cover was Lutrasil® (Freudenberg, Weinheim, Germany), 17 g/m^2 in weight. At the time of planting, covers were placed on the soil surface and removed 50 and 30 days after germination in the first and second year, respectively. Whole potato tubers (diameter 28–35 mm) were planted at a distance of 0.75 m between and 0.35 m within rows (4.08 plants/m^2). Only the two middle rows were used for sampling.

2.2. Soil Properties and Agricultural Management Practice

At Pula, the experiments were carried out on Terra rossa, a type of red clay soil produced by the weathering of limestone and dolomites, with a soil texture of 40.7% sand, 26.4% silt, and 30.9% clay. Other soil characteristics were as follows: pH 7.2, organic matter 1.5%, P_2O_5 45.8 mg/100 g soil, and K_2O 16.2 mg/100 g soil. The experiment in Split was conducted on an anthropogenic, carbonate, deep clay-loamy soil on sandstone and marly soil, with a soil texture of 28% sand, 32.6% silt and 39.4% clay. Other soil characteristics were as follows: pH 7.6, organic matter 2.1%, P_2O_5 6.0 mg/100 g soil, and K_2O 31.8 mg/100 g soil. The average air temperatures in Pula and Split during both seasons are presented in Figure 1.

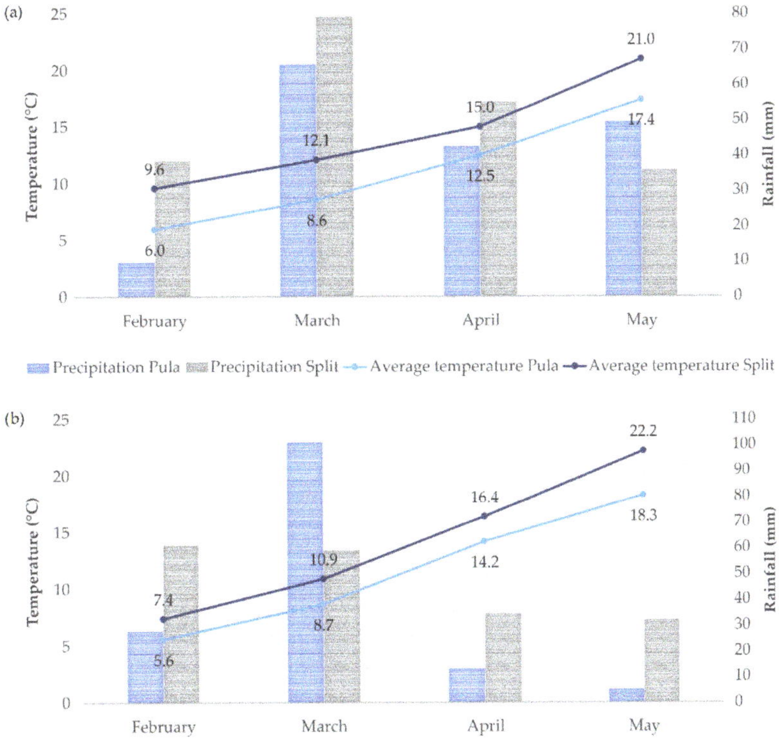

Figure 1. Average air temperatures and rainfall in the (**a**) first and (**b**) second year at the Pula and Split locations.

The experimental fields in both seasons were plowed to a depth of 30 cm, and 735 kg ha^{-1} of mineral fertilizer 7N-6.1P-17.4K was incorporated into the soil. Subsequently, herbicide (Sencor WG70 at 1 kg ha^{-1}, active substance metribuzin) was applied to the soil with a rotary cultivator. The inter rows were hoed once before plants reached 15 cm in height and the plants were mound before canopy closure. A fungicide (Ridomil Gold at 2.5 kg ha^{-1}, active substance metalaxyl m and mancozeb) and insecticide (Actar 25 WG 0.07 kg ha^{-1}, active substance thiamethoxam) were applied according to crop monitoring once or twice during a season. The crop was irrigated by drip irrigation (T-tape; T-Systems International, San Diego, CA, USA), with an emitter spacing at 20 cm (capacity of 1 L h^{-1}) as needed.

Tubers were harvested manually, 60 days after germination of covered plants, which in Pula was on June 2nd and May 25th, and in Split on May 27th and May 25th in the first and second seasons, respectively.

Only healthy, well-shaped tubers of marketable size (diameter 35–80 mm) were used for chemical analysis and sensory evaluation.

2.3. Tuber Mineral Composition

The tubers were washed, dried with tissue paper, weighed, and diced. Finally, a portion was oven-dried at 65 °C until constant weight. The dehydrated material was ground, passed through a 1 mm sieve, and subsequently used for the determination of minerals. In this study, five mineral elements (N, P, K, Mg, Ca) were analyzed. Approximately 0.5 g of the oven-dried material was mineralized in a muffle furnace at 550 °C for 12 h. After cooling, the resulting ash was dissolved using hydrochloric acid (HCl). Phosphorus was determined with the vanadate-molybdate yellow color method using a UV–visible spectrophotometer (Cary 50 Scan, Varian, Palo Alto, CA, USA) at 420 nm. Potassium was determined using flame photometry (model 410 flame photometer, Sherwood Scientific, Cambridge, UK). The other minerals (Ca and Mg) were determined by atomic absorption spectrometry (Spectraa 220, Varian, Palo Alto, CA, USA). The nitrogen content was determined by Kjeldahl digestion with a Kjeltec System 1026 (Tecator, Höganas, Sweden). The tuber protein content was obtained by multiplying the obtained results with the factor 6.25. Quantification of individual minerals in the samples was performed using calibration curves. Data are expressed as g or mg kg^{-1} of dry weight (DW).

2.4. Soluble Carbohydrates and Starch Analysis

The content of soluble carbohydrates was analyzed separately by high-performance liquid chromatography (HPLC). The procedure was fully described by Štampar et al. [22]. Briefly, samples for sugar determination (glucose, fructose, and sucrose) were prepared from 1.5 kg of fresh tubers and then divided into three subsamples. They were converted into pulp individually by a mixer and homogenized with Ultra-Turrax T-25 (Ika-Labortechnik, Staufen, Germany). The tuber puree (10 g) was diluted to 50 mL with ultra-pure water and centrifuged at 6000× g for 15 min. The extract was filtered through 0.45 μm Millipore filters prior to analysis. Sugars analyses were performed by injecting 20 μL of the sample extract on an Aminex HPX-87C column held at 85 °C at a flow rate of 0.6 mL/min, with ultra-pure water as eluent. Sugars present in each sample were identified and quantified against retention times and peak areas of analytical standards, respectively. Standards for sucrose, glucose and fructose were obtained from Fluka Chemical (New York, NY, USA).

The percentage of starch was calculated by specific weight. Briefly, the weight of 10 potato tubers was determined both in air and after their immersion in water. The specific weight was calculated according to Vasanthan et al. [23].

2.5. Vitamin C

The concentration of L-ascorbic acid was measured as reported by Tausz et al. [24]. Briefly, L-ascorbic acid was extracted from 200 mg of lyophilized tuber tissue with 5 mL of cold 1.5% (*w*/*v*) metaphosphoric acid containing 1mM EDTA. The mixture was homogenized (T-25 Ultra-Turrax) for 50 s in an ice bath. All extraction procedures were performed in dim light. After filtration through 0.2 μm Minisart SRP 15 filter, extracts were subjected to isocratic HPLC analysis on a Spectra-Physics HPLC system equipped with Spectra Focus UV-VIS detector and a Lichrosorb RP-8 (250 × 4.6 mm) column with an Lichrosorb RP-8 (50 × 4.6 mm) precolumn (Alltech Associaties, Inc., Deerfield, MA, USA) using methanol/water (1/3, *v*/*v*) containing 1 mM hexadecylammoniumbromide and 0.05% (*w*/*v*) sodium dihydrogen phosphate monohydrate (pH 3.6) as solvent, at a flow rate of 1 mL/min, with a run time of 20 min, and photometric detection at 248 nm. Identification of ascorbic acid was achieved by comparing the retention time as well as by the addition of a standard. The concentration of ascorbic acid was obtained with the external standard method.

2.6. Sensory Evaluation

Potato tubers of uniform shape and size were used for sensory evaluation within 5 days after harvest. The tubers were peeled of, cut into strips, and fried in hot sunflower oil at 180 °C for 15 min. The fried strips were drained for 30 to 60 s, salted, and served hot to panelists.

Sensory attributes were assessed by a trained sensory panel (15 trained panelists). The panels consisted of 60% women and 40% men, aged 30 to 48 years. Panelists were trained in flavor and texture attributes of fried potatoes. Samples were coded and served to panelists randomly in white shallow ceramic plates. Between samples, panelist were served with room temperature tap water.

The sensory attributes were assessed on a 5-point numerical scale with overall impression being the only exception (10 point). Sensory descriptors were as follows: odor intensity (1—no odor, 5—excellent), crispness (1—too crispy or not crispy, 5—optimal), texture (1—bad, 5—optimal), overall taste (1—unacceptable, 5—excellent), overall appearance (1—bad, 5—excellent), and overall impression (1—bad, 10—excellent).

2.7. Data Analysis

Factorial analysis of variance (ANOVA) was applied on all data and the data set was split according to harvest year and analyzed separately. Means were separated by Tukey's high significant difference (HSD) test where the F-test was significant. Additionally, partial least square (PLS) classification was performed on the data. Statistica version 13.4 (Tibco Inc., Palo Alto, CA, USA) was used for all statistical analyses.

3. Results

3.1. Yield and Dry Matter

Total yield was found to be significantly different according to planting location, where higher yields were obtained in Pula compared to Split in both harvest years (Tables 1 and 2). The application of the plant cover was not significantly different for the total tuber yield in both years (Tables 1 and 2). In the first year, yield was significantly higher for Berber and Red Scarlet compared to Jerla, while in the second year, Berber and Red Scarlet had significantly higher yield compared to Jerla and Vivaldi (Tables 1 and 2).

Dry matter content was found to be significantly different between planting locations only in the second harvest year, where tubers grown in Split had significantly higher dry matter content than tubers grown in Pula (Tables 1 and 2). The interaction between cultivar and location was significant for dry matter content in the second year of the study (Table 3), where only Vivaldi achieved the significantly higher dry matter content in Split compared to Pula (Table 4). Similarly, the effect of the plant cover was significant only in the second harvest year, where plants grown with the cover yielded tubers with higher dry matter content (Table 2). In both harvest years, dry matter content of tubers was influenced by cultivars, where Berber exhibited the highest dry matter content followed by Adora and Jerla in the first year, and Vivaldi and Jerla in the second year, with slight differences (Tables 1 and 2).

Table 1. Nutritional properties of potato tubers affected by location, plant cover and cultivar in the first year of growing.

Nutritional Parameters	Location			Cover †			Cultivar				
	Split	Pula	p-Value	Yes	No	p-Value	Adora	Berber	Jerla	Red Scarlet	p-Value
Yield (t/ha)	30.0 ± 7.8	33.1 ± 7.2	<0.001 *	15.3 ± 8.5	16.1 ± 7.3	0.134	29.6 ± 4.4 ab	34.5 ± 8.1 b	28.0 ± 5.0 a	34.0 ± 9.8 b	<0.001 *
Dry matter (%)	20.9 ± 1.1	20.5 ± 1.2	0.281	21.1 ± 1.2	20.3 ± 1.0	0.102	20.7 ± 1.0 ab	21.5 ± 1.1 b	20.5 ± 1.4 ab	20.1 ± 0.6 a	0.003 *
Starch (%)	13.8 ± 1.0	13.3 ± 1.2	0.286	14.0 ± 1.1	13.2 ± 1.0	0.121	13.6 ± 1.0 ab	14.3 ± 1.1 b	13.4 ± 1.4 ab	13.0 ± 0.6 a	0.003 *
Sucrose (g/100 g DW)	5.3 ± 1.2	5.0 ± 1.6	0.082	5.9 ± 1.4	4.5 ± 1.2	0.001 *	4.5 ± 1.4 a	6.0 ± 1.4 b	4.3 ± 0.9 a	6.0 ± 1.1 b	<0.001 *
Glucose (g/100 g DW)	0.93 ± 0.13	0.91 ± 0.12	0.672	0.94 ± 0.15	0.89 ± 0.09	0.393	0.94 ± 0.10 ab	0.97 ± 0.11 b	0.83 ± 0.10 a	0.93 ± 0.15 ab	0.016 *
Fructose (g/100 g DW)	0.82 ± 0.07	0.83 ± 0.11	0.815	0.83 ± 0.08	0.82 ± 0.10	0.788	0.84 ± 0.10	0.81 ± 0.10	0.82 ± 0.08	0.82 ± 0.10	0.921
Vitamin C (mmol/kg DW)	7.4 ± 1.6	7.5 ± 2.0	0.958	6.8 ± 1.5	8.1 ± 1.8	0.063	9.0 ± 1.4 c	7.0 ± 1.9 ab	7.7 ± 1.6 bc	6.1 ± 0.8 a	<0.001 *
Protein content (g/100 g DW)	7.9 ± 2.1	7.5 ± 1.4	0.348	7.8 ± 1.9	7.6 ± 1.7	0.558	7.2 ± 2.0	7.2 ± 1.0	8.1 ± 1.1	8.4 ± 2.5	0.244
Ca (mg/kg DW)	858 ± 405	566 ± 229	0.014 *	696 ± 359	728 ± 362	0.672	842 ± 443 b	773 ± 422 b	750 ± 209 b	485 ± 215 a	<0.001 *
Mg (mg/kg DW)	808 ± 348	771 ± 175	0.048 *	778 ± 280	801 ± 272	0.146	814 ± 257 b	855 ± 168 b	921 ± 98 b	567 ± 370 a	<0.001 *
K (g/kg DW)	19.9 ± 2.4	16.9 ± 2.2	0.037 *	18.6 ± 3.3	18.2 ± 2.1	0.700	17.2 ± 4.0 a	17.6 ± 1.8 ab	19.1 ± 1.6 ab	19.7 ± 2.1 b	0.004 *
P (g/kg DW)	2.4 ± 0.3	1.8 ± 0.7	0.013 *	2.1 ± 0.7	2.1 ± 0.5	0.876	1.6 ± 0.9 a	2.0 ± 0.4 ab	2.3 ± 0.4 b	2.5 ± 0.4 c	<0.001 *

* $p < 0.05$; Different letters indicate homogenous groups in Tukey's HSD test ($p < 0.05$); Data are presented as mean ± standard deviation (N = 3); † Cover—Lutrasil® (Freudenberg, Weinheim, Germany), 17 g/m^2 in weight.

Table 2. Nutritional properties of potato tubers affected by location, plant cover and cultivar in the second year.

Nutritional Parameters	Location			Cover †			Cultivar				
	Split	Pula	*p*-Value	Yes	No	*p*-Value	Berber	Jerla	Red Scarlet	Vivaldi	*p*-Value
Yield (t/ha)	11.4 ± 4.4	20.1 ± 7.6	0.040 *	15.3 ± 8.5	16.1 ± 7.3	0.524	19.3 ± 8.5 b	12.9 ± 4.5 a	17.8 ± 6.6 b	12.9 ± 9.4 a	<0.001 *
Dry matter (%)	21.0 ± 0.7	20.7 ± 0.7	0.032 *	21.4 ± 0.4	20.3 ± 0.5	<0.001 *	21.1 ± 0.5 b	20.8 ± 0.7 b	20.4 ± 0.6 a	21.0 ± 0.8 b	<0.001 *
Starch (%)	13.8 ± 0.6	13.6 ± 0.6	0.092	14.1 ± 0.4	13.2 ± 0.5	0.001 *	14.0 ± 0.6 b	13.6 ± 0.6 ab	13.2 ± 0.5 a	13.9 ± 0.7 b	<0.001 *
Sucrose (g/100 g DW)	5.7 ± 1.5	6.1 ± 1.2	0.217	6.5 ± 1.5	5.3 ± 0.9	0.011 *	7.1 ± 1.3 b	5.8 ± 1.2 ab	5.8 ± 1.0 ab	5.0 ± 1.1 a	<0.001 *
Glucose (g/100 g DW)	0.95 ± 0.10	0.90 ± 0.08	<0.001 *	0.96 ± 0.07	0.88 ± 0.10	<0.001 *	0.88 ± 0.11 a	0.93 ± 0.09 ab	0.99 ± 0.07 b	0.89 ± 0.08 a	0.008
Fructose (g/100 g DW)	0.88 ± 0.13	0.90 ± 0.07	0.220	0.93 ± 0.09	0.85 ± 0.10	0.003 *	0.86 ± 0.11 a	0.87 ± 0.09 ab	0.96 ± 0.10 b	0.88 ± 0.08 ab	0.044
Vitamin C (mmol/kg DW)	7.0 ± 1.6	7.5 ± 1.7	0.202	6.5 ± 1.3	8.1 ± 1.5	0.009 *	7.7 ± 2.0 b	7.4 ± 1.5 ab	6.2 ± 1.1 a	7.8 ± 1.6 b	0.044 *
Protein content (g/100 g DW)	5.9 ± 1.2	8.0 ± 1.9	0.007 *	7.1 ± 2.1	6.8 ± 1.6	0.561	6.0 ± 1.7 a	8.0 ± 2.5 b	7.3 ± 1.0 ab	6.4 ± 1.4 ab	0.009 *
Ca (mg/kg DW)	1021 ± 588	952 ± 681	0.783	1100 ± 746	873 ± 479	0.383	1035 ± 784	1202 ± 727	884 ± 325	825 ± 596	0.282
Mg (mg/kg DW)	838 ± 125	955 ± 136	0.041 *	907 ± 143	886 ± 144	0.629	877 ± 116 a	888 ± 180 ab	1012 ± 124 b	810 ± 44 a	0.002 *
K (g/kg DW)	12.4 ± 4.9	17.0 ± 2.8	0.046 *	16.0 ± 3.5	13.4 ± 5.2	0.180	14.7 ± 3.9 a	13.3 ± 5.0 a	16.6 ± 5.3 b	14.0 ± 3.8 a	0.017 *
P (g/kg DW)	2.3 ± 0.4	2.7 ± 0.4	0.018 *	2.6 ± 0.4	2.4 ± 0.5	0.102	2.3 ± 0.4 a	2.6 ± 0.5 ab	2.7 ± 0.4 b	2.4 ± 0.4 ab	0.021 *

* $p < 0.05$; Different letters indicate homogenous groups in Tukey's HSD test ($p < 0.05$); Data are presented as mean ± standard deviation (N = 3); † Cover—Lutrasil® (Freudenberg, Weinheim, Germany), 17 g/m^2 in weight.

Table 3. Interaction between location, plant cover and cultivar effects on nutritional parameters in the first and second harvest years.

Year	Interaction	Nutritional Parameters											
		Ca (mg/kg DW)	Mg (mg/kg DW)	K (g/100 g DW)	P (g/100 g DW)	Glucose (g/100 g DW)	Fructose (g/100 g DW)	Sucrose (g/100 g DW)	Vitamin C (mmol/kg DW)	Dry Matter (%)	Starch (%)	Protein Content (g/100 g DW)	Yield (t/ha)
First	Location × Cover	ns	ns	ns	ns	ns	ns	ns	ns	ns	ns	ns	ns
	Cover × Cultivar	ns	ns	ns	ns	ns	ns	ns	*	ns	ns	ns	ns
	Location × Cultivar	***	***	*	***	ns	ns	***	*	ns	ns	ns	ns
	Location × Cover × Cultivar	ns	ns	ns	ns	ns	ns	ns	ns	ns	ns	ns	ns
Second	Location × Cover	ns	ns	ns	ns	ns	ns	ns	ns	ns	ns	ns	ns
	Cover × Cultivar	ns	ns	ns	ns	ns	ns	ns	ns	ns	ns	ns	ns
	Location × Cultivar	*	ns	ns	ns	ns	ns	ns	ns	*	*	ns	ns
	Location × Cover × Cultivar	ns	ns	ns	ns	ns	ns	ns	ns	ns	ns	ns	ns

* $p < 0.05$, *** $p < 0.001$, ns—not significant.

Table 4. Nutritional properties of potato tubers affected by the interaction between location and cultivar.

Year	Location	Cultivar	Ca (mg/kg DW)	Mg (mg/kg DW)	K (g/kg DW)	P (g/kg DW)	Sucrose (g/100 g DW)	Vitamin C (mmol/kg DW)	Dry matter (%)	Starch (%)
First	Split	Adora	1184 ± 258 c	1028 ± 70 d	20.0 ± 3.5 bc	2.3 ± 0.2 bc	5.6 ± 0.8 bc	8.5 ± 1.5 bc	21.1 ± 0.5	13.9 ± 0.4
		Berber	1132 ± 205 c	1000 ± 57 d	18.3 ± 2.2 bc	2.3 ± 0.2 bc	5.6 ± 0.7 bc	6.8 ± 1.3 ab	21.6 ± 1.5	14.5 ± 1.4
		Jerla	801 ± 219 bc	977 ± 80 d	20.3 ± 0.9 bc	2.5 ± 0.4 c	4.0 ± 1.0 a	8.5 ± 1.2 bc	20.7 ± 1.3	13.6 ± 1.2
		Red Scarlet	315 ± 145 a	226 ± 11 a	20.8 ± 2.1 c	2.5 ± 0.5 c	6.2 ± 1.3 c	6.0 ± 0.5 a	20.4 ± 0.5	13.3 ± 0.4
	Pula	Adora	499 ± 290 ab	601 ± 174 b	14.3 ± 1.9 a	0.8 ± 0.7 a	3.3 ± 0.7 a	9.5 ± 1.3 c	20.4 ± 1.3	13.3 ± 1.2
		Berber	413 ± 200 ab	710 ± 93 bc	16.9 ± 0.9 ab	1.7 ± 0.2 b	6.4 ± 1.9 c	7.1 ± 2.5 ab	21.4 ± 0.5	14.1 ± 0.7
		Jerla	698 ± 203 ab	866 ± 85 cd	17.9 ± 1.1 abc	2.1 ± 0.1 bc	4.6 ± 0.7 ab	6.9 ± 1.7 ab	20.3 ± 1.6	13.2 ± 1.6
		Red Scarlet	654 ± 108 ab	908 ± 150 d	18.6 ± 1.6 bc	2.4 ± 0.2 c	5.8 ± 1.0 bc	6.3 ± 1.0 ab	19.8 ± 0.6	12.7 ± 0.6
Second	Split	Berber	651 ± 350 a	799 ± 67	12.6 ± 4.4	2.0 ± 0.4	7.0 ± 1.7	6.9 ± 1.8	21.2 ± 0.6 cd	14.1 ± 0.6 cd
		Jerla	1442 ± 866 a	770 ± 89	10.7 ± 5.3	2.4 ± 0.3	6.1 ± 0.9	7.8 ± 1.3	20.8 ± 0.7 abcd	13.6 ± 0.6 abcd
		Red Scarlet	1060 ± 353 a	981 ± 154	13.6 ± 6.2	2.4 ± 0.3	5.5 ± 1.0	6.1 ± 1.2	20.5 ± 0.5 ab	13.3 ± 0.4 ab
		Vivaldi	928 ± 450 a	801 ± 52	12.6 ± 4.3	2.4 ± 0.4	4.3 ± 0.6	7.3 ± 1.7	21.5 ± 0.7 d	14.2 ± 0.6 d
	Pula	Berber	1419 ± 936 a	955 ± 103	16.9 ± 1.5	2.5 ± 0.3	7.1 ± 0.9	8.6 ± 1.8	21.1 ± 0.5 bcd	13.9 ± 0.7 bcd
		Jerla	961 ± 523 a	1005 ± 173	16.0 ± 3.2	2.8 ± 0.6	5.6 ± 1.4	6.9 ± 1.7	20.8 ± 0.7 abc	13.7 ± 0.6 abcd
		Red Scarlet	707 ± 183 a	1043 ± 88	19.6 ± 1.4	3.0 ± 0.3	6.1 ± 1.0	6.3 ± 1.0	20.3 ± 0.6 a	13.2 ± 0.7 a
		Vivaldi	722 ± 743 a	818 ± 36	15.5 ± 3.0	2.4 ± 0.4	5.6 ± 1.2	8.3 ± 1.5	20.6 ± 0.8 abc	13.5 ± 0.6 abc

Different letters in columns indicate homogenous groups in Tukey's HSD test ($p < 0.05$); Data are presented as mean ± standard deviation (N = 3).

3.2. Carbohydrates

Location had no significant impact on potato tuber starch content in both harvest years (Tables 1 and 2). The effect of the plant cover was found to be significant in the second year, where plants with the cover yielded tubers with higher starch levels compared to plants grown without the cover (Table 2). In the first year, the difference was not statistically significant (Table 1). Similarly to dry matter, Berber had the highest starch content in the first year, closely followed by Adora, Jerla and finally Red Scarlet (Table 1). In the second year of the investigation, Berber and Vivaldi exhibited the highest starch content, followed by Jerla and Red Scarlet (Table 2).

The interaction between cultivar and location was found to be significant for the starch content in the second year (Table 3). Vivaldi achieved significantly higher starch content in Split compared to Pula (Table 4). Red Scarlet had similar starch content at both locations (Table 4). Tuber sucrose content did not differ between planting locations in both harvest years (Tables 1 and 2). The application of the plant cover had significant influence on sucrose content in both harvest years, where tubers grown with the plant cover had higher sucrose content than those grown without (Tables 1 and 2). Sucrose content also differed between cultivars in both harvest years (Tables 1 and 2). In the first year, Berber and Red Scarlet had significantly higher sucrose content compared to Jerla and Adora (Table 1) while in the second year, Berber had significantly higher sucrose content compared to the Vivaldi cultivar (Table 2). Significant interactions were observed between planting location and potato cultivar in the first year (Table 3), where the Adora cultivar showed significantly higher sucrose content in Split compared to Pula (Table 4).

Potato tuber glucose content was found to be significantly higher in tubers grown in Split in the second harvest year (Table 2). Significant interactions for sucrose content were observed between locations and cultivars in the first year (Table 3) where the sucrose content was significantly higher for Adora in Split compared to Pula (Table 4). In the first year, although higher on average in Split, there were no significant differences between planting locations (Table 1). Plants that were covered yielded tubers with significantly higher glucose levels in the second harvest year (Table 2). Significant differences in glucose content between cultivars were detected in both harvest years (Tables 1 and 2). In the first year, Berber had significantly higher glucose content compared to Jerla, and it was comparable to the other investigated cultivars (Table 1). In the second year, Red Scarlet achieved significantly higher glucose content compared to Berber (Table 2).

Fructose content in potato tubers was not influenced by planting location in both harvest years (Tables 1 and 2). Significantly higher fructose content was observed in tubers that were grown with the plant cover in the second harvest year (Table 2). In the first year, there were no significant differences in fructose content between cultivars nor the application of the plant cover (Table 1). In the second harvest year, similarly to glucose, significantly higher fructose content was detected in Red Scarlet compared to the Berber cultivar (Table 2).

3.3. Vitamin C

There were no observed differences in vitamin C content between potato tubers grown on different locations in both years (Tables 1 and 2), but significant interactions were observed between the application of the plant cover and location, as well as between location and cultivar in the first year (Table 3). In Split, Adora and Jerla exhibited significantly higher vitamin C content compared to Red Scarlet, while in Pula, Adora was most abundant in vitamin C among the investigated cultivars (Table 4). The interaction between cultivar and plant cover was highly expressed with Berber, where the use of the planting cover significantly decreased the amount of vitamin C present in its tuber compared to uncovered plants (data not shown). In the second year, significant differences were observed between tubers grown with or without plant cover, where tubers grown without the cover exhibited higher vitamin C content (Table 2). In the first year, Adora had significantly higher vitamin

C content compared to Berber and Red Scarlet cultivars (Table 1). In the second year, Berber and Vivaldi had significantly higher vitamin C content compared to Red Scarlet (Table 2).

3.4. Protein Content

There were no significant differences observed in potato tubers' protein content in the first harvest year either according to planting location, plant cover or cultivar (Table 1). Higher protein content was observed in the second year in tubers grown in Pula, while there was no statistically significant difference observed between potato tubers grown with or without plant cover (Table 2). In the second year, Jerla exhibited significantly higher protein content compared to Berber (Table 2).

3.5. Mineral Content

Potato tubers' mineral content was found to be significantly different between planting locations in both years, except for Ca in the second year (Tables 1 and 2). Significant interactions between location and potato cultivar were observed for Ca, Mg, K and P in the first year as well as for Ca in the second year (Table 3). The interaction shows that all cultivars, except Jerla, had higher Ca concentrations in Split compared to Pula in the first year, while in the second year, Tukey's post hoc test revealed no differences among cultivars and locations (Table 4). In the first year, Adora, Berber, and Jerla exhibited significantly higher Mg levels in Split compared to in Pula, while the opposite was observed for Red Scarlet (Table 4). The K and P levels were significantly higher in Split compared to in Pula only for Berber in the first year (Table 4). In the second harvest year, potato tubers grown in Split had higher Mg, K and P concentrations (Table 2). Plant cover had no influence on any of the observed tuber mineral levels in both harvest years.

In both harvest years, potato tubers' Mg, K and P content was significantly different between cultivars, while Ca content was found to be significantly different in the first harvest year (Tables 1 and 2). In the first year, the Ca and Mg contents were lowest in Red Scarlet, whic attained significantly higher Mg levels in the second year compared to Berber and Vivaldi (Tables 1 and 2). In the first year, Red Scarlet had significantly higher K content compared to Adora, while in the second year, Red Scarlet had the highest K content among the investigated cultivars (Tables 1 and 2). The P content was highest in Red Scarlet in the first year, while in the second year, Red Scarlet had significantly higher P content only compared to Berber (Tables 1 and 2).

3.6. Sensory Analysis of Oil-Fried "Early" Potatoes

In the first year, all sensory characteristics of oil-fried "early" potatoes were significantly influenced by planting location, where potato tubers grown in Split were consistently graded higher than those grown in Pula (Table 5). In the second year, none of the observed sensory characteristics were found to be significantly different between potato planting locations (Table 6).

Table 5. Sensory characteristics of potato tubers affected by location, plant cover and cultivar in the first year.

Sensory Characteristics	Location			Cover †			Cultivar				
	Split	Pula	*p*-Value	Yes	No	*p*-Value	Adora	Berber	Jerla	Red Scarlet	*p*-Value
Overall appearance	3.8 ± 0.8	3.3 ± 0.9	<0.001 *	3.5 ± 0.9	3.6 ± 0.9	0.496	3.4 ± 0.9 a	3.8 ± 1.0 b	3.5 ± 0.9 ab	3.6 ± 0.9 ab	0.050 *
Odor intensity	3.4 ± 0.8	3.1 ± 0.9	0.047 *	3.2 ± 0.8	3.3 ± 0.9	0.458	3.1 ± 1.0	3.3 ± 0.9	3.3 ± 0.7	3.2 ± 0.8	0.468
Overall taste	3.5 ± 0.9	3.0 ± 0.9	0.003 *	3.2 ± 0.9	3.3 ± 1.0	0.086	3.1 ± 1.0 a	3.6 ± 0.9 b	3.1 ± 0.8 a	3.3 ± 0.8 ab	0.003 *
Crispness	3.3 ± 0.9	2.8 ± 1.0	<0.001 *	3.1 ± 1.0	3.0 ± 1.0	0.938	3.1 ± 1.1 ab	3.4 ± 1.0 b	3.0 ± 0.8 ab	2.7 ± 0.9 a	0.002 *
Texture	3.7 ± 0.8	3.0 ± 0.9	<0.001 *	3.4 ± 0.8	3.4 ± 1.0	0.877	3.4 ± 1.0	3.6 ± 1.0	3.3 ± 0.8	3.2 ± 0.8	0.057
Overall impression	6.4 ± 1.6	5.3 ± 1.6	<0.001 *	5.8 ± 1.6	5.9 ± 1.8	0.059	5.7 ± 1.8	6.3 ± 1.9	5.7 ± 1.6	5.8 ± 1.5	0.793

* $p < 0.05$, Different letters in potato cultivars indicate homogenous groups in Tukey's HSD test ($p < 0.05$); Data are presented as mean ± standard deviation (N = 15); † Cover—Lutrasil® (Freudenberg, Weinheim, Germany), 17 g/m^2 in weight.

Table 6. Sensory characteristics of potato tubers affected by location, plant cover and cultivar in the second year.

Sensory Characteristics	Location			Cover †			Cultivar				
	Split	Pula	*p*-Value	Yes	No	*p*-Value	Berber	Jerla	Red Scarlet	Vivaldi	*p*-Value
Overall appearance	3.5 ± 0.9	3.6 ± 0.8	0.293	3.5 ± 0.9	3.6 ± 0.8	0.451	3.6 ± 0.8	3.7 ± 0.8	3.7 ± 0.9	3.3 ± 0.9	0.080
Odor intensity	3.2 ± 0.8	3.2 ± 0.8	0.779	3.1 ± 0.9	3.2 ± 0.8	0.352	3.1 ± 0.8	3.3 ± 0.8	3.2 ± 0.8	3.2 ± 0.9	0.461
Overall taste	3.3 ± 0.9	3.2 ± 0.9	0.663	3.2 ± 0.9	3.3 ± 0.9	0.220	3.3 ± 0.9	3.3 ± 0.8	3.4 ± 0.9	3.0 ± 0.8	0.161
Crispness	3.2 ± 0.9	3.2 ± 0.9	0.937	3.2 ± 0.9	3.3 ± 1.0	0.212	3.6 ± 0.8 b	3.1 ± 0.9 a	3.3 ± 0.9 ab	2.9 ± 0.9 a	<0.001 *
Texture	3.4 ± 0.9	3.4 ± 0.9	0.920	3.4 ± 0.8	3.5 ± 0.9	0.256	3.6 ± 0.8	3.4 ± 0.9	3.5 ± 0.9	3.3 ± 0.8	0.392
Overall impression	5.5 ± 1.9	5.5 ± 1.7	0.683	5.3 ± 1.8	5.7 ± 1.7	0.083	5.8 ± 1.7 b	5.6 ± 1.7 ab	5.8 ± 1.7 b	4.8 ± 1.8 a	0.011 *

* $p < 0.05$; Different letters in potato cultivars indicate homogenous groups in Tukey's HSD test ($p < 0.05$); Data are presented as mean ± standard deviation (N = 15); † Cover—Lutrasil® (Freudenberg, Weinheim, Germany), 17 g/m^2 in weight.

Plant cover had no influence on the observed sensory characteristics in both harvest years (Tables 5 and 6). A significant difference in both harvest years was observed in the crispness characteristic of oil-fried potatoes between potato cultivars. Significantly higher scores were observed for Berber compared to Jerla in the first year, and Berber compared to Jerla and Vivaldi in the second year (Tables 5 and 6). Overall impression was significantly higher for Berber and Red Scarlet compared to Vivaldi in the second year (Table 6).

3.7. PLS Analysis of Obtained Data

To further reveal differences between the obtained data, multivariate analysis was performed on potato tubers' physicochemical parameters, as well as on sensory characteristics. More precisely, a PLS model was developed for each dependent effect—potato cultivar, planting location and plant cover. The results for the first harvest year are presented in Figure 2 and for the second harvest year in Figure 3. In the first harvest year, the developed PLS model based on potato tubers' physicochemical parameters successfully differentiated between potato cultivars. The Berber cultivar was characterized by high dry matter, starch, sucrose, glucose, and Mg content combined with low protein and Vitamin C content, while Adora potato tubers represented the opposite (Figure 2a).

The obtained PLS model differed between potato tubers grown with or without plant cover. Tubers grown with the plant cover exhibited lower vitamin C concentrations and higher K, P, and protein content than potato tubers grown without the cover (Figure 2b). However, the model was not successful in differentiating between potato tubers' planting location (Figure 2c). The developed PLS model based on oil-fried potato sensory scores could not differentiate between potato cultivars (Figure 2d). Differences could be observed between potato tubers grown with and without the plant cover, where oil-fried potatoes grown with the plant cover were graded higher for overall appearance, impression, texture, and crispness over the potatoes which were grown without the cover (Figure 2e). Moreover, French fries from potatoes grown in Split were graded higher in the same sensory characteristics as those grown with the plant cover, i.e., overall appearance, impression, texture, and crispness, compared to those grown in Pula or without the plant cover (Figure 2f).

The obtained PLS model based on physicochemical parameters for the second harvest year showed clear differences between potato tubers grown with or without plant cover. Tubers grown with the plant cover had higher dry matter and starch content, higher soluble sugar content (sucrose, glucose, and fructose), higher Ca content, and lower vitamin C content than those grown without the cover. The plant cover had no impact on protein, K, P and Mg content (Figure 3c).

Differences could also be observed between potato tubers grown in Pula and Split. According to the developed PLS model, potato tubers grown in Split had higher dry matter and starch content combined with lower protein, K, P and Mg concentrations. However, planting location had no impact on soluble sugar and Ca content (Figure 3b).

There were no differences between potato cultivars in the second harvest year (Figure 3a). The developed PLS model based on oil-fried potato sensory scores could not differentiate among potato cultivars (Figure 3d), whether potatoes were grown with or without the plant cover (Figure 3f) or between planting locations (Figure 3e).

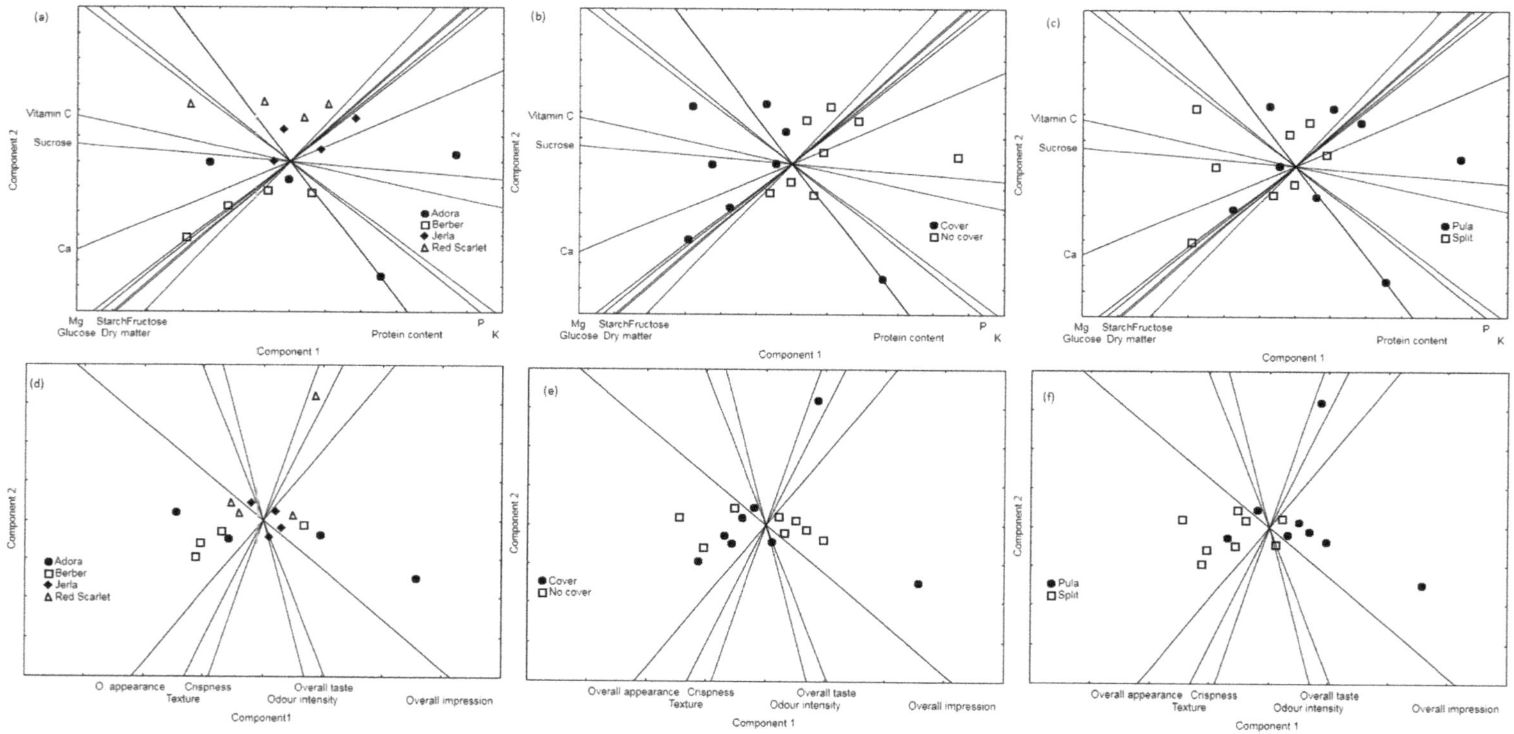

Figure 2. PLS models of the effect of (**a**) cultivar, (**b**) plant cover, (**c**) location on "early" potato physicochemical parameters and the effect of (**d**) cultivar, (**e**) plant cover, (**f**) location on oil-fried "early" potato sensory characteristics in the first harvest year.

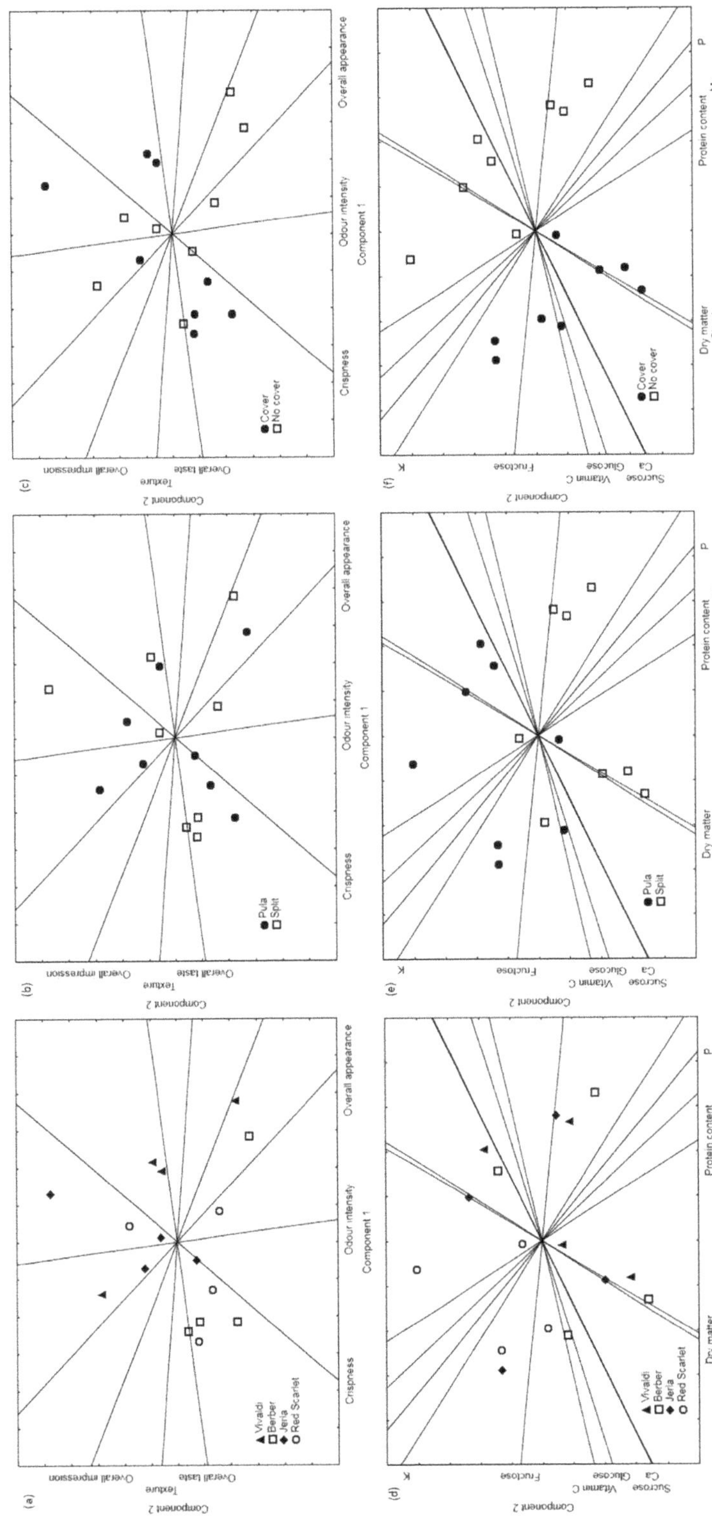

Figure 3. PLS models of the effect of (**a**) cultivar, (**b**) location, (**c**) plant cover on "early" potato physicochemical parameters and the effect of (**d**) cultivar, (**e**) location, (**f**) plant cover on oil-fried "early" potato sensory characteristics in the second harvest year.

4. Discussion

Our results showed that the planting location had a significant influence on potato yield, dry matter, glucose, and protein content in the second harvest year. The climate in Split is on average 2 °C warmer, which can stimulate the "earliness" of potato cultivars and therefore favor quicker dry matter accumulation in the tubers, as was observed previously [24]. Temperature influences dry matter accumulation directly by affecting the daily growth rate [25]. The total dry matter production is determined by the length of the growth cycle, which is divided into four phases critical for dry matter accumulation [25]. When the plant is subjected to temperate conditions, the tuberization and subsequent growth rate increases with temperatures from 15 °C up to the optimal 22 °C [25]. The latter temperature range induced earlier transition to crop senescence [25] which could explain the higher potato dry matter in Split compared to Pula in the second year. Our results are similar those of the study published by Lombardo et al. [24], where the authors reported that among three planting locations in Sicily (Italy), the highest dry matter content was observed in "early" tubers grown at the location with the highest minimum temperature. The same authors published another study, where they found that the dry matter accumulation was higher in tubers at the warmer planting location (the minimum temperature was higher by +2.6 °C on average), but only in the case of the conventional farming system [26].

In our experiment, starch content was not significantly different between locations, except in the second year and only for Vivaldi, which had higher starch content in Split compared to Pula. Fructose content did not differ between the investigated planting locations in both years, while glucose was significantly higher in Split but only in the second year. Sucrose content was significantly higher only for Adora in the first year in Split compared to Pula. Chung et al. [27] observed significant differences in potato starch and glucose content between two planting locations in Canada, where they attributed the discrepancies to environmental conditions such as total rainfall, temperature, and soil type. Simkova et al. [28] investigated the effect of the planting location on potato starch content, where the authors observed significant differences among planting locations with different above-sea altitudes. In our case, the absence of a more pronounced planting location effect could be attributed to the similarity of Pula and Split, where both are located closely to the Mediterranean coast and with similar climatic conditions. Contrary to Lombardo et al. [24] and Lombardo et al. [26], where significant differences were observed in vitamin C content between different planting locations, our research showed that the planting location did not influence vitamin C content. Protein content was found to be significantly higher in Pula compared to Split only in the first year. Lombardo et al. [26] did not observe differences in potato protein content between planting locations.

The effect of planting location was more pronounced in potato mineral content, where significant differences were observed in both harvest years, which could be influenced by differences in soil types. According to the literature, potato tubers' mineral content is influenced by soil mineral composition [29], by either conventional or organic production [30], or cultivar [30,31]. The same conclusions are presented in the PLS models, where distinctive locations can be observed in the second harvest year but not in the first. Lombardo et al. [24] and Wekesa et al. [29] reported that the tuber mineral content was significantly influenced by the planting location, which is in line with our findings, especially in the second year. In the first year, significant interactions in the tuber mineral content were observed for all investigated minerals, where significant differences were observed only for the Adora cultivar, where higher K and P content was found in Split compared to Pula.

In our study, the yield was not significantly affected by the application of the plant cover, which is opposite to the findings of Jabłońska-Ceglarek and Wadas [32], where the application of a non-woven polypropylene cover increased the average yield by 23.34% 60 days after planting. Probably, the effect of plant cover is less pronounced in less severe climates (higher early spring temperatures), as were present in our study.

Our results show that in the first year there were no differences in dry matter content between covered and non-covered plants, but in the second year, the use of the plant cover stimulated the accumulation of dry matter in potato tubers. The study by Jabłońska-Ceglarek and Wadas [32] showed that the plant cover did not influence tuber dry matter and starch content until 75 days after planting, suggesting that it is dependent on the physiological age of the tuber. So, it seems that the chemical composition of the tuber is dependent on the environmental conditions during potato vegetation, the potato cultivation method as well as on the crop maturity.

The plant covering investigated in this experiment had a similar effect on the tuber's nutritional parameters as the planting location in Split in the second year. Nutritional parameters which correlate with potato tuber physiological maturity—dry matter and carbohydrate content, including starch, sucrose, glucose, and fructose—were significantly higher in the second harvest year, while the vitamin C content was significantly lower in tubers grown under the plant cover. This effect can be explained by earlier crop transition to later developmental phases, as well as plant senescence, which stimulates earlier dry matter accumulation and tuber maturity [25] at higher temperatures under the plant cover. At the same time, Cho et al. [33] demonstrated that as the maturity level of potato tubers increases, the vitamin C content decreases, which could explain the observed decrease in vitamin C content in tubers grown under the plant cover. Additionally, Njoku et al. [34] showed a drop in vitamin C content in citrus fruits grown at warmer planting locations. Cantore et al. [35] suggested that the use of the non-woven covers raise air and soil temperature around the plant by 0.8 to 1.8 °C, thus increasing net assimilation and water use efficiency, plant weight and yield, which could explain the effect of earlier tuber initiation and growth under the cover. Additionally, Wadas and Kosterna [36] further corroborated the faster growing rate after plant emergence of potato cultivars under the plant cover compared to bare soil. Jabłońska-Ceglarek and Wadas [32] also observed a decrease in tuber vitamin C content in plants grown under the cover compared to no crop cover. In a study by Michalik [37], the use of non-woven covers contributed to dry matter accumulation and yield in some sweet pepper cultivars, but plants grown without a protective cover contained higher L-ascorbic acid content. Additionally, the study by Luthra et al. [38] found that processing potatoes had lower tuber yield in comparison to table cultivars, but had high ascorbic acid and low sucrose content. A reduction in ascorbic acid content and elevated sucrose content in tomato fruit was observed when mulching with black PE film was applied [39], while some authors suggest that soil covering with plastic film promotes plant productivity and reduces the fruit quality (acidity, pH, carotenoids, ascorbic acid, and potassium), except for total soluble solids of some tomato cultivars [40]. Other research also suggests that using both perforated foil or non-woven polypropylene plant covers elevates soil temperature by 1 to 2 °C, forcing plant emergence, shortening the tuber setting period and accelerating plant growth [36,41].

Our results showed that the plant cover did not influence the tubers' mineral content in both years. The results obtained by Wadas et al. [42] show differences between the application of different potato plant covers in the tubers' P and Mg content, where higher content of the minerals was found in tubers grown under the plant cover.

The choice of potato cultivar had a significant influence on all investigated parameters, except for fructose and protein content in the first year and Ca content in the second year. The strong effect of genotype has been confirmed by many researchers. Lombardo et al. [24] investigated the effect of eight different cultivars on "early" potato dry matter, vitamin C, and mineral content, and found significant differences among the investigated cultivars. Simkova et al. [28] investigated the differences in starch and other nutritional parameters among 16 different potato cultivars and detected significant differences.

The obtained values for dry matter, starch and protein content in our study are in line with the values obtained in "early" potato cultivars by Lombardo et al. [26]. The results obtained for sucrose are lower, while the glucose and fructose content is higher compared to the results from the same study [26]. Additionally, the obtained vitamin C content in our investigated cultivars is higher compared to the values reported in the literature [24,26]. Our results showed that the most abundant mineral in potato tubers is K, followed by P.

The obtained values for K, P, Ca, and Mg are in line with the results published by Wekesa et al. [29], as well as the USDA nutritional database [43].

The planting location in Split in the first year had a positive effect on the observed sensory characteristics of oil-fried early potatoes, while in the second harvest year, the differences were not significant. Agblor and Scanlon [44] observed differences in fry texture and color between two planting locations, where better texture properties were determined from fried potatoes made from tubers with higher Ca content. In our study, higher Ca content was observed in tubers grown in Split compared to Pula in the first year, while no significant differences were observed in the second year between planting locations. Two key factors were important in obtaining deep-fried potato products of good quality—high dry matter content and low reducing sugar content [20]. A subtle positive effect on sensory characteristics was observed with the use of plant cover, where differences could be observed through the developed PLS model in the first year. The effect of the plant cover was not significant on the fried potato sensory characteristics in both harvest years, which implies that the potato "earliness" induced by the plant cover did not cause any negative effects on the perceived sensory attributes of "early" fried potatoes. The effect of cultivar was significant on several sensory attributes in our study, including overall appearance, overall taste, and crispness in the first year, and crispness and overall impression in the second year. Among the investigated cultivars, Berber was among the highest rated fried potatoes in every sensory attribute, with significant differences between cultivars.

The application of a plant cover significantly increased the potato tubers' dry matter, starch, and sugar content in the second year. Sensory analysis of the oil-fried "early" potatoes revealed no differences between potatoes grown with or without the plant cover. We also observed significantly higher dry matter content in potatoes grown at the Split location in the second year, while no differences in the sensory scores between oil-fried potatoes grown at the investigated locations were observed. By employing a plant cover or by choosing a warmer planting location, the desired potato maturity level could be reached in less time, and one could more effectively exploit the "early" potato market.

Author Contributions: Conceptualization, S.G.B. and D.B.; Data curation, N.M. and S.G.B.; Formal analysis, J.P., D.Ž., A.S.I.P., M.O., G.D. and B.U.; Funding acquisition, D.B.; Investigation, J.P., A.S.I.P., M.O., G.D., B.U. and D.B.; Methodology, S.G.B. and D.B.; Supervision, S.G.B. and D.B.; Visualization, N.M.; Writing—original draft, N.M., S.G.B. and J.P.; Writing—review and editing, N.M., S.G.B., J.P., D.Ž., A.S.I.P., M.O., G.D., B.U. and D.B. All authors have read and agreed to the published version of the manuscript.

Funding: This research received no external funding.

Institutional Review Board Statement: Not applicable.

Informed Consent Statement: Not applicable.

Data Availability Statement: The data are available within the article.

Conflicts of Interest: The authors declare no conflict of interest.

References

1. FAOSTAT. Food and Agriculture Organization Corporate Statistical Database. Available online: https://www.fao.org/faostat/en/#data (accessed on 22 February 2022).
2. King, J.C.; Slavin, J.L. White Potatoes, Human Health, and Dietary Guidance. *Adv. Nutr. Int. Rev. J.* **2013**, *4*, 393S–401S. [CrossRef] [PubMed]
3. Kreutzmann, S.; Bassompierre, M.; Thybo, A.K.; Buch, L.; Engelsen, S.B. Exploratory Study of Potato Cultivar Differences in Sensory and Hedonistic Applicability Tests. *Potato Res.* **2011**, *54*, 13–28. [CrossRef]
4. UNECE. *Standard FFV-52 Concerning the Marketing and Commercial Quality Control of Early and Ware Potatoes*; UNECE: Geneva, Switzerland, 2017.
5. Buono, V.; Paradiso, A.; Serio, F.; Gonnella, M.; De Gara, L.; Santamaria, P. Tuber Quality and Nutritional Components of "Early" Potato Subjected to Chemical Haulm Desiccation. *J. Food Compos. Anal.* **2009**, *22*, 556–562. [CrossRef]
6. Wijesinha-Bettoni, R.; Mouillé, B. The Contribution of Potatoes to Global Food Security, Nutrition and Healthy Diets. *Am. J. Potato Res.* **2019**, *96*, 139–149. [CrossRef]

7. Burlingame, B.; Mouillé, B.; Charrondière, R. Nutrients, Bioactive Non-Nutrients and Anti-Nutrients in Potatoes. *J. Food Compos. Anal.* **2009**, *22*, 494–502. [CrossRef]
8. Navarre, D.A.; Shakya, R.; Hellmann, H. Chapter 6—Vitamins, Phytonutrients, and Minerals in Potato. In *Advances in Potato Chemistry and Technology*; Academic Press: Cambridge, MA, USA, 2016; pp. 117–166, ISBN 9780128000021.
9. FAO. *International Year of the Potato—2008*; FAO: Rome, Italy, 2008.
10. Jansky, S.H. Potato Flavor. *Am. J. Potato Res.* **2010**, *87*, 209–217. [CrossRef]
11. Caracciolo, F.; Cembalo, L. Traceability and Demand Sensitiveness: Evidences From Italian Fresh Potatoes Consumption. *Int. J. Food Syst. Dyn.* **2010**, *1*, 352–365. [CrossRef]
12. Kim, Y.-U.; Seo, B.-S.; Choi, D.-H.; Ban, H.-Y.; Lee, B.-W. Impact of High Temperatures on the Marketable Tuber Yield and Related Traits of Potato. *Eur. J. Agron.* **2017**, *89*, 46–52. [CrossRef]
13. Ban, D.; Vrtačić, M.; Goreta Ban, S.; Dumičić, G.; Oplanić, M.; Horvat, J.; Žnidarčič, D. Effect of Variety, Direct Covering and Date of Harvest on the Early Potato Growth and Yield. In Proceedings of the 46th Croatian and 6th International Symposium on Agriculture, Opatija, Croatia, 14–18 February 2011; Pospišil, M., Ed.; University of Zagreb, Faculty of Agriculture: Zagreb, Croatia, 2011; pp. 496–500.
14. Stark, J.C.; Novy, R.G.; Whitworth, J.L.; Knowles, N.R.; Pavek, M.J.; Thornton, M.; Spear, R.; Brown, C.R.; Charlton, B.A.; Sathuvalli, V.; et al. Mountain Gem Russet: A Potato Variety with High Early and Full Season Yield Potential and Excellent Fresh Market and Early Processing Characteristics. *Am. J. Potato Res.* **2016**, *93*, 158–171. [CrossRef]
15. Hamouz, K.; Lachman, J.; Dvořák, P.; Trnková, E. Influence of Non-Woven Fleece on the Yield Formation of Early Potatoes. *Plant Soil Environ.* **2011**, *52*, 289–294. [CrossRef]
16. Hirai, G. The Effect of Non-Woven Fabric Floating Row Covers on the Emergence, Growth, and Bulb Yield of Direct-Seeded Onions (*Allium Cepa* L.) in a Subarctic Area. *Hortic. J.* **2019**, *88*, 67–75. [CrossRef]
17. Kazimierczak, R.; Srednicka-Tober, D.; Hallmann, E.; Kopczynska, K.; Zarzynska, K. The Impact of Organic vs. Conventional Agricultural Practices on Selected Quality Features of Eight Potato Cultivars. *Agronomy* **2019**, *9*, 799. [CrossRef]
18. McKenzie, M.; Corrigan, V. Chapter 12—Potato Flavor. In *Advances in Potato Chemistry and Technology*; Academic Press: Cambridge, MA, USA, 2016; pp. 339–368, ISBN 9780128000021.
19. Pedreschi, F.; Mariotti, M.S.; Cortés, P. Chapter 15—Fried and Dehydrated Potato Products. In *Advances in Potato Chemistry and Technology*; Academic Press: Cambridge, MA, USA, 2016; pp. 459–474, ISBN 9780128000021.
20. Pedreschi, F. Frying of Potatoes: Physical, Chemical, and Microstructural Changes. *Dry. Technol.* **2012**, *30*, 707–725. [CrossRef]
21. Štampar, F.; Usenik, V.; Dolenc-Šturm, K. Evaluating of Some Quality Parameters of Different Apricot Cultivars Using Hplc Method. *Acta Aliment.* **1999**, *28*, 297–309. [CrossRef]
22. Vasanthan, T.; Bergthaller, W.; Driedger, D.; Yeung, J.; Sporns, P. Starch from Alberta Potatoes: Wet-Isolation and Some Physicochemical Properties. *Food Res. Int.* **1999**, *32*, 355–365. [CrossRef]
23. Tausz, M.; Wonisch, A.; Grill, D.; Morales, D.; Soledad Jiménez, M. Measuring Antioxidants in Tree Species in the Natural Environment: From Sampling to Data Evaluation. *J. Exp. Bot.* **2003**, *54*, 1505–1510. [CrossRef]
24. Lombardo, S.; Pandino, G.; Mauromicale, G. The Influence of Growing Environment on the Antioxidant and Mineral Content of "Early" Crop Potato. *J. Food Compos. Anal.* **2013**, *32*, 28–35. [CrossRef]
25. Kooman, P.L.; Haverkort, A.J. Modelling Development and Growth of the Potato Crop Influenced by Temperature and Daylength: LINTUL-POTATO. In *Potato Ecology and Modelling of Crops under Conditions Limiting Growth*; Springer: Heidelberg, Germany, 1995; pp. 41–59. [CrossRef]
26. Lombardo, S.; Pandino, G.; Mauromicale, G. The Effect on Tuber Quality of an Organic versus a Conventional Cultivation System in the Early Crop Potato. *J. Food Compos. Anal.* **2017**, *62*, 189–196. [CrossRef]
27. Chung, H.J.; Li, X.Q.; Kalinga, D.; Lim, S.T.; Yada, R.; Liu, Q. Physicochemical Properties of Dry Matter and Isolated Starch from Potatoes Grown in Different Locations in Canada. *Food Res. Int.* **2014**, *57*, 89–94. [CrossRef]
28. Simkova, D.; Lachman, J.; Hamouz, K.; Vokal, B. Effect of Cultivar, Location and Year on Total Starch, Amylose, Phosphorus Content and Starch Grain Size of High Starch Potato Cultivars for Food and Industrial Processing. *Food Chem.* **2013**, *141*, 3872–3880. [CrossRef]
29. Wekesa, M.N.; Okoth, M.W.; Abong', G.O.; Muthoni, J.; Kabira, J.N. Effect of Soil Characteristics on Potato Tuber Minerals Composition of Selected Kenyan Varieties. *J. Agric. Sci.* **2014**, *6*, 163. [CrossRef]
30. Lombardo, S.; Pandino, G.; Mauromicale, G. The Mineral Profile in Organically and Conventionally Grown "Early" Crop Potato Tubers. *Sci. Hortic.* **2014**, *167*, 169–173. [CrossRef]
31. Tamasi, G.; Cambi, M.; Gaggelli, N.; Autino, A.; Cresti, M.; Cini, R. The Content of Selected Minerals and Vitamin C for Potatoes (*Solanum Tuberosum* L.) from the High Tiber Valley Area, Southeast Tuscany. *J. Food Compos. Anal.* **2015**, *41*, 157–164. [CrossRef]
32. Jabłońska-Ceglarek, R.; Wadas, W. Effect of Nonwoven Polypropylene Covers on Early Tuber Yield of Potato Crops. *Plant Soil Environ.* **2005**, *51*, 226–231. [CrossRef]
33. Cho, K.S.; Jeong, H.J.; Cho, J.H.; Park, Y.E.; Hong, S.Y.; Won, H.S.; Kim, H.J. Vitamin C Content of Potato Clones from Korean Breeding Lines and Compositional Changes during Growth and after Storage. *Hortic. Environ. Biotechnol.* **2013**, *54*, 70–75. [CrossRef]
34. Njoku, P.C.; Ayuk, A.A.; Okoye, C.V. Temperature Effects on Vitamin C Content in Citrus Fruits. *Pak. J. Nutr.* **2011**, *10*, 1168–1169. [CrossRef]

35. Cantore, V.; Pace, B.; Calabrese, N.; Boari, F.; Schiattone, M.I. Effects of Non-Woven Fabric and Fertilizer on Air and Soil Temperature, Leaf Gas Exchange, Yield and Quality of Wild Rocket Grown in Organic Farming. *Acta Hortic.* **2013**, *1005*, 479–486. [CrossRef]
36. Wadas, W.; Kosterna, E. Effect of Perforated Foil and Polypropylene Fibre Covers on Development of Early Potato Cultivars. *Plant Soil Environ.* **2007**, *53*, 136–141. [CrossRef]
37. Michalik, L. The Effect of Non-Woven PP Fabric Covers on the Yielding and the Fruit Quality of Field-Grown Sweet Peppers. *Acta Sci. Pol.-Hortorum Cultus* **2010**, *9*, 25–32.
38. Luthra, S.; Gupta, V.; Kaundal, B.; Tiwari, J. Genetic Analysis of Tuber Yield, Processing and Nutritional Traits in Potato (*Solanum Tuberosum*). *Indian J. Agric. Sci.* **2018**, *88*, 1214–1221.
39. Horvat, J.; Ban, D.; Goreta Ban, S.; Oplanić, M.; Žnidarčič, D. Fertigation and Mulching Effect Bioactive and Nutritive Compounds of Tomato Fruit (*Lycopersicon Esculentum* Mill.). In Proceedings of the 45th Croatian & 5th International Symposium on Agriculture, Opatija, Croatia, 15–19 February 2010; Marić, S., Lončarić, Z., Eds.; Poljoprivredi fakultet Sveučilišta Josipa Jurja Strossmayera u Osijeku: Opatija, Croatia, 2010; pp. 634–638.
40. Caliman, F.; da Silva, D.; Stringheta, P.; Fontes, P.; Moreira, G.; Mattedi, A.; Naher, L.; Ismail, A. Relation between Plant Yield and Fruit Quality Characteristics of Tomato. *Biosci. J.* **2008**, *24*, 46–52.
41. Wadas, W. Using Non-Woven Polypropylene Covers in Potato Production: A Review. *J. Cent. Eur. Agric.* **2016**, *17*, 734–748. [CrossRef]
42. Wadas, W.; Jabloriska-Ceglarek, R.; Kurowska, A. Effect of Using Covers in Early Crop Potato Culture on the Content of Phosphprus and Magnesium in Tubers. *J. Elem.* **2008**, *13*, 275–280.
43. USDA. *National Nutrient Database for Standard Reference*; USDA: Washington, DC, USA, 2018.
44. Agblor, A.; Scanlon, M.G. Effect of Storage Period, Cultivar and Two Growing Locations on the Processing Quality of French Fried Potatoes. *Am. J. Potato Res.* **2002**, *79*, 167–172. [CrossRef]

horticulturae

Article

On-Farm Supplemental Irrigation of 'Roja Lisa' Cactus Pear: Pre- and Postharvest Effects

Valentín Melero-Meraz [1], Jorge A. Zegbe [1,*], Mayra Denise Herrera [1], Salvado H. Guzmán-Maldonado [2], Guillermo Medina-García [1], Blanca I. Sánchez-Toledano [1], Raquel K. Cruz-Bravo [1] and Miguel Servín-Palestina [1]

1 Campo Experimental Zacatecas, Instituto Nacional de Investigaciones Forestales, Agrícolas y Pecuarias, Km 24.5 Carretera Zacatecas-Fresnillo, Calera de V.R., Zacatecas 98500, Mexico; melero.valentin@inifap.gob.mx (V.M.-M.); herrera.mayra@inifap.gob.mx (M.D.H.); medina.guillermo@inifap.gob.mx (G.M.-G.); sanchez.blanca@inifap.gob.mx (B.I.S.-T.); cruz.raquel@inifap.gob.mx (R.K.C.-B.); servin.miguel@inifap.gob.mx (M.S.-P.)

2 Campo Experimental Bajío, Instituto Nacional de Investigaciones Forestales, Agrícolas y Pecuarias, Km 6.5 Carretera Celaya-San Miguel de Allende, Celaya 38110, Mexico; guzman.horacio@inifap.gob.mx

* Correspondence: zegbe.jorge@inifap.gob.mx

Abstract: *Opuntia* species grow in arid and semiarid lands, where water for irrigation is scarce. However, supplemental irrigation can be a feasible strategy for commercial cactus pear orchards. From 2018 to 2020, a commercial cactus pear orchard was managed to validate the effect of supplemental irrigation on fruit yield, crop water use efficiency, fruit quality, and storability of 'Roja Lisa' cactus pear grown in the semiarid region of Mexico. The irrigation treatments were no irrigation and supplemental irrigation, with four replications. Crop water use was less and, therefore, water productivity greater in non-irrigated plants than in plants with supplemental irrigation. Mean fruit yield, mean fruit mass, and proportion of commercial fruit increased with supplemental irrigation. These differences were more pronounced in growing seasons with less rainfall. Fruit quality at harvest or after room temperature or cold storage was examined. Fruit mass loss rate was reduced in fruit receiving supplemental irrigation in both storage conditions. In addition, supplemental irrigation was consistent with water savings and food security programs in marginal areas: this irrigation strategy improved both pre- and postharvest some quality components of cactus pear fruit. Therefore, this irrigation strategy is suggested for cactus pear growers, depending on the availability of water for irrigation.

Keywords: *Opuntia ficus-indica*; crop water use; fruit quality; fruit yield; storability

Citation: Melero-Meraz, V.; Zegbe, J.A.; Herrera, M.D.; Guzmán-Maldonado, S.H.; Medina-García, G.; Sánchez-Toledano, B.I.; Cruz-Bravo, R.K.; Servín-Palestina, M. On-Farm Supplemental Irrigation of 'Roja Lisa' Cactus Pear: Pre- and Postharvest Effects. *Horticulturae* **2022**, *8*, 483. https://doi.org/10.3390/horticulturae8060483

Academic Editors: Yang Bi, Yongcai Li and Di Gong

Received: 7 May 2022
Accepted: 25 May 2022
Published: 28 May 2022

1. Introduction

Global warming is damaging ecosystems around the world. This is more noticeable in arid and semiarid regions where, now, the temperatures are more extreme, rainfall patterns are changing, soil erosion is increasing, and water scarcity is reflected in poverty and human migration [1]. In addition, water shortage drastically reduces both livestock and agricultural activities in these agricultural lands [2]. However, combining available sustainable technologies and appropriate crops can enhance both crop productivity and grower incomes.

Cactus pear is a succulent, crassulacean acid metabolism plant that is used in semiarid lands to restore degraded lands, as animal fodder, to grow tender cladodes for human consumption [3], for biogas production [4], and as a fruit crop [5,6]. In most producing countries, cactus pear for fruit consumption is cultivated mainly under rainfed conditions. Nevertheless, in some Mediterranean and American countries, this fruit crop is drip-irrigated to increase fruit productivity [5,7–9]. However, groundwater for irrigation is limited in all cactus-pear-growing areas [10], including north-central México, due to the overexploitation of aquifers [11]. Therefore, water conservation and savings, and water productivity enhancement through new irrigation strategies are crucial to both food production and security [12]. Supplemental irrigation may be an alternative for crops cultivated under rainfed conditions [13]. This irrigation strategy has been applied successfully in chickpeas, fava beans, lentils [14], common beans [15], wheat [16], sugarcane [17], sorghum [18], and maize [19]. In all these crops, yield and yield components were significantly improved with supplemental irrigation. In the case of maize, the application of this irrigation technic resulted in economic advantages for growers. Regarding fruit trees, supplemental irrigation also improved yields in peach [13] and olive [20] trees. In the latter fruit crop, water use efficiency was improved without affecting oil content. Unlike the study of Van Der Merwe [21], supplemental irrigation was tested successfully in commercial and non-commercial cactus pear varieties in a semiarid area of north-central Mexico [18]. After three consecutive growing seasons, compared to fully irrigated plants, supplemental irrigation enhanced water use efficiencies and water productivity in favor of water savings by 52%. Cactus pear plants had fruit yields and marketable fruit sizes similar to fully irrigated plants, but some fruit quality parameters were enhanced and the fruit storage life of all cultivars was extended at both room and cold temperatures, which is important for transportation, marketing, and consumers [22].

To our best knowledge, except experimentally, supplemental irrigation has not been tested previously at the level of commercial cactus pear orchards. This study uses a commercial cactus pear orchard to validate the effects of supplemental irrigation on the fruit yield, crop water use, fruit quality, and storability of 'Roja Lisa' cactus pear grown in a semiarid region, where weather conditions are more extreme than the places where previous trials on supplemental irrigation were performed [22]. Therefore, we hypothesize that this irrigation strategy can be replicated at a commercial scale, with similar effects to those reported previously [22,23].

2. Materials and Methods

2.1. Experimental Site, Vegetative Material, and Orchard Management

The experiment was conducted in a commercial cactus pear orchard located in 'La Victoria', Pinos, Zacatecas, Mexico (lat. 22°22′ N, long. 101°67′ W, elevation 2161 m) from 2018 to 2020. Fifteen-year-old cactus pear plants of the cultivar 'Roja Lisa' (*O. ficus-indica* (L.) Mill.) were used. This cultivar can be used as fodder, vegetable (tender cladodes), and fruit. This essay is oriented to its consumption as fresh fruit. The fruit peel and pulp are red, medium size (8.5 and 5.7 cm of equatorial and polar diameters, respectively), fruit mass (136 g; 60% pulp and 40% peel) that has juicy pulp with 14.5° Brix of sugar content at the ready-to-eat stage. The plants were trained to an in-row-oriented system and spaced at 5.6 × 2.7 m between rows and plants, respectively. The orchard's management program included cladode pruning, fruit thinning, drip irrigation, mineral and organic fertilization, and weed and pest control, as required.

2.2. Treatments and Experimental Design

The field experiment had two irrigation treatments: no irrigation (NI, as control) and supplemental irrigation (SI). A section of the orchard was drip-irrigated three times during each dry season (April–June) of the 2018 and 2019 growing seasons and just once in 2020 due to malfunctioning of the irrigation pumping system. The irrigation was applied according to the grower's criterion; nevertheless, the irrigation water applied was calculated using three-quarters of the plant area, and volume was estimated as the product of the number of drippers, dripper flow rate (4 L h^{-1}), and time used for each irrigation event. Irrigation water applied was calculated by adding all irrigation events. The physical soil parameters for permanent wilting point and field capacity were established at 0.14 and 0.27 cm^3 cm^{-3}, respectively. From bloom until harvest, soil water content (θ) was monitored in both treatments using time-domain reflectometry (TDR, Mini-Trase System, Soil Moisture Equipment Corp., Santa Barbara, CA, USA). Probes were placed 20 cm deep

MDPI

in the soil at 10 and 70 cm from emitters and plants, respectively. Climate information was recorded by an automated weather station placed 600 m from the experiment. This information, in part, was used to estimate the reference evapotranspiration by the Penman–Monteith method [9] and to track air temperature and precipitation (Figure 1). The year 2019 was the driest growing season. The experiment was conducted in a completely random design with four replicates, each comprising two uniform plants whose crop load was adjusted by fruit thinning [24] to minimize, in part, the variation among replicates. One plant was used for destructive sampling, while the other one was used to assess fruit yield, fruit sizing distribution, fruit quality, and storability.

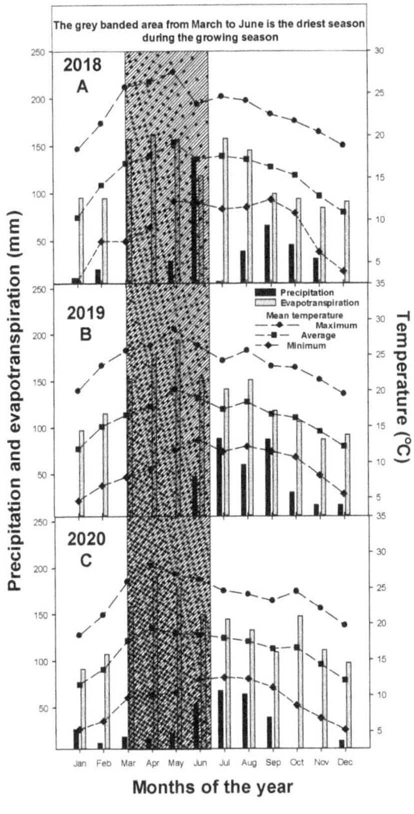

Figure 1. Climate conditions prevailed at the experimental site from 2018 to 2020. La Victoria, Pinos, Zacatecas, México.

2.3. *Preharvest Determinations*

2.3.1. Relative Water Content (RWC)

RWC was determined from two fruiting cladodes per replication as follows: between 12:00 and 13:00 h on every sampling date, two stem segments were taken out with a cork borer (17 mm internal diameter). The tissue samples were placed and sealed in Eppendorf tubes. The tissue specimens were individually weighed to determine the fresh mass (FM), turgid mass (TM), and dry mass (DM). The tissue samples were placed into the Eppendorf tubes and hydrated to full turgidity for 3 h to determine TM. DM was determined by oven-drying the samples at 65 °C to constant mass. RWC was calculated as RWC (%) = ((FM − DM)/(TM − DM)) × 100.

2.3.2. Crop Production

Cactus pear productivity was measured as fruit yield, mean fruit mass, and fruit size distribution. The harvest took place from June 28 to August 30 in 2018, from Jun 26 to August 5 in 2019, and from July 13 to August 4 in 2020 when the fruit peel color was reddish-green. Fruit from each plant was harvested separately and graded from Grade 1 to Grade 4 using the equatorial diameter (cm). Grade 1 fruit had a diameter between 6.0 and 7.0 cm. The corresponding values for Grade 2, Grade 3, and Grade 4 were 5.0 to 5.9 cm, 4.0 to 4.9 cm, and shorter than 4.0 cm, respectively. Fruit from each plant was counted and weighed, and the total mass was measured as gross yield. The mean fresh mass of fruit was calculated by dividing the gross yield by the number of fruits per plant. Water productivity was measured here by dividing the gross yield (kg) and commercial fruit yield (fruit of Grade 1 and 2) by the volume of applied irrigation water (m^3).

2.4. Postharvest Determinations

Fruit Quality

To assess the fruit quality (FQ) of the third harvest event, fruit at color break (green to red) was harvested randomly from the outer and middle parts of cactus pear plants. In the postharvest laboratory, three lots each of 24 Grade 1 fruits (6 fruit per replicate per treatment) were formed. One group was used to evaluate FQ at harvest; the other two lots were stored in a cold room or at room temperature. The storage periods ceased when, on average, fruit lost 8% of its original mass, and FQ was measured. The cold room temperature (T) and relative humidity (RH) values ($\pm$ standard deviation) were set at 7 °C and 90 $\pm$ 4%, respectively. At room temperature, T and RH values were 24 $\pm$ 2 °C and 40 $\pm$ 10% in 2018, 25 $\pm$ 2 °C and 37 $\pm$ 7% in 2019, and 25 $\pm$ 2 °C and 38 $\pm$ 7% in 2020. Room T and RH were determined every two hours with a data logger (model 42276, ExTech, Instruments, Waltham, MA, USA). Before storing, the spines were removed from the fruit, and the fruit was treated with a solution of copper sulfate (2.5 mL L^{-1}) and chlorine (1%).

FQ evaluations at harvest or after storage followed the same protocol. The mass of each fruit was recorded with a precision scale (VE-303, Velab, Ciudad de Mexico, Mexico). After removing the fruit skin, two flesh firmness determinations were made on two opposite sides, in the equatorial part of each fruit, using a press-mounted Wagner penetrometer (model FT 327, Wagner Instruments, Greenwich, CT, USA) with an 11.1 mm head. For each flesh firmness determination, some juice drops were mixed and the soluble solids concentration was measured with a digital refractometer with automatic temperature compensation (model PR-32α, Atago, CO., Ltd., Tokyo, Japan). Each fruit was split into peel and pulp, and each tissue was weighed separately with a precision scale (VE-303, Velab, Ciudad de Mexico, Mexico). The dry mass of fruit pulp was determined from 25 g composite samples of fresh pulp tissue (including seeds) from three fruits. Samples were oven-dried at 60 °C for one week to constant mass. Fruit mass loss was evaluated weekly by weighing the fruit individually at both storage conditions, taking into account the initial mass to calculate this parameter.

2.5. Data Analysis

The pre- and postharvest data were analyzed with a completely randomized model using the general linear model procedure of Statistical Analysis System software (SAS Institute ver. 9.4, 2002–2010, Cary, NC, USA). As fruit size distribution was expressed in percentage, it was arcsine-transformed and de-transformed for its presentation. Treatment means were separated by Fisher's least significant difference test at $p \leq 0.05$. Fruit decay incidence was analyzed by the chi-square test.

3. Results

3.1. Preharvest Results

Soil Water Content (θ) and Relative Water Content (RWC)

The θ of plants under no irrigation was below the permanent wilting point (PWP), but when plants received supplemental irrigation plus rainfall, it was between field capacity (FC) and PWP in 2019 and 2020. In contrast, except for one occasion at 204 days in 2019, despite the early rainfall in the 2020 growing season (Figure 1), θ remained below the PWP in non-irrigated plants (Figure 2c,d). The RWC was sensitive to changes in θ in both treatments, but plants under supplemental irrigation had the greatest RWC in both growing seasons (Figure 2a,b). The θ and RWC were not measured in 2018, but it is assumed that they were similar to 2019 and 2020.

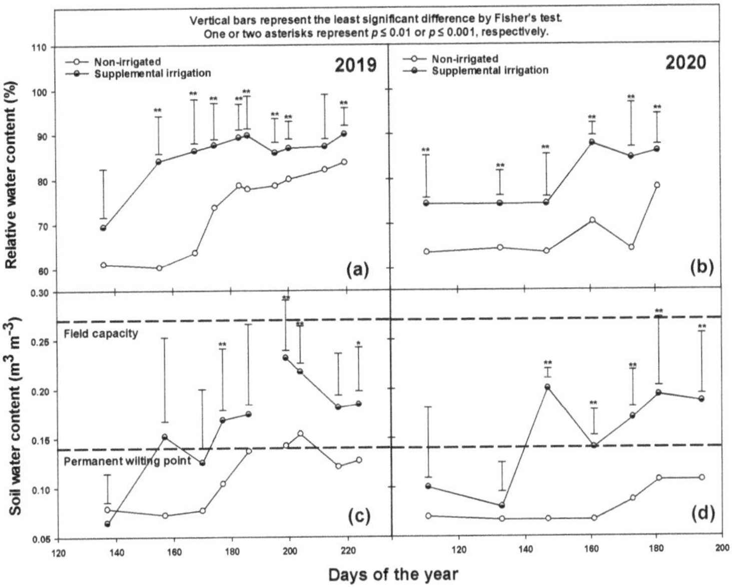

Figure 2. Changes in relative water content of cladodes (**a**,**b**) and soil water content (**c**,**d**) in experimental plots of 'Roja Lisa' cactus pear plants undergoing no irrigation or supplemental irrigation in 2019 and 2020. The data were collected in the commercial orchard 'Las 100 Hectáreas'. La Victoria, Pinos, Zacatecas, Mexico.

Based on the irrigation water applied, crop water uses on the plants given supplemental irrigation was, over a three-year average, 2.5-fold greater than in non-irrigated plants (Table 1). Except for in 2018, this was reflected in increased total fruit yield, commercial fruit yield, and mean mass of fresh fruit of irrigated plants, with a corresponding reduction in water productivity in the first two growing seasons (Table 2). Fruit size distribution was enhanced in plants undergoing supplemental irrigation, particularly in the fruit of the first category (the most marketable fruit) in the second and third growing seasons (Table 3).

Table 1. Accumulated reference evapotranspiration (ETo), irrigation water applied (IWA), and crop water use (CWU) of 'Roja Lisa' cactus pear under non-irrigated and supplemental irrigation in 2018, 2019, and 2020. Data were collected from a commercial orchard, 'Las 100 Hectáreas'. La Victoria, Pinos, Zacatecas, Mexico.

Year/Irrigation Treatment	ETo (mm)	IWA (mm)	CWU (mm)
2018			
Non-irrigated	588	0	155
Supplemental irrigation	588	219	374
2019			
Non-irrigated	657	0	50
Supplemental irrigation	657	247	297
2020			
Non-irrigated	522	0	113
Supplemental irrigation	522	24	137

Table 2. Influence of supplemental irrigation on total fruit yield (TFY), commercial fruit yield (CFY, Grade 1 + Grade 2), and mean mass of fresh fruit (MMFF) and water productivity of 'Roja Lisa' cactus pear. Fruit were harvested from the commercial orchard 'Las 100 Hectáreas'. La Victoria, Pinos, Zacatecas, Mexico.

Year/Irrigation Treatment	TFY (t ha^{-1})	CFY (t ha^{-1})	MMFF (g)	Water Productivity (kg m^{-3})	
				CFY	TFY
2018					
Non-irrigated	17.3a *	14.0a	113.4a	9.1a	11.1a
Supplemental irrigation	18.4a	16.7a	117.7a	4.5b	4.9b
Least significant difference	9.3	9.6	29.6	4.1	3.0
Significance ($p > F$)	0.77	0.52	0.74	0.03	0.002
Coefficient of variation (%)	30.1	36.1	14.8	35.0	21.3
2019					
Non-irrigated	17.7a	13.7b	109.3b	27.4a	35.4a
Supplemental irrigation	32.3a	28.5a	126.9a	9.6b	10.9b
Least significant difference	15.1	14.5	14.3	9.4	10.6
Significance ($p > F$)	0.05	0.05	0.02	0.004	0.001
Coefficient of variation (%)	35.0	39.6	7.0	29.4	26.4
2020					
Non-irrigated	23.8a	19.4b	103.8a	17.2a	21.1a
Supplemental irrigation	29.1a	26.3a	133.0b	19.3a	21.3a
Least significant difference	5.6	5.4	15.3	4.0	4.1
Significance ($p > F$)	0.06	0.02	0.004	0.25	0.93
Coefficient of variation (%)	12.2	13.6	7.5	12.6	11.2

* Within columns per year, means followed by the same letter were not significantly different by Fisher's test at $p \leq 0.05$.

Table 3. Influence of supplemental irrigation on the fruit size distribution of 'Roja Lisa' cactus pear. Fruit were harvested from the commercial orchard 'Las 100 Hectáreas'. La Victoria, Pinos, Zacatecas, Mexico.

	Fruit Size Distribution (%)			
	Grades (Equatorial Diameter, cm)			
Year/Irrigation Treatment	**1 (7.0–6.0)**	**2 (5.9–5.0)**	**3 (4.9–4.0)**	**4 (3.9–3.5)**
2018				
Non-irrigated	24.0a *	56.0a	19.1a	0.9a
Supplemental irrigation	33.4a	57.7a	7.8a	1.11a
Least significant difference	34.3	17.7	21.3	1.6
Significance ($p > F$)	0.53	0.82	0.24	0.72
Coefficient of variation (%)	69.2	18.0	91.5	96.4
2019				
Non-irrigated	7.2b	70.2a	21.7a	0.8a
Supplemental irrigation	22.2a	64.6a	12.9a	0.3a
Least significant difference	9.1	10.2	11.1	1.6
Significance ($p > F$)	0.01	0.23	0.10	0.48
Coefficient of variation (%)	36.0	8.8	37.0	166.7
2020				
Non-irrigated	8.3b	73.1a	17.0a	1.6a
Supplemental irrigation	19.7a	70.7a	9.6b	0.0b
Least significant difference	5.8	5.5	4.5	0.9
Significance ($p > F$)	0.00	0.32	0.01	0.01
Coefficient of variation (%)	23.8	4.4	19.6	69.2

* Within columns per year, means followed by the same letter were not significantly different by Fisher's test at $p \leq 0.05$.

3.2. *Postharvest Results*

3.2.1. Fruit Quality

At harvest, irrigated plants averaged a more fleshy mass of fruit and its components (peel and pulp) over the three growing seasons, although the differences were not always significant. These results remained consistent after room temperature storage for 13 weeks in 2018 and 9 weeks in 2019 and 2020 and after cold storage for 14, 11, and 10 weeks in 2018, 2019, and 2020, respectively (Table 4).

At harvest, plants under both treatments produced fruit with similar firmness in all years and after storage at either temperature, except in 2019, when fruits from irrigated plants and stored in a cold room were maintained the firmest (Table 5).

In 2018, at harvest, soluble solids concentration showed no measurable differences between non-irrigated and supplementally irrigated fruit. The same was true after fruit storage at room temperature or in a cold room. In contrast, soluble solids concentration was reduced in supplementary irrigated fruit at harvest in 2019 and 2020. This last trend was also consistent after storage at room temperature or in a cold room. The pulp dry matter acted similarly to the soluble solids concentration (Table 5).

Fruit from non-irrigated plants had the fastest fruit mass loss (FML) rate and, therefore, reached the threshold of 8% of the FML rate established for this non-climacteric fruit. Alternatively, fruit deterioration was observed before reaching this threshold. The opposite was observed in supplementally irrigated fruit, which had a long storage life (Figure 3a–c). Similar behavior was observed in fruit stored in a cold room. In both fruit types, the FML rate was lower with cold room storage than at room temperature. Nevertheless, the assay ended before reaching the FML threshold because signs of fruit deterioration were observed (Figure 3d–f). No measurable differences were found in 2020 (Figure 3f).

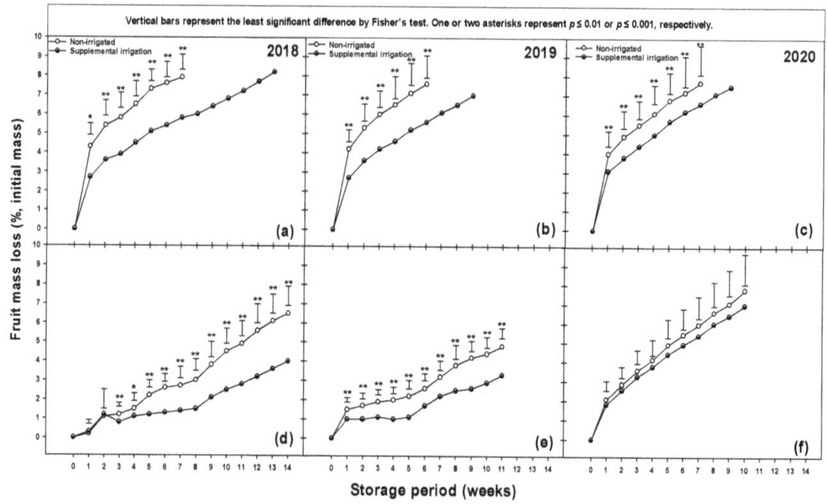

Figure 3. Changes in fruit mass loss of 'Roja Lisa' cactus pear, as influenced by supplemental irrigation in three consecutive growing seasons. Fruit were harvested from the commercial orchard 'Las 100 Hectárias'. La Victoria, Pinos, Zacatecas, Mexico.

Table 4. Influence of supplemental irrigation on fruit mass attributes of 'Roja Lisa' cactus pear. Fruit were harvested from the commercial orchard 'Las 100 Hectárias'. La Victoria, Pinos, Zacatecas, Mexico.

Irrigation/Fruit Condition	Fresh Mass of Fruit (g)			Fresh Mass of Peel (g)			Fresh Mass of Pulp (g)		
	2018	2019	2020	2018	2019	2020	2018	2019	2020
Harvest									
Non-irrigation	122.7b *	114.2b	136.4a	59.6b	52.1b	54.2b	63.1b	63.1a	82.2a
Supplemental irrigation	150.5a	133.7a	160.1a	70.2a	65.6a	73.0a	80.3a	68.2a	87.1a
Least significant difference	14.6	18.4	25.1	5.1	9.8	13.6	11.4	10.4	13.6
Significance ($p > F$)	0.003	0.04	0.06	0.002	0.01	0.02	0.01	0.28	0.41
Coefcient of variation (%)	11.8	7.9	12.7	12.8	8.6	13.6	13.9	11.3	15.5
Room temperature **									
Non-irrigation	116.2b	121.2b	129.6b	41.0b	41.5b	41.6b	75.2a	79.7a	88.0a
Supplemental irrigation	138.8a	143.4a	152.4a	52.8a	55.7a	61.4a	85.9a	87.7a	91.0a
Least significant difference	14.4	15.6	16.0	6.1	6.7	7.5	14.5	14.3	12.6
Significance ($p > F$)	0.007	0.01	0.01	0.002	0.003	0.001	0.125	0.25	0.61
Coefcient of variation (%)	10.3	11.0	10.3	14.1	12.6	13.4	11.7	14.7	13.2
Cold room ***									
Non-irrigation	139.3b	119.1b	126.6b	50.9b	47.6b	41.4b	88.4a	71.4b	84.2a
Supplemental irrigation	168.6a	161.9a	146.7a	65.2a	64.9a	54.9a	103.3a	96.9a	91.8a
Least significant difference	22.7	19.9	12.2	9.4	8.1	5.6	17.7	13.0	8.1
Significance ($p > F$)	0.02	0.001	0.01	0.01	0.002	0.002	0.09	0.003	0.06
Coefcient of variation (%)	8.5	11.4	12.2	11.7	11.7	16.8	8.8	14.1	12.7

* Within columns, per fruit conditions means followed by the same letter were not significantly different by Fisher's test at $p \leq 0.05$. ** Room temperature conditions were set at 24 ± 2 °C and 40 ± 10% in 2018; 25 ± 2 °C and 37 ± 7% in 2019; and 25 ± 2 °C and 38 ± 7% in 2020. *** Cold storage conditions were set at 7 °C and 90 ± 4%.

Table 5. Influence of supplemental irrigation on some fruit quality attributes of 'Roja Lisa' cactus pear. Fruit were harvested from the commercial orchard 'Las 100 Hectárias'. La Victoria, Pinos, Zacatecas, Mexico.

Irrigation/Fruit Condition	Fresh Firmness (Newtons)			Soluble Solids Concentration (Brix)			Dry Mass of Pulp (mg g^{-1} FW)		
	2018	2019	2020	2018	2019	2020	2018	2019	2020
Harvest									
Non-irrigation	34.6a *	28.3a	28.7a	12.0a	13.9a	13.3a	145.8a	219.3a	143.9a
Supplemental irrigation	34.2a	32.5a	28.2a	12.2a	11.9b	12.6a	148.0a	210.4b	131.2a
Least significant difference	4.1	3.8	5.6	0.8	0.08	0.9	6.3	4.1	15.5
Significance ($p > F$)	0.85	0.08	0.86	0.5	0.001	0.11	0.50	0.000	0.09
Coefficient of variation (%)	18.3	14.2	18.1	6.8	6.9	6.8	5.7	1.8	10.9
Room temperature **									
Non-irrigation	22.2a	23.1a	23.5a	11.4a	12.8a	13.3a	139.7a	187.1a	151.1a
Supplemental irrigation	22.3a	22.5a	21.7a	10.8a	11.0b	11.3b	141.4a	173.8b	127.7b
Least significant difference	4.9	3.8	3.7	0.8	0.7	0.4	17.5	5.3	9.5
Significance ($p > F$)	0.99	0.8	0.30	0.07	0.001	0.0001	0.8	0.001	0.001
Coefficient of variation (%)	17.5	18.5	15.3	5.0	6.6	5.6	10.8	3.1	12.3
Cold room ***									
Non-irrigation	33.2a	29.9b	31.7a	11.6a	12.2a	13.6a	129.3a	191.7a	150.9a
Supplemental irrigation	33.6a	33.3a	30.8a	11.4a	11.4b	13.0b	134.3a	187.0a	147.5a
Least significant difference	4.4	2.6	3.1	0.4	0.7	0.5	18.6	12.5	8.9
Significance ($p > F$)	0.85	0.04	0.52	0.60	0.04	0.02	0.54	0.4	0.38
Coefficient of variation (%)	10.6	13.1	10.3	5.6	5.8	8.1	21.7	5.5	10.4

* Within columns per fruit conditions means followed by the same letter were not significantly different by Fisher's test at $p \leq 0.05$. ** Room temperature conditions were 24 ± 2 °C and 40 ± 10% in 2018; 25 ± 2 °C and 37 ± 7% in 2019, and 25 ± 2 °C and 38 ± 7% in 2020. *** Cold storage conditions were set at 7 °C and 90 ± 4%.

3.2.2. Fruit Decay Incidence

The chi-square (χ^2) test of independence concluded that irrigation treatment and storage condition are independent of each other for fruit decay presence (χ^2 = 1.32, p = 0.25; Fisher's exact test p = 0.262). The highest incidence of decay occurred in the fruit from plants given supplemental irrigation that were then stored in the cold room in 2018 or 2019. Decay was minimal or absent in 2020 (Table 6).

Table 6. Number and percentage (in brackets) of fruit with decay from 'Roja Lisa' cactus pear plants under different irrigation treatments and storage conditions in three consecutive growing seasons.

Year/Irrigation Treatments	Storage Conditions	
	Room Temperature	Cold Room
2018		
No irrigation	1 (4.2)	2 (8.3)
Supplemental irrigation	2 (8.3)	4 (16.7)
2019		
No irrigation	2 (8.3)	0 (0)
Supplemental irrigation	2 (8.3)	7 (29.2)
2020		
No irrigation	0 (0)	0 (0)
Supplemental irrigation	1 (4.2)	0 (0)

4. Discussion

Supplemental irrigation could be a feasible alternative to full irrigation to enhance the productivity of cactus pear orchards in semiarid agricultural lands [22,23]. This experiment examined its effect on fruit yield, crop water use, fruit quality, and storability of 'Roja Lisa' cactus pear grown in a commercial orchard in the semiarid region of Mexico.

Similar to other fruit trees [25–27], the relative water content of cactus pear cladodes acted as an indicator of plant water status and was sensitive to changes in soil water content in both irrigation treatments. This occurred due to water inputs (irrigation or rainfall) and created measurable differences between non-irrigated and supplementally irrigated plants (Figures 1 and 2). The permanent wilting point for this kind of soil was established at 0.14 m^3 m^{-3}, but the soil water content in non-irrigated cactus pear plants was as low as 0.06 m^3 m^{-3} in 2020 (Figure 2). However, after rainfall, these plants not only recovered but also produced a commercial yield (Table 2). This shows that the permanent wilting point for this plant must be established because this soil physical parameter is the product of the plant (succulent)–soil–atmosphere continuum [28]. This determination was not made during this study, and it was not the purpose of this study either.

Because non-irrigated plants received water input only from rain, supplementally irrigated plants had greater crop water use and, therefore, lower water productivity than those under no irrigation (Tables 1 and 2). The latter conclusion is supported by previous irrigation research on cactus pear plants that received various irrigation techniques [9], including a comparison of full and supplemental irrigation [22]. Similar results were seen in peach and olive orchards grown under water scarcity [29], similar to the weather conditions of the place where this experiment was conducted. In 2020, water productivity, measured as total fruit yield and commercial fruit yield, was similar between irrigation treatments (Table 2). This was attributed to both treatments receiving a similar amount of water, mainly due to rain (Table 1). The reduced water applied to supplementally irrigated plants was due to the malfunctioning of the irrigation system. However, this season was rainy, which could have masked the effect of both treatments on total fruit yield. This suggests that in a rainy growing season, irrigation may be suspended, at least in succulent plants such as cactus pear [21].

This experiment was expected to enhance cactus pear fruit yield by applying supplemental irrigation in a semiarid region facing water scarcity for agricultural activities. As seen at the start of a mineral fertilization program for this plant species [30], in 2018, the irrigation treatments made no measurable difference to total fruit yield, commercial fruit yield, and mean fruit mass (Table 2). This may be attributed to the plants' adjustment to irrigation [9] because in the two following growing seasons, commercial fruit yield, mean fruit mass, and fruit size distribution were greater in supplementally irrigated plants than in non-irrigated plants. The cladodes of the latter had the lowest relative water content (Figure 2) and a shriveled appearance by the middle of the growing season, indicating a water stress effect. Hence, it is likely that these plants had limited photosynthetic activity and, therefore, a low rate of photo-assimilate remobilization from source tissues into the reproductive sink organs [31]. The data also indicate that in a rainy season, which is atypical in this region, supplemental irrigation must be applied judiciously to produce marketable fruit with lower irrigation costs (Table 3).

The influence of preharvest management on the postharvest quality maintenance and storability of cactus pear fruit has received little attention [23]. We assessed the influence of no irrigation and supplemental irrigation on cactus pear fruit at harvest and their effect on postharvest life during storage at room temperature and in cold storage. Cactus pear plants receiving supplemental irrigation produced fruit with significantly greater fruit and peel mass and occasionally significantly more pulp than cactus pear plants grown without irrigation (Table 4). This suggests that the non-irrigated fruit would have had an insufficient supply of carbohydrates from the source cladodes (leaves as a source were limiting) or that the fruit (sink tissues) was unable to utilize the available carbohydrates from the source cladodes fully (sink-limited) [32]. Except for this measure, irrigation treatments did not affect flesh firmness at harvest or after cold or room temperature storage (Table 5). Although this study did not examine fruits at the cellular level, neither irrigation treatments nor storage conditions modified fruit cell density, as occurred in other fruits [33,34], including fruit from various cactus pear cultivars [23]. The soluble solids concentration and pulp dry matter were usually greater in non-irrigated fruit than in supplementally irrigated fruit, at harvest and after storage at either temperature (Table 5). The reduced soluble solids concentration and dry mass of pulp in irrigated fruit are probably due to dilution. The larger cells of irrigated fruits can store more water than the presumably smaller ones in non-irrigated fruits or under deficit irrigation [9,23,33,34]. The association of fresh mass of fruit with soluble solids concentration ($r = -0.18$; $p = 0.0002$) and dry mass of pulp ($r = -0.05$; $p = 0.54$) was weak and negative.

A mass loss of 5% causes many perishable commodities to appear wilted and shriveled, among other postharvest physiological disorders [35]. For white cactus pear varieties, an 8% mass loss during storage creates visible changes in fruit appearance (wilting and shriveling) and fruit texture (flaccidity and softening) [36]. In this experiment, supplementally irrigated fruit minimized fruit mass loss rates and prolonged fruit storage life under both storage conditions, as reported previously for cactus pear undergoing deficit irrigation [9] or supplemental irrigation [23]. After harvest, fruit mass loss occurs mainly via transpiration [37,38]. Therefore, supplemental irrigation is likely to promote positive changes at the levels of the epidermis [23,39] that minimize fruit mass loss to the surrounding atmosphere [35]. This finding has important implications for the local and long-distance transportation, marketing, and final consumers of this exotic fruit.

In cactus pear plants, fruit decay incidence depends on the cultivar, fruit maturation stage, and storage conditions. In this experiment, all fruit were collected at the same fruit maturity (color break) and were spineless, curated with a solution of copper sulfate and chlorine, and stored, theoretically, under the same conditions. Temperature variation could be a key factor in postharvest disease occurrence. Cactus pear fruit rot is caused by *Alternaria alternata*, *Chlamydomyces* spp., *Fusarium* spp., *Macrophomina phaseolina*, and *Penicillium polonicum* [40,41]. Although this experiment was not designed to study cactus pear fruit diseases, both storage temperatures and causal agents deserve special attention in order to minimize fruit postharvest losses.

5. Conclusions

This study was conducted in a commercial cactus pear orchard to validate the effects of supplemental irrigation on the fruit yield, crop water use, fruit quality, and storability of 'Roja Lisa' cactus pear grown in the semiarid region of Mexico. This irrigation strategy is easily implemented on a farm scale. Although crop water use and water productivity were low in supplementally irrigated plants, total fruit yield, mean fruit mass, and commercial fruit yield were greater than in non-irrigated plants. These differences were more pronounced in growing seasons with low rainfall and vice versa. Fruit quality, measured as flesh firmness, was the same at harvest and maintained under both storage conditions (room temperature or cold room storage), but soluble solids concentration and dry mass of pulp were consistently low in supplementally irrigated fruit at harvest and under both storage conditions. An important finding was that the fruit mass loss rate was the lowest in those fruit receiving supplemental irrigation. This is very important for fruit storage, transportation, marketing, and final consumption. Therefore, this irrigation strategy is suggested for cactus pear growers with limited water availability for irrigation, and it corroborates our hypothesis. Cactus pear growers need more knowledge and training to efficiently manage supplemental irrigation applications.

Author Contributions: Conceptualization, J.A.Z. and V.M.-M.; Methodology, V.M.-M., J.A.Z., B.I.S.-T. and M.S.-P.; Formal analysis, J.A.Z., B.I.S.-T. and G.M.-G.; Investigation, V.M.-M., J.A.Z., R.K.C.-B., M.D.H., S.H.G.-M., B.I.S.-T., G.M.-G. and M.S.-P.; Resources, J.A.Z.; Data curation, V.M.-M., J.A.Z., R.K.C.-B., M.D.H., S.H.G.-M., B.I.S.-T., G.M.-G. and M.S.-P.; Writing—original draft preparation, V.M.-M. and J.A.Z.; Writing—review and editing, J.A.Z.; Supervision, J.A.Z.; Project administration, J.A.Z.; Funding acquisition, J.A.Z. All authors have read and agreed to the published version of the manuscript.

Funding: This research work was funding by the Instituto Nacional de Investigaciones Forestales, Agrícolas y Pecuarias (INIFAP) through funding research project No. 8403134459.

Data Availability Statement: The datasets generated and/or analyzed during the current study are not publicly available because this information belongs to our sponsor (INIFAP), but they are available from the corresponding author upon reasonable request.

Acknowledgments: We highly appreciate the enthusiastic cooperation of Jesús Álvarez Gómez, the owner of the cactus pear orchard. We also thank Mary Lou Mendum (University California, Davis) for improving the final presentation of this document. We appreciate the valuable comments and suggestions from three reviewers.

Conflicts of Interest: The authors declare no conflict of interest.

References

1. Nhemachena, C.; Nhamo, L.; Matchaya, G.; Nhemachena, C.R.; Muchara, B.; Karuaihe, S.T.; Mpandeli, S. Climate change impacts on water and agriculture sectors in Southern Africa: Threats and opportunities for sustainable development. *Water* **2020**, *12*, 2673. [CrossRef]
2. Pereira, L.S. Water and agriculture: Facing water scarcity and environmental challenges. Invited overview paper. *CIGR J. Agric. Eng. Int.* **2005**, *VII*, 1–26.
3. Louhaich, M.; Nefzaoui, A.; Guevara, J.C. Cactus ecosystem goods and services. In *Crop Ecology, Cultivation and Uses of Cactus Pear*; Inglese, P., Mondragon, C., Nefzaoui, A., Sáenz, C., Eds.; The Food and Agriculture Organization of the United Nations and the International Centre for Agricultural Research in the Dry Areas: Rome, Italy, 2017; pp. 159–169.
4. Varnero, M.T.; Homer, I. Biogas production. In *Crop Ecology, Cultivation and Uses of Cactus Pear*; Inglese, P., Mondragon, C., Nefzaoui, A., Sáenz, C., Eds.; The Food and Agriculture Organization of the United Nations and the International Centre for Agricultural Research in the Dry Areas: Rome, Italy, 2017; pp. 187–193.
5. Potgieter, J.; D'Aquino, S. Fruit production and post-harvest management. In *Crop Ecology, Cultivation and Uses of Cactus Pear*; Inglese, P., Mondragon, C., Nefzaoui, A., Sáenz, C., Eds.; The Food and Agriculture Organization of the United Nations and the International Centre for Agricultural Research in the Dry Areas: Rome, Italy, 2017; pp. 51–71.
6. Prisa, D. *Opuntia ficus-indica* the key plant in climate change: Characteristics, cultivation and uses. *GSC Biol. Pharm. Sci.* **2021**, *17*, 94–105. [CrossRef]
7. Garcia de Cortázar, V.; Nobel, P.S. Biomass and fruit production for the prickly pear cactus, *Opuntia ficus-indica*. *J. Am. Soc. Hortic. Sci.* **1992**, *117*, 558–562. [CrossRef]
8. Nerd, A.; Karady, A.; Mizrahi, Y. Out-of-season prickly pear: Fruit characteristics and effect of fertilization and short droughts on productivity. *HortScience* **1991**, *26*, 527–529. [CrossRef]
9. Zegbe, J.A.; Serna-Pérez, A. Irrigation options to save water while enhancing export-size fruit and storability of 'Smooth Red' cactus pear. *J. Sci. Food Agric.* **2018**, *98*, 5503–5508. [CrossRef]
10. Konikow, L.F.; Kendy, E. Groundwater depletion: A global problem. *Hydrogeol. J.* **2005**, *13*, 317–320. [CrossRef]
11. Moreno-Vázquez, J.L.; Marañón-Pimentel, B.; López-Córdova, D. Los acuíferos sobrexplotados: Origen, crisis y gestión social. In *El Agua en México: Cauces y Encauces*; Jiménez-Cisneros, B., Torregorosa y Armenta, M.T., Arboites-Aguiliar, L., Eds.; Academia Mexicana de Ciencia y Consejo Nacional del Agua: Ciudad de México, México, 2010; pp. 79–115.
12. Postel, S.L. Entering an era of water scarcity: The challenges ahead. *Ecol. Appl.* **2000**, *10*, 941–948. [CrossRef]
13. Oweis, T.; Hachum, A. *Supplemental Irrigation: A Highly Efficient Water-Use Practice*; International Centre for Agricultural Research in the Dry Areas (ICARDA): Alepo, Syria, 2012; 16p.
14. Oweis, T.; Hachum, A. Improving water productivity in the dry areas of West Asia and North Africa. In *Water Productivity in Agriculture: Limits and Opportunities for Improvement*; Kijne, W.J., Barker, R., Molden, D., Eds.; CAB International: Wallingford, UK, 2003; Volume 1, pp. 179–197.
15. Acosta-Díaz, E.; Acosta-Gallegos, J.A.; Amador-Ramírez, M.D.; Padilla-Ramírez, J.S. Effect of supplemental irrigation on biomass and seed yield of rainfed dry bean in highlands of Mexico. *Agric. Téc. Méx.* **2009**, *35*, 157–167.
16. Abderrazzak, B.; Daoui, K.; Kajji, A.; Ibriz, M.; Dahan, R. Effects of supplemental irrigation and nitrogen applied on yield and yield components of bread wheat at the saïs region of Morocco. *J. Exp. Agric. Int.* **2013**, *3*, 904–913. [CrossRef]
17. Campos, P.F.; Júnior, J.A.; Casaroli, D.; Fontoura, P.R.; Evangelista, A.W.P. Varieties of sugarcane to supplemental irrigation in Brazilian savanna. *Eng. Agríc. Jaboticabal* **2014**, *34*, 1139–1149. [CrossRef]
18. Wale, A.; Sebnie, W.; Girmay, G.; Beza, G. Evaluation of the potentials of supplementary irrigation for improvement of sorghum yield in Wag-Himra, North Eastern, Amhara, Ethiopia. *Cogent Food Agric.* **2019**, *5*, 1664203. [CrossRef]
19. Cavalcante, E.S.; Feitosa de Lacerda, C.; Távora-Costa, R.N.; Gheyi, H.R.; Luzia-Pinho, L.; Sérvulo-Bezerra, F.M.; Cruz de Oliveira, A.; Florentino-Canjá, J. Supplemental irrigation using brackish water on maize in tropical semi-arid regions of Brazil: Yield and economic analysis. *Sci. Agric.* **2021**, *78* (Suppl. 1), e20200151. [CrossRef]
20. Razouk, R.; Ibijbijen, J.; Kajji, A. Optimal time of supplemental irrigation during fruit development of rainfed olive tree (*Olea europaea*, cv. Picholine Marocaine) in Morocco. *Am. J. Exp. Agric.* **2013**, *3*, 685–697. [CrossRef]
21. Van Der Merwe, L.L.; Wessels, A.B.; Ferreira, D.I. Supplemental irrigation for spineless cactus pear. *Acta Hortic.* **1997**, *438*, 77–82. [CrossRef]

22. Zegbe, J.A.; Servín-Palestina, M. Supplemental irrigation to save water while growing cactus pear in semi-arid regions. *Irrig. Drain.* **2021**, *70*, 269–280. [CrossRef]
23. Zegbe, J.A. Quality and storability of cactus pear fruit as improved by supplemental irrigation. *J. Prof. Assoc. Cactus Dev.* **2020**, *22*, 1–17.
24. Zegbe, J.A.; Mena-Covarrubias, J. Two reproductive bud thinning alternatives for cactus pear. *HortTechnology* **2010**, *20*, 202–205. [CrossRef]
25. Diaz-Pérez, J.C.; Shackel, K.A.; Sutter, E.G. Relative water content and water potential of tissue cultured apple shoots under water deficits. *J. Exp. Bot.* **1995**, *46*, 111–118. [CrossRef]
26. Pérez-Pérez, J.G.; Syvertsen, J.P.; Botía, P.; García-Sánchez, F. Leaf water relations and net gas exchange responses of salinized carrizo citrange seedlings during drought stress and recovery. *Ann. Bot.* **2007**, *100*, 335–345. [CrossRef]
27. Popescu, M.; Popescu, G.C. Diurnal changes in leaf photosynthesis and relative water content of grapevine. *Curr. Trends Nat. Sci.* **2014**, *3*, 74–81.
28. Tolk, J.A. Soils, permanent wilting points. In *Encyclopedia of Water Science*; Stewart, B.A., Howell, T.A., Eds.; Marcel Dekket Pub: New York, NY, USA, 2003; pp. 927–929.
29. Ghrab, M.; Masmoudi, M.M.; Ben Mechlia, N. Water productivity in fruit trees orchards under water scarcity. *Acta Hortic.* **2017**, *1150*, 317–322. [CrossRef]
30. Zegbe, J.A.; Serna-Pérez, A.; Mena-Covarrubias, J. Mineral nutrition enhances yield and affects fruit quality of 'Cristalina' cactus pear. *Sci. Hortic.* **2014**, *167*, 63–70. [CrossRef]
31. Smith, M.R.; Rao, I.M.; Merchant, A. Source-sink relationships in crop plants and their influence on yield development and nutritional quality. *Front. Plant Sci.* **2018**, *9*, 1889. [CrossRef] [PubMed]
32. Heldt, H.-W.; Piechulla, B. Phloem transport distributes photoassimilates to various sites of consumption and storage. In *Plant Biochemistry*, 5th ed.; Heldt, H.-W., Piechulla, B., Eds.; Academic Press: Cambridge, MA, USA, 2021; pp. 323–333.
33. Crisosto, C.H.; Johnson, R.S.; Luza, J.G.; Crisosto, G.M. Irrigation regimes affect fruit soluble solids concentration and rate of water loss of 'O'Henry' peaches. *HortScience* **1994**, *29*, 1169–1171. [CrossRef]
34. Mpelasoka, B.S.; Behboudian, M.H.; Dixon, J.; Neal, S.M.; Caspari, H.W. Improvement of fruit quality and storage potential of 'Braeburn' apple through deficit irrigation. *J. Hortic. Sci. Biotechnol.* **2000**, *75*, 615–621. [CrossRef]
35. Lufu, R.; Ambaw, A.; Opara, U.L. Water loss of fresh fruit: Influencing pre-harvest, harvest and postharvest factors. *Sci. Hortic.* **2020**, *272*, 109519. [CrossRef]
36. Cantwell, M. Post-harvest management of fruits and vegetable stems. In *Agro-Ecology, Cultivation, and Uses of Cactus Pear*; Barbera, G., Inglese, P., Pimienta-Barrios, E., Eds.; Plant Production and Protection 132; Food and Agriculture Organization of the United Nations: Rome, Italy, 1995; pp. 120–136.
37. Corrales García, J.; Hernández Silva, J.L. Cambios en la calidad postcosecha de variedades de tuna con y sin semilla. *Rev. Fitotec. Mex.* **2005**, *28*, 9–16.
38. Corrales Garcia, J.; Franco Moreno, F.; Rodriguez Campos, J. Fruit characterization of twenty accessions of cactus pear (*Opuntia* spp.) II. Changes in post-harvest. *Acta Hortic.* **2006**, *728*, 205–209. [CrossRef]
39. Maguire, K.M.; Lang, A.; Banks, N.H.; Hall, A.; Hopcroft, D.; Bennett, R. Relationship between water vapour permeance of apples and micro-cracking of the cuticle. *Postharvest Biol. Technol.* **1999**, *17*, 89–96. [CrossRef]
40. Faedda, R.; Daquino, S.; Granata, G.; Pane, A.; Palma, A.; Sanzani, S.M.; Schena, L.; Cacciola, S.O. Postharvest fungal diseases of cactus pear fruit in southern Italy. *Acta Hortic.* **2016**, *1144*, 215–218. [CrossRef]
41. Granata, G.; Faedda, R.; Ochoa, M.J. Diseases of cactus pear. In *Crop Ecology, Cultivation and Uses of Cactus Pear*; Inglese, P., Mondragon, C., Nefzaoui, A., Sáenz, C., Eds.; The Food and Agriculture Organization of the United Nations and the International Centre for Agricultural Research in the Dry Areas: Rome, Italy, 2017; pp. 115–123.

horticulturae

MDPI

Article

On-Tree Fruit Bagging and Cold Storage Maintain the Postharvest Quality of Mango Fruit

Atif Nadeem [1], Zienab Fawzy Reiad Ahmed [2,*], Syed Bilal Hussain [3,*], Alaa El-Din K. Omar [4,5], Muhammad Amin [6], Saqib Javed [1], Amjad Ali [1], Sami Ullah [1], Kashif Razzaq [1], Ishtiaq A. Rajwana [1], Shafa Nayab [1], Vasileios Ziogas [7], Shamel M. Alam-Eldein [8] and Amany M. Mira [8]

1 Department of Horticulture, Muhammad Nawaz Shareef University of Agriculture, Multan 600000, Pakistan
2 Department of Integrative Agriculture, College of Agriculture and Veterinary Medicine, United Arab Emirates University, Al Ain 15551, United Arab Emirates
3 Citrus Research and Education Center, University of Florida, 700 Experiment Station Rd., Lake Alfred, FL 33850, USA
4 Institute of Research and Consulting, King Faisal University, Al-Hassa 31982, Saudi Arabia
5 Horticulture Department, Faculty of Agriculture, Kafrelsheikh University, Kafr El-Sheikh 33516, Egypt
6 Department of Horticultural Sciences, Faculty of Agriculture & Environment, The Islamia University of Bahawalpur, Bahawalpur 63100, Pakistan
7 Institute of Olive Tree Subtropical Crops and Viticulture, ELGO-DIMITRA, 73134 Chania, Greece
8 Department of Horticulture, Faculty of Agriculture, Tanta University, Tanta 31527, Egypt
* Correspondence: zienab.ahmed@uaeu.ac.ae (Z.F.R.A.); syedbilalhussain@ufl.edu (S.B.H.)

Abstract: The present study investigates the influence of on-tree fruit bagging on the quality and shelf life of mango (*Mangifera indica* L. cv. 'Samar Bahisht Chaunsa') during cold storage (12 °C ± 1; 85–90% RH) for 0, 10 and 20 days (d) and subsequent shelf storage under ambient conditions (25 °C ± 1; 60–65% RH). Fruits were covered with brown paper bags at the marble stage and then harvested at commercial maturity. Results showed that 0 d and 10 d cold-stored fruits, irrespective of bagging treatments, retained eatable quality and shelf-life up to 7 d and 5 d during ambient storage, respectively. However, bagged fruits had better postharvest performance compared with non-bagged fruits by exhibiting slower weight loss, higher fruit firmness, more total soluble solids, vitamin C and total phenolic content and higher activities of catalase and superoxide dismutase during cold storage and ambient shelf storage. On the other hand, 20 d cold-stored fruits, both bagged and non-bagged, were decayed when kept under ambient conditions. It is proposed that mango fruit bagging could be a potential cultural practice to preserve postharvest quality up to 10 d of cold storage, followed by 5 d under ambient conditions.

Keywords: *Mangifera indica*; brown paper bag; cold storage; postharvest performance; antioxidants; phenolics

Citation: Nadeem, A.; Ahmed, Z.F.R.; Hussain, S.B.; Omar, A.E.-D.K.; Amin, M.; Javed, S.; Ali, A.; Ullah, S.; Razzaq, K.; Rajwana, I.A.; et al. On-Tree Fruit Bagging and Cold Storage Maintain the Postharvest Quality of Mango Fruit. *Horticulturae* **2022**, *8*, 814. https://doi.org/10.3390/horticulturae8090814

Academic Editors: Yang Bi, Yongcai Li and Di Gong

Received: 1 August 2022
Accepted: 1 September 2022
Published: 6 September 2022

1. Introduction

Mango (*Mangifera indica* L.), also known as the 'King of Fruits', is one of the most famous tropical fruits which has high nutritional value with excellent fragrance and delicious taste [1]. Pakistan is the world's 5th largest mango producer, accounting for about 5% of global demand [2]. In Pakistan, it is grown on an area of 158,000 hectares with an annual production of around 1.723 million tons [3]. Notably, the 'Samar Bahisht Chaunsa' mango, having a unique flavour and fragrance, is popular in domestic and foreign markets such as Saudi Arabia, the United Arab Emirates, Iran, the United Kingdom, and Kuwait [4–6].

Good agricultural practices and environmental factors during the fruit growing season largely determine the postharvest performance of the produce [7]. Among them, on-tree fruit bagging is an emerging agricultural practice that involves the covering of fruit with a cloth or paper bag, which is widely used for pear [8], apple [9], grape [10], and mango [11,12]. Technically, it alters the microenvironment of the fruit, which facilitates the reduction of pest infestations, wounds, bruises, sunburn, blemishes, and scars [13]. For example, in the Konkan region (Maharashtra-India), prolonged winter and fog are unfavourable for mango production, and it damages the external appearance of the fruit. Therefore, the bagging practice is used to minimize the negative effect of unfavourable environmental conditions [14]. Bagging can also improve the internal quality of fruits by promoting pigmentation at harvest [15]. Apart from improving fruit physical appearance, bagging encourages fruit production without the application of pesticides [16]. Currently, this practice is commercially used in Japan, China, Australia, and the United States of America [17–19]. It is interesting to mention that countries such as Mexico, Chile and Argentina do not import fruits unless they were previously bagged [20].

Due to climacteric ripening behaviour and poor postharvest management practices, mango fruits have a short postharvest life span, these factors can restrict their export potential. Generally, fruit storage at low-temperature delays the ripening process, however, the optimal temperature ranges vary among fruit species. In mango, ambient storage temperature for 21 days (d) causes significantly higher physiological weight loss (7.31%) as compared to low-temperature storage at 10 °C (2.15%) [21]. With reduced fruit weight loss and respiration rate, low-temperature storage can prolong the postharvest shelf-life [22–24]. Hence, the inclusion of pre-harvest and post-harvest management practices in the production system can improve fruit quality and shelf life.

Pakistan has a number of mango varieties that have better adaptability under the prevailing agro-climatic conditions and can produce high-quality fruits. In recent times, the demand for Pakistani mango fruits is receiving more importance and commercial interest. To fulfil local and international demand, the production of high-quality fruit is a prerequisite. So far, mango is largely produced without bagging for domestic as well as the export market. Considering the above-mentioned benefits of fruit bagging and cold storage, it can be hypothesized that on-tree fruit bagging along with cold storage can further improve the postharvest performance of mango fruits. In this study, postharvest quality and shelf life under ambient and cold storage conditions were compared between the bagged and non-bagged fruits of the 'Samar Bahisht Chaunsa' mango.

2. Materials and Methods

2.1. Fruit and Bagging Material

For this study, the experimental fruits of the 'Samar Bahist Chaunsa' mango were grown in a commercial mango orchard in Multan, Pakistan (30°13′05′′ N 71°34′15′′ E). At the marble stage (45 d after the fruit set), the fruits were bagged using brown paper bags (25 × 20 cm) and at the same time, some fruits were tagged and labelled as control treatment—fruit not bagged.

2.2. Treatments

A total of 135 physiologically mature and healthy fruits (twenty-seven fruits from each tree) were harvested for each bagged and non-bagged treatment, and fruits were treated with Amistar fungicide (8 mL L^{-1} water) for 3 min to overcome sap burn injury. Afterwards, fruits were divided into three groups. One group was evaluated at harvest (0 d at cold storage) and the second and the third group were, respectively, evaluated after 10 d and 20 d of cold storage (12 °C ± 1, 85–90% RH). After each storage interval, fruit samples were sent to the Postharvest Science and Technology Lab, MNS University of Agriculture, Multan, Pakistan for shelf life and quality analysis at ambient conditions (25 °C ± 1, 60–65% RH). In each treatment, there were three replications and each replication had 15 fruits.

2.3. Measurements

The physiological parameters (ethylene production rate, and CO_2 production rate) and physical parameters (fruit peel colour and weight loss) were recorded daily until fruits started to decay whereas destructive analyses were performed on the first day and last day of shelf storage.

2.3.1. Respiration Rate and Ethylene Production

The respiration rate and ethylene production rate were determined by the release of CO_2 and ethylene, respectively. Four fruit derived from each replicate per treatment were placed in a 2.25 L plastic container and sealed at room temperature. After 60 min of incubation, a portable gas analyzer (F-940 Gas Analyzer, Felix, Camas, WA, USA) was used to detect the release of CO_2 and ethylene in a static system. The respiration rate and ethylene production were expressed as μmol CO_2 Kg^{-1} h^{-1} and μmol C_2H_4 Kg^{-1} h^{-1}, respectively.

2.3.2. Peel Colour

Three sites close to the equatorial part of each mango were chosen to determine the peel colour with a colourimeter (CR-400, Konica Minolta, Tokyo, Japan), which was expressed as L^* value (L^* = 100 is white and L^* = 0 is black), a^* value [green (−) to red (+) axis], and b^* value [blue (−) to yellow (+) axis].

2.3.3. Fruit Firmness

Fruit firmness was determined using a handheld penetrometer (FR-5120, Lutron Electronics Enterprises, Taiwan) equipped with an 8 mm diameter flat probe. Two readings were obtained from 2 equidistant points on the equatorial axis of each fruit, after the removal of the skin. Firmness was expressed as a force in Newton (N).

2.3.4. Weight Loss

Fruit weight loss was calculated by taking initial weight on the 0 d and final weight on the sampling day until fruits started to decay.

$$\text{Weight Loss (\%)} = (\text{WO} - \text{WS})/\text{WS} \times 100$$

where WO and WS represent the initial and the final weights of mangoes, respectively.

2.3.5. Soluble Solids Content

Soluble solids content (SSC) was measured using a digital refractometer (PAL-1, Atago Co., Tokyo, Japan) and expressed in percentage (%).

2.3.6. Titratable Acidity and Juice pH

Titratable acidity (TA) was determined by taking 10 mL juice in a 100 mL volumetric flask, followed by the addition of 2–3 drops of phenolphthalein to the juice sample. After that, the sample was titrated against 0.1 N NaOH solution until the endpoint light pink was obtained. TA was expressed as %. Whereas, juice pH was calculated by using a digital pH meter (Starter 3100 OHAUS Corporation, Parsippany, NJ, USA).

2.3.7. Ascorbic Acid Content

The content of ascorbic acid was determined as suggested by Ruck [25] and modified by Xylia, Chrysargyris, Ahmed and Tzortzakis [24]. Briefly, a 10 mL homogenized juice sample was placed in a 100 mL volumetric flask and the final volume was achieved by adding 0.4% oxalic acid. After that, the mixture was filtered through Whatman No. 41 filter paper. About 5 mL of the filtrate aliquot was taken in a flask and titrated against the dye (2,6-Dichlorophenolindophenol) until the light pink colour was achieved. The content was expressed as mg 100 mL^{-1} juice.

2.3.8. Total Phenolic Content and Total Antioxidant Activity

Briefly, 1 g pulp tissue was homogenized with a 5 mL cold extraction mixture [HCL: acetone:methanol (2:8:90)]. After that, samples were centrifuged at 9000 rpm for 5 min at 4 °C and supernatant was collected and used for total antioxidant activity (TAC) and total phenolic content (TPC) determination.

The TAC was assessed by following a method described by Saleem et al. [26]. 50 µL extract was homogenized with 5 µL of 2,2-diphenylpicrylhydrazyl solution (0.004%) in a test tube, followed by 30 min of dark incubation at room temperature. After incubation, 200 µL of each sample was collected in a microplate, and samples were run through a spectrophotometer (Epoch Eliza reader, Winooski, VT, USA) at a wavelength of 517 nm and expressed as % inhibition.

The amount of TPC was measured by following the Folin-Ciocalteu assay. In test tubes, 100 µL of the supernatant was mixed with 200 µL of Folin-Ciocalteu reagent (10%) and vortexed for 10 s. After that, 800 µL of sodium carbonate solution (700 mM) was added and vortexed for another 10 s. Then, the mixture was subjected to dark incubation at room temperature for 2 h. After incubation, 200 µL of the mixture was placed on a microplate, and samples were read at a wavelength of 765 nm. The amount of TPC was expressed as mg Gallic Acid Equivalents (GAE) in a 100 g FW sample (GAE mg 100 g^{-1}).

2.3.9. Determination of Catalase, Peroxidase and Superoxide Dismutase Enzyme Activity

Briefly, 1 g pulp tissue was homogenized with 2 mL of phosphate buffer (pH 7.0–7.8) by using a mortar and pestle. After that, samples were centrifuged at 9000 rpm for 5 min at 4 °C and the supernatant was collected and used for the determination of catalase (CAT), peroxidase (POD) and superoxide dismutase (SOD) enzyme activity. The enzyme activities were expressed as Unit mg^{-1} protein [27]

Catalase Activity

The extracted sample (100 µL) was mixed with a 100 µL solution of hydrogen peroxide (5.9 mM). To evaluate CAT activity, sample absorbance was noted at a wavelength of 240 nm by using a spectrophotometer (Epoch Eliza reader, Winooski, VT, USA).

Peroxidase Activity

100 µL of hydrogen peroxide (40 mM), 100 µL guaiacol, and 800 µL phosphate buffer (50 mM, pH 5) were used in an 8:1:1 to make the POD reaction mixture. Afterwards, 100 µL of enzyme extract was homogenized with 100 µL of the reaction mixture and then subjected to spectrophotometric (Epoch Eliza reader, Winooski, VT, USA) examination at 470 nm.

Superoxide Dismutase Activity

100 µL of enzyme extract was homogenized with 200 µL of methionine, 500 µL of phosphate buffer (50 mM, pH 5), 200 µL of Triton X, 100 µL of Nitro blue tetrazolium, and 800 µL of distilled water. The mixture was then exposed to UV light for 15 min, followed by the addition of 100 µL of riboflavin. Then the samples were exposed to spectrophotometric (Epoch Eliza reader, Winooski, VT, USA) observation at 560 nm to determine the absorbance values.

2.4. Statistical Analysis

The experiment was designed using a randomised complete block design with a two-factor factorial layout (bagging treatment and storage interval). Using Statistix 9® software, the acquired data was statistically analysed (ANOVA and LSD test at $p \leq 0.05$ level). The figures were generated using Sigma Plot 10.0 (SPSS, Chicago, IL, USA).

3. Results

3.1. Ethylene Production, Respiration Rate and Weight Loss

The change in ethylene production between the bagged and non-bagged treatments was not significant at each cold storage interval. However, irrespective of bagging treatment, there was a rise in ethylene production at 10 d of storage, followed by no significant change on 20 d of cold storage (Figure 1A). During shelf-life storage, ethylene production in 0 d cold-stored fruits followed an increasing trend, irrespective of bagging treatments

(Figure 1B). However, ethylene production fluctuated in 10 d cold-stored fruits throughout the ambient storage (Figure 1C).

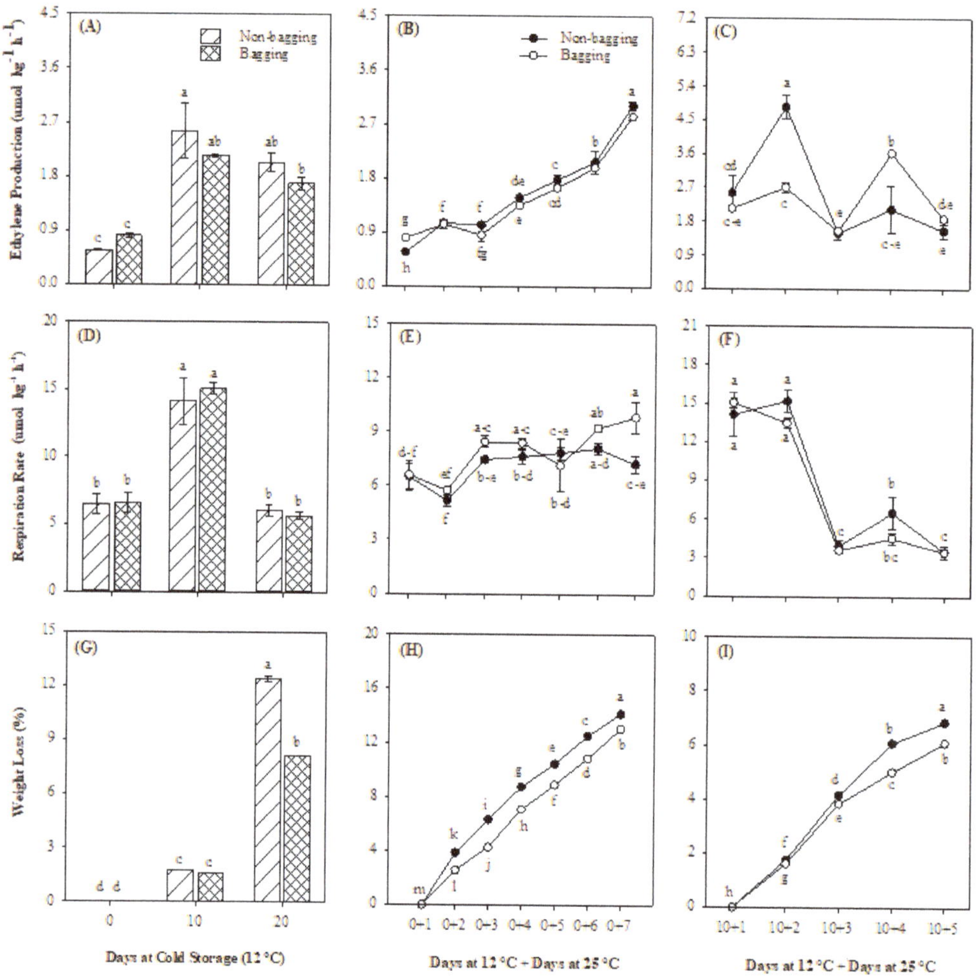

Figure 1. Effect of on-tree fruit bagging treatments on post-harvest ethylene production (**A–C**), respiration rate (**D–F**) and weight loss (**G–I**) in mango fruit during cold storage intervals and under subsequent ambient conditions. Error bars show standard errors of the means ($n = 3$). Different letters indicate significant differences according to the LSD test at $p \leq 0.05$. When 20-day cold-stored fruits were kept under ambient conditions, fruits were decayed within 24 h hence no tests were performed.

Similar to ethylene production, the effect of bagging treatments on respiration rate was found non-significant at each storage interval. However, the respiration rate was higher in 10 d of cold storage fruits as compared to 0 d and 20 d of cold storage (Figure 1D). Under ambient storage, 0 d stored fruits had no changes in respiration rate on each shelf day, except significantly more respiration rate in bagged fruits on the seventh day (7 d) of ambient storage (Figure 1E). Moreover, 10 d cold-stored fruits showed a sudden decline in respiration rate on the third day (3 d) of ambient storage, however, the difference was non-significant between the bagging treatments (Figure 1F).

There was an increase in weight loss with the passage of storage intervals, however, the difference between the bagged and non-bagged fruits was non-significant on each interval, except on 20 d of cold storage with significantly less weight loss in bagged fruits as compared to non-bagged fruit (Figure 1G). During ambient storage, 0 d stored fruits (both bagged and non-bagged) exhibited an increased weight loss with the advancement in shelf storage. However, more weight loss was observed in non-bagged fruits starting from the second day (2 d) to the seventh day (7 d) of ambient storage (Figure 1H). Similarly, weight loss changes were observed in 10 d cold-stored fruits under ambient storage (Figure 1I).

3.2. Fruit Firmness

A significant effect of bagging treatments was found on fruit firmness at different cold storage intervals. Except on 20 d of cold storage, bagged fruits maintained higher values of fruit firmness as compared to non-bagged fruits on 0 d and 10 d of cold storage intervals. Overall, there was a significant decline in fruit firmness with the passage of cold storage interval (Table 1). During ambient storage, 0 day and 10 d cold-stored fruits exhibited more fruit firmness in bagged fruits as compared to non-bagged fruits. Moreover, fruit firmness was significantly decreased to the last day of their shelf life which was 7 d and 5 d in 0 d and 10 d cold-stored fruits, respectively (Table 1).

Table 1. Effect of on-tree fruit bagging treatments on mango fruit firmness during cold storage intervals and under ambient conditions. Values are means of triplicates ± SD. Different letters in superscripts indicate significant differences according to LSD test at $p \leq 0.05$. D, decayed fruit hence no tests were performed.

Treatment	Fruit Firmness (N)	
	Non-Bagging	Bagging
	a. Days at 12 °C	
0	142.67 ± 0.33^{b}	150.00 ± 0.58^{a}
10	79.33 ± 0.88^{d}	94.23 ± 0.50^{c}
20	39.00 ± 0.58^{e}	37.17 ± 0.73^{f}
	b. Days at 12 °C + Days at 25 °C	
0 + 0	142.67 ± 0.33^{b}	150.00 ± 0.58^{a}
0 + 7	27.30 ± 1.37^{d}	94.80 ± 0.79^{c}
10 + 0	79.33 ± 0.88^{b}	94.23 ± 0.50^{a}
10 + 5	38.47 ± 0.36^{d}	46.93 ± 1.90^{c}
20 + 0	39.00 ± 0.58^{b}	37.17 ± 0.73^{a}
20 + 1	D	D

3.3. Fruit Peel Colour (L^*, a^* and b^*)

Except in 10 d cold-stored fruits, bagging treatments did not affect the L^* value (Figure 2A). During ambient storage, 0 d cold-stored fruits exhibited no significant change throughout the storage, except a significantly high value of L^* in bagged fruits on the seventh day (7 d) (Figure 2B). Moreover, when 10 d stored fruits were kept under ambient storage conditions, bagged fruits exhibited a higher value of L^* as compared to control (Figure 2C). Similar to the effect of bagging treatments on L^* value, a* value was also significantly changed in 10 d cold-stored fruits (Figure 2D). During ambient storage, 0 d cold-stored non-bagged fruits exhibited an increasing profile from the second day (2 d) to the seventh day (7 d) while bagged fruits had decreasing profile from the second day (2 d) to the sixth day (6 d) with non-significant changes followed by a significant increase (Figure 2E). During ambient storage, 10 d cold-stored bagged fruits had significantly higher values of a^* on the second day (2 d) and fourth day (4 d) (Figure 2F). As for the b^* value is concerned, only 10 d cold-stored fruits exhibited significant changes due to bagging treatments (Figure 2G). During ambient storage, 0 d cold-stored non-bagged fruits had a significantly higher value of b^* on the fifth day (5 d) and seventh day (7 d) as compared to bagged fruits (Figure 2H). While 10 d cold-stored bagged fruits showed higher values of b^*

on the first day (1 d), second day (2 d) and fifth day (5 d) as compared to non-bagged fruits (Figure 2I).

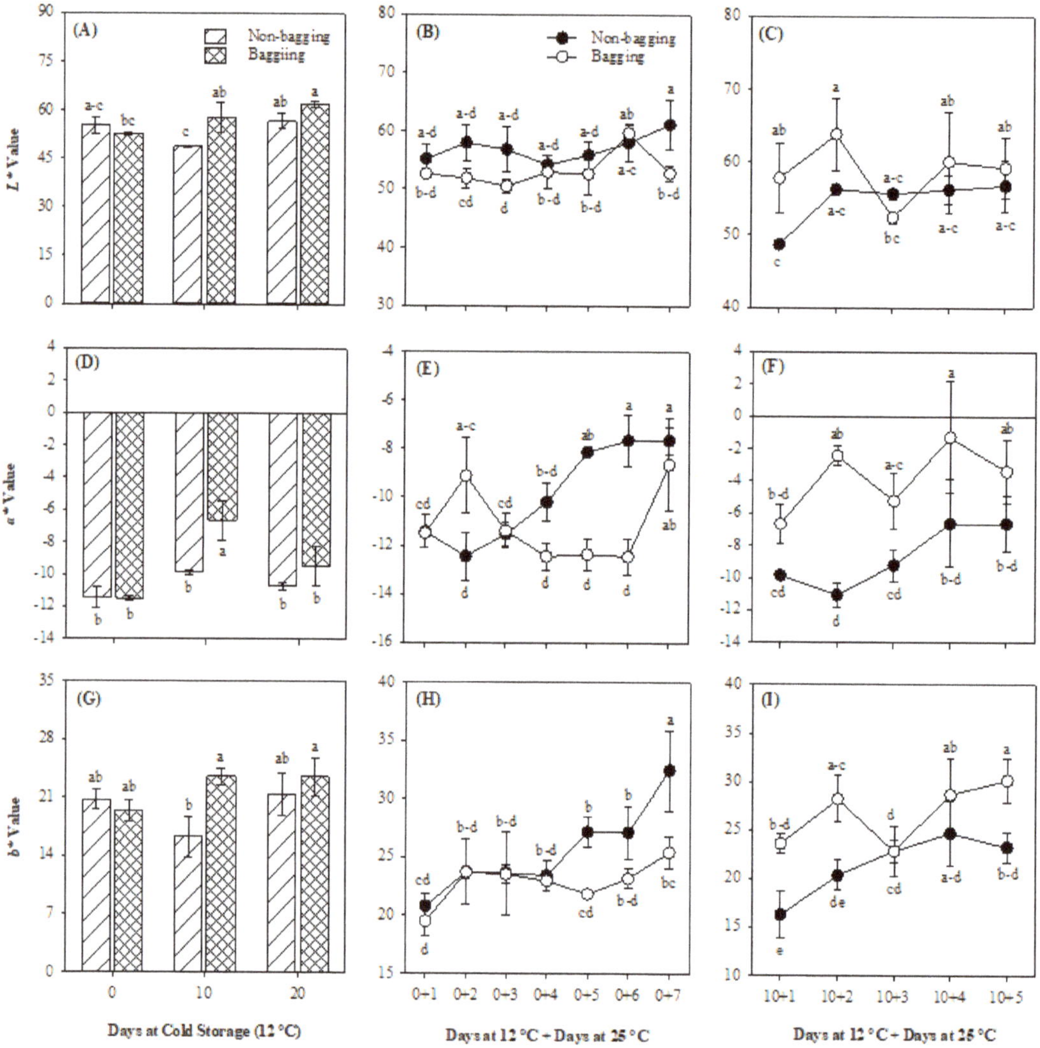

Figure 2. Effect of on-tree fruit bagging treatments on L^* value (**A–C**), a^* value (**D–F**) and b^* value (**G–I**) in mango fruit during cold storage intervals and under subsequent ambient conditions. Error bars show standard errors of the means (n = 3). Different letters indicate significant differences according to the LSD at $p \leq 0.05$. When 20-day cold-stored fruits were kept under ambient conditions, fruits were decayed within 24 h hence no tests were performed.

3.4. *Soluble Solids Content, Titratable Acidity and Juice pH*

Except on 0 d of cold storage, bagging treatments had a significant effect on SSC. There was increased SSC content in bagged fruits than in non-bagged fruits on 10 d and 20 d of cold storage. Overall, there was an increasing profile of SSC from 0 d to 20 d of cold storage (Table 2). Moreover, 0 d and 10 d cold-stored bagged fruits yielded more SSC than non-bagged fruits. Also, a significant increase in SSC was found during ambient storage (Table 2). There was a significant difference in TA content due to bagging treatments on each storage interval, except for 20 d storage (Table 2). Overall, the TA content followed the decreasing trend with the passage of cold storage intervals. During ambient storage of 0 d and 10 d cold-stored fruit, TA contents were significantly reduced with the passage of days, irrespective of bagging treatments (Table 2). Bagging treatments significantly changed the juice pH on each storage interval (Table 2). Bagged fruits had lower juice pH than non-bagged fruits. During the ambient storage of 0 d cold-stored fruit, the change in juice pH was non-significant between the first day (1 d) and seventh day (7 d); however, a significant increase in juice pH was observed in both bagged and non-bagged fruits on the fifth day (5 d) as compared with the first day (1 d) (Table 2).

Table 2. Effect of on-tree fruit bagging treatments on soluble solids contents (SSC), titratable acidity (TA) and juice pH in mango fruit during cold storage intervals and under ambient conditions. Values are means of triplicates ± SD. Different letters in superscripts indicate significant differences according to LSD test at $p \leq 0.05$. D, decayed fruit hence no tests were performed.

Treatment	SSC (%)		TA (%)		Juice pH	
	Non-Bagging	Bagging	Non-Bagging	Bagging	Non-Bagging	Bagging
a. Days at 12 °C						
0	7.55 ± 0.38 ^{e}	8.33 ± 0.06 ^{e}	0.62 ± 0.01 ^{a}	0.45 ± 0.02 ^{b}	4.63 ± 0.09 ^{c}	4.27 ± 0.06 ^{d}
10	14.53 ± 0.24 ^{d}	15.63 ± 0.35 ^{c}	0.59 ± 0.00 ^{a}	0.41 ± 0.02 ^{bc}	5.44 ± 0.09 ^{a}	4.65 ± 0.06 ^{c}
20	23.03 ± 0.33 ^{b}	25.50 ± 0.55 ^{a}	0.38 ± 0.00 ^{cd}	0.35 ± 0.03 ^{d}	5.19 ± 0.02 ^{b}	4.27 ± 0.0 ^{d}
b. Days at 12 °C + Days at 25 °C						
0 + 0	7.55 ± 0.38 ^{d}	8.33 ± 0.06 ^{c}	0.62 ± 0.01 ^{a}	0.45 ± 0.02 ^{b}	4.63 ± 0.09 ^{b}	4.27 ± 0.06 ^{b}
0 + 7	25.42 ± 0.42 ^{b}	27.03 ± 0.27 ^{a}	0.29 ± 0.04 ^{c}	0.19 ± 0.01 ^{d}	5.42 ± 0.08 ^{a}	5.74 ± 0.16 ^{a}
10 + 0	14.53 ± 0.24 ^{c}	15.63 ± 0.35 ^{c}	0.59 ± 0.00 ^{a}	0.41 ± 0.02 ^{b}	5.44 ± 0.09 ^{a}	4.65 ± 0.06 ^{c}
10 + 5	22.29 ± 0.35 ^{b}	25.10 ± 0.36 ^{a}	0.40 ± 0.01 ^{b}	0.26 ± 0.02 ^{c}	4.64 ± 0.01 ^{c}	4.86 ± 0.03 ^{b}
20 + 0	23.03 ± 0.33 ^{b}	25.50 ± 0.55 ^{a}	0.38 ± 0.00 ^{a}	0.35 ± 0.03 ^{a}	5.19 ± 0.02 ^{a}	4.27 ± 0.0 ^{b}
20 + 1	D	D	D	D	D	D

3.5. Vitamin C, Total Antioxidant Activity and Total Phenolic Contents

A significantly more vitamin C content was found in bagged fruits as compared to non-bagged fruits during each cold-storage interval. However, there was a significant decrease in non-bagged fruit with the passage of cold storage time (Table 3). During the ambient storage of 0 d cold-stored fruit, the change in vitamin C content was non-significant between the first day (1 d) and the seventh day (7 d). However, a significant decline in vitamin C content was observed in both bagged and non-bagged fruits on the fifth day (5 d) as compared with the first day (1 d) (Table 3). The effect of bagging treatment was found non-significant on TAC during cold-storage intervals and also during ambient storage (Table 3). However, there was a significant difference in TPC due to bagging treatments on each storage interval. Bagged fruits exhibited more TPC as compared to non-bagged fruits (Table 3). During ambient storage, 0 d cold-stored fruits had no significant change in TPC, irrespective of bagging treatments. While, 10 d cold-stored fruits presented an increased profile of TPC from the first day (1 d) to the fifth day (5 d), irrespective of bagging treatments (Table 3).

Table 3. Effect of on-tree fruit bagging treatments on vitamin C, total antioxidant activity (TAC) and total phenolic content (TPC) in mango fruits during cold storage intervals and under ambient conditions. Values are means of triplicates ± SD. Different letters in superscripts indicate significant differences according to LSD test at $p \leq 0.05$. D, decayed fruits hence no tests were performed.

Treatment	Vitamin C (mg 100 mL^{-1} Juice)		TAC (% Inhibition)		TPC (mg GAE 100 g^{-1})	
	Non-Bagging	Bagging	Non-Bagging	Bagging	Non-Bagging	Bagging
			a. Days at 12 °C			
0	32.98 ± 0.42 ^{b}	38.40 ± 0.67 ^{a}	83.97 ± 0.17 ^{a}	87.68 ± 0.50 ^{a}	72.26 ± 0.67 ^{bc}	83.03 ± 3.53 ^{a}
10	21.23 ± 0.18 ^{d}	27.80 ± 0.26 ^{c}	85.71 ± 1.05 ^{a}	83.83 ± 3.44 ^{a}	67.71 ± 0.66 ^{c}	76.16 ± 2.48 ^{b}
20	12.30 ± 0.46 ^{f}	15.70 ± 0.26 ^{e}	81.83 ± 2.76 ^{a}	82.31 ± 1.23 ^{a}	74.96 ± 1.15 ^{b}	85.55 ± 1.40 ^{a}
			b. Days at 12 °C + Days at 25 °C			
0 + 0	32.98 ± 0.42 ^{c}	38.40 ± 0.67 ^{a}	83.97 ± 0.17 ^{b}	87.68 ± 0.50 ^{a}	72.26 ± 0.67 ^{bc}	83.03 ± 3.53 ^{a}
0 + 7	34.70 ± 1.21 ^{bc}	36.88 ± 0.48 ^{ab}	84.49 ± 0.78 ^{b}	86.98 ± 0.54 ^{a}	67.44 ± 0.74 ^{c}	80.32 ± 3.98 ^{ab}
10 + 0	21.23 ± 0.18 ^{c}	27.80 ± 0.26 ^{a}	85.71 ± 1.05 ^{a}	83.83 ± 3.44 ^{a}	67.71 ± 0.66 ^{c}	76.16 ± 2.48 ^{b}
10 + 5	15.77 ± 0.26 ^{d}	26.67 ± 0.42 ^{b}	86.98 ± 0.17 ^{a}	88.95 ± 0.77 ^{a}	78.77 ± 2.29 ^{ab}	85.13 ± 1.66 ^{a}
20 + 0	12.30 ± 0.46 ^{b}	15.70 ± 0.26 ^{a}	81.83 ± 2.76 ^{a}	82.31 ± 1.23 ^{a}	74.96 ± 1.15 ^{b}	85.55 ± 1.40 ^{a}
20 + 1	D	D	D	D	D	D

3.6. *Antioxidative Enzyme Activities (CAT, POD and SOD)*

Bagged fruits had significantly higher CAT activity than non-bagged fruits, except for 20 d of cold storage (Table 4). Moreover, a non-significant change in CAT activity was found during shelf storage of 0 d and 10 d cold-stored fruits (Table 4). Except for 0 d of cold storage, POD activity had a non-significant difference during cold storage (Table 4). Moreover, a non-significant change in POD activity was found during ambient storage of 0 d and 10 d cold-stored fruits (Table 4). There was a significant difference in SOD activity between bagged and non-bagged fruits on each storage interval, except for 20 d of cold storage. Non-bagged fruits improved the SOD activity on 10 d of cold storage as compared to 0 d of cold storage. However, bagged fruits did not show any significant change during cold storage (Table 4). During ambient storage, 0 d cold-stored non-bagged fruits improved their SOD activity. However, on the seventh day (7 d), no change was observed in bagged fruit's SOD activity (Table 4). A significant decline in SOD activity was observed in both 10 d cold-stored bagged and non-bagged fruits under shelf storage (Table 4).

Table 4. Effect of on-tree fruit bagging treatments on catalase (CAT), peroxidase (POD) and superoxide dismutase (SOD) activities in mango fruit during cold storage intervals and under ambient conditions. Values are means of triplicates ± SD. Different letters in superscripts indicate significant differences according to the LSD test at $p \leq 0.05$. D, decayed fruit hence no tests were performed.

Treatment	CAT (U mg^{-1} Protein)		POD (U mg^{-1} Protein)		SOD (U mg^{-1} Protein)	
	Non-Bagging	Bagging	Non-Bagging	Bagging	Non-Bagging	Bagging
			a. Days at 12 °C			
0	3.97 ± 0.19 ^{c}	8.39 ± 1.24 ^{a}	0.16 ± 0.01 ^{b}	2.79 ± 0.48 ^{a}	53.41 ± 1.63 ^{c}	86.78 ± 3.23 ^{a}
10	4.18 ± 0.70 ^{c}	7.29 ± 0.28 ^{ab}	0.59 ± 0.19 ^{b}	0.59 ± 0.36 ^{b}	75.25 ± 2.25 ^{b}	86.61 ± 2.60 ^{a}
20	3.62 ± 0.40 ^{c}	5.27 ± 1.12 ^{bc}	0.73 ± 0.43 ^{b}	0.93 ± 0.36 ^{b}	81.32 ± 0.76 ^{ab}	87.06 ± 3.06 ^{a}
			b. Days at 12 °C + Days at 25 °C			
0 + 0	3.97 ± 0.19 ^{c}	8.39 ± 1.24 ^{a}	0.16 ± 0.0 ^{c}	2.79 ± 0.48 ^{a}	53.41 ± 1.63 ^{c}	86.78 ± 3.23 ^{a}
0 + 7	6.66 ± 1.19 ^{ab}	9.86 ± 1.23 ^{a}	1.23 ± 0.48 ^{bc}	2.05 ± 0.45 ^{ab}	69.54 ± 2.92 ^{b}	83.88 ± 1.64 ^{a}
10 + 0	4.18 ± 0.70 ^{b}	7.29 ± 0.28 ^{a}	0.59 ± 0.19 ^{a}	0.59 ± 0.36 ^{a}	75.25 ± 2.25 ^{b}	86.61 ± 2.60 ^{a}
10 + 5	3.05 ± 0.15 ^{b}	8.12 ± 1.10 ^{a}	0.66 ± 0.08 ^{a}	1.04 ± 0.10 ^{a}	70.13 ± 1.74 ^{c}	74.84 ± 2.79 ^{bc}
20 + 0	3.62 ± 0.40 ^{a}	5.27 ± 1.12 ^{a}	0.73 ± 0.43 ^{a}	0.93 ± 0.36 ^{a}	81.32 ± 0.76 ^{a}	87.06 ± 3.06 ^{a}
20 + 1	D	D	D	D	D	D

each storage interval, except for 20 d storage (Table 2). Overall, the TA content followed the decreasing trend with the passage of cold storage intervals. During ambient storage of 0 d and 10 d cold-stored fruit, TA contents were significantly reduced with the passage of days, irrespective of bagging treatments (Table 2). Bagging treatments significantly changed the juice pH on each storage interval (Table 2). Bagged fruits had lower juice pH than non-bagged fruits. During the ambient storage of 0 d cold-stored fruit, the change in juice pH was non-significant between the first day (1 d) and seventh day (7 d); however, a significant increase in juice pH was observed in both bagged and non-bagged fruits on the fifth day (5 d) as compared with the first day (1 d) (Table 2).

Table 2. Effect of on-tree fruit bagging treatments on soluble solids contents (SSC), titratable acidity (TA) and juice pH in mango fruit during cold storage intervals and under ambient conditions. Values are means of triplicates ± SD. Different letters in superscripts indicate significant differences according to LSD test at $p \leq 0.05$. D, decayed fruit hence no tests were performed.

Treatment	SSC (%)		TA (%)		Juice pH	
	Non-Bagging	Bagging	Non-Bagging	Bagging	Non-Bagging	Bagging
			a. Days at 12 °C			
0	7.55 ± 0.38^{e}	8.33 ± 0.06^{e}	0.62 ± 0.01^{a}	0.45 ± 0.02^{b}	4.63 ± 0.09^{c}	4.27 ± 0.06^{d}
10	14.53 ± 0.24^{d}	15.63 ± 0.35^{c}	0.59 ± 0.00^{a}	0.41 ± 0.02^{bc}	5.44 ± 0.09^{a}	4.65 ± 0.06^{c}
20	23.03 ± 0.33^{b}	25.50 ± 0.55^{a}	0.38 ± 0.00^{cd}	0.35 ± 0.03^{d}	5.19 ± 0.02^{b}	4.27 ± 0.0^{d}
			b. Days at 12 °C + Days at 25 °C			
0 + 0	7.55 ± 0.38^{d}	8.33 ± 0.06^{c}	0.62 ± 0.01^{a}	0.45 ± 0.02^{b}	4.63 ± 0.09^{b}	4.27 ± 0.06^{b}
0 + 7	25.42 ± 0.42^{b}	27.03 ± 0.27^{a}	0.29 ± 0.04^{c}	0.19 ± 0.01^{d}	5.42 ± 0.08^{a}	5.74 ± 0.16^{a}
10 + 0	14.53 ± 0.24^{c}	15.63 ± 0.35^{c}	0.59 ± 0.00^{a}	0.41 ± 0.02^{b}	5.44 ± 0.09^{a}	4.65 ± 0.06^{c}
10 + 5	22.29 ± 0.35^{b}	25.10 ± 0.36^{a}	0.40 ± 0.01^{b}	0.26 ± 0.02^{c}	4.64 ± 0.01^{c}	4.86 ± 0.03^{b}
20 + 0	23.03 ± 0.33^{b}	25.50 ± 0.55^{a}	0.38 ± 0.00^{a}	0.35 ± 0.03^{a}	5.19 ± 0.02^{a}	4.27 ± 0.0^{b}
20 + 1	D	D	D	D	D	D

3.5. Vitamin C, Total Antioxidant Activity and Total Phenolic Contents

A significantly more vitamin C content was found in bagged fruits as compared to non-bagged fruits during each cold-storage interval. However, there was a significant decrease in non-bagged fruit with the passage of cold storage time (Table 3). During the ambient storage of 0 d cold-stored fruit, the change in vitamin C content was non-significant between the first day (1 d) and the seventh day (7 d). However, a significant decline in vitamin C content was observed in both bagged and non-bagged fruits on the fifth day (5 d) as compared with the first day (1 d) (Table 3). The effect of bagging treatment was found non-significant on TAC during cold-storage intervals and also during ambient storage (Table 3). However, there was a significant difference in TPC due to bagging treatments on each storage interval. Bagged fruits exhibited more TPC as compared to non-bagged fruits (Table 3). During ambient storage, 0 d cold-stored fruits had no significant change in TPC, irrespective of bagging treatments. While, 10 d cold-stored fruits presented an increased profile of TPC from the first day (1 d) to the fifth day (5 d), irrespective of bagging treatments (Table 3).

Table 3. Effect of on-tree fruit bagging treatments on vitamin C, total antioxidant activity (TAC) and total phenolic content (TPC) in mango fruits during cold storage intervals and under ambient conditions. Values are means of triplicates ± SD. Different letters in superscripts indicate significant differences according to LSD test at $p \leq 0.05$. D, decayed fruits hence no tests were performed.

Treatment	Vitamin C (mg 100 mL^{-1} Juice)		TAC (% Inhibition)		TPC (mg GAE 100 g^{-1})	
	Non-Bagging	Bagging	Non-Bagging	Bagging	Non-Bagging	Bagging
			a. Days at 12 °C			
0	32.98 ± 0.42 ^{b}	38.40 ± 0.67 ^{a}	83.97 ± 0.17 ^{a}	87.68 ± 0.50 ^{a}	72.26 ± 0.67 ^{bc}	83.03 ± 3.53 ^{a}
10	21.23 ± 0.18 ^{d}	27.80 ± 0.26 ^{c}	85.71 ± 1.05 ^{a}	83.83 ± 3.44 ^{a}	67.71 ± 0.66 ^{c}	76.16 ± 2.48 ^{b}
20	12.30 ± 0.46 ^{f}	15.70 ± 0.26 ^{e}	81.83 ± 2.76 ^{a}	82.31 ± 1.23 ^{a}	74.96 ± 1.15 ^{b}	85.55 ± 1.40 ^{a}
			b. Days at 12 °C + Days at 25 °C			
0 + 0	32.98 ± 0.42 ^{c}	38.40 ± 0.67 ^{a}	83.97 ± 0.17 ^{b}	87.68 ± 0.50 ^{a}	72.26 ± 0.67 ^{bc}	83.03 ± 3.53 ^{a}
0 + 7	34.70 ± 1.21 ^{bc}	36.88 ± 0.48 ^{ab}	84.49 ± 0.78 ^{b}	86.98 ± 0.54 ^{a}	67.44 ± 0.74 ^{c}	80.32 ± 3.98 ^{ab}
10 + 0	21.23 ± 0.18 ^{c}	27.80 ± 0.26 ^{a}	85.71 ± 1.05 ^{a}	83.83 ± 3.44 ^{a}	67.71 ± 0.66 ^{c}	76.16 ± 2.48 ^{b}
10 + 5	15.77 ± 0.26 ^{d}	26.67 ± 0.42 ^{b}	86.98 ± 0.17 ^{a}	88.95 ± 0.77 ^{a}	78.77 ± 2.29 ^{ab}	85.13 ± 1.66 ^{a}
20 + 0	12.30 ± 0.46 ^{b}	15.70 ± 0.26 ^{a}	81.83 ± 2.76 ^{a}	82.31 ± 1.23 ^{a}	74.96 ± 1.15 ^{b}	85.55 ± 1.40 ^{a}
20 + 1	D	D	D	D	D	D

3.6. *Antioxidative Enzyme Activities (CAT, POD and SOD)*

Bagged fruits had significantly higher CAT activity than non-bagged fruits, except for 20 d of cold storage (Table 4). Moreover, a non-significant change in CAT activity was found during shelf storage of 0 d and 10 d cold-stored fruits (Table 4). Except for 0 d of cold storage, POD activity had a non-significant difference during cold storage (Table 4). Moreover, a non-significant change in POD activity was found during ambient storage of 0 d and 10 d cold-stored fruits (Table 4). There was a significant difference in SOD activity between bagged and non-bagged fruits on each storage interval, except for 20 d of cold storage. Non-bagged fruits improved the SOD activity on 10 d of cold storage as compared to 0 d of cold storage. However, bagged fruits did not show any significant change during cold storage (Table 4). During ambient storage, 0 d cold-stored non-bagged fruits improved their SOD activity. However, on the seventh day (7 d), no change was observed in bagged fruit's SOD activity (Table 4). A significant decline in SOD activity was observed in both 10 d cold-stored bagged and non-bagged fruits under shelf storage (Table 4).

Table 4. Effect of on-tree fruit bagging treatments on catalase (CAT), peroxidase (POD) and superoxide dismutase (SOD) activities in mango fruit during cold storage intervals and under ambient conditions. Values are means of triplicates ± SD. Different letters in superscripts indicate significant differences according to the LSD test at $p \leq 0.05$. D, decayed fruit hence no tests were performed.

Treatment	CAT (U mg^{-1} Protein)		POD (U mg^{-1} Protein)		SOD (U mg^{-1} Protein)	
	Non-Bagging	Bagging	Non-Bagging	Bagging	Non-Bagging	Bagging
			a. Days at 12 °C			
0	3.97 ± 0.19 ^{c}	8.39 ± 1.24 ^{a}	0.16 ± 0.01 ^{b}	2.79 ± 0.48 ^{a}	53.41 ± 1.63 ^{c}	86.78 ± 3.23 ^{a}
10	4.18 ± 0.70 ^{c}	7.29 ± 0.28 ^{ab}	0.59 ± 0.19 ^{b}	0.59 ± 0.36 ^{b}	75.25 ± 2.25 ^{b}	86.61 ± 2.60 ^{a}
20	3.62 ± 0.40 ^{c}	5.27 ± 1.12 ^{bc}	0.73 ± 0.43 ^{b}	0.93 ± 0.36 ^{b}	81.32 ± 0.76 ^{ab}	87.06 ± 3.06 ^{a}
			b. Days at 12 °C + Days at 25 °C			
0 + 0	3.97 ± 0.19 ^{c}	8.39 ± 1.24 ^{a}	0.16 ± 0.0 ^{c}	2.79 ± 0.48 ^{a}	53.41 ± 1.63 ^{c}	86.78 ± 3.23 ^{a}
0 + 7	6.66 ± 1.19 ^{ab}	9.86 ± 1.23 ^{a}	1.23 ± 0.48 ^{bc}	2.05 ± 0.45 ^{ab}	69.54 ± 2.92 ^{b}	83.88 ± 1.64 ^{a}
10 + 0	4.18 ± 0.70 ^{b}	7.29 ± 0.28 ^{a}	0.59 ± 0.19 ^{a}	0.59 ± 0.36 ^{a}	75.25 ± 2.25 ^{b}	86.61 ± 2.60 ^{a}
10 + 5	3.05 ± 0.15 ^{b}	8.12 ± 1.10 ^{a}	0.66 ± 0.08 ^{a}	1.04 ± 0.10 ^{a}	70.13 ± 1.74 ^{c}	74.84 ± 2.79 ^{bc}
20 + 0	3.62 ± 0.40 ^{a}	5.27 ± 1.12 ^{a}	0.73 ± 0.43 ^{a}	0.93 ± 0.36 ^{a}	81.32 ± 0.76 ^{a}	87.06 ± 3.06 ^{a}
20 + 1	D	D	D	D	D	D

4. Discussion

Accelerated ripening and softening are major factors shortening the shelf life and restricting the supply chains of mango fruit. Ripening of the mango fruit is a genetically complex process triggered by an increase in ethylene production and respiration rate [28]. In the current study, both bagged and non-bagged mango fruits exhibited a significant increase in ethylene production and respiration rate up to 10 d of cold storage, suggesting the initiation of the ripening process. After 20 d of cold storage, mango fruits reduced the respiration rate than those in 10 d cold-stored fruits. Montalvo et al. [29] found the highest rate of respiration in 'Ataulfo' mango during ripening which then decreased at the advanced stages of the fruit ripening process. Therefore, the reduction in the respiration rate observed in our study might be related to the fruit ripening process. Moreover, the differences in ethylene production and respiration rate between bagged and non-bagged mango fruits were not significant. These data are in accordance with other climacteric fruits, such as pears that exhibited similar results [30].

The marketing of fleshy fruits is seriously affected by weight loss and firmness loss during postharvest handling. Increased values of fruit firmness are linked with the increased potential of the fruit to travel long distances without quality deterioration. According to our results, when fruits were kept under ambient conditions after 0 d and 10 d of cold storage, weight loss increased progressively; however, bagged fruits exhibited slow weight loss which suggests that bagging treatment could slow down the metabolic process. Sharma et al. [31] also reported that when 'Delicious' apple fruits were bagged 30 d before harvest, a reduction in postharvest weight loss was observed as compared to non-bagged fruits. Furthermore, fruits exhibit more weight loss under ambient conditions with rapid ethylene production during ripening [32]; thereby, fruits become shrivelled and mushy and ultimately reduce their market value and quality [33]. Fruit firmness is reduced via the disintegration of cell walls which involves numerous enzymes such as polygalacturonate, galactosidases, and pectin methylesterase, as well as fruit respiration [34,35]. During ripening, mango fruits lose their firmness over time [36]. In the present study, a significant reduction in fruit firmness was observed over time, either during cold storage or ambient conditions; however, increased values of fruit firmness were recorded in bagged fruits. Our findings are in line with the results of Sharma, Pal, Asrey, Sagar, Dhiman and Rana [31] who noticed that the fruit firmness in bagged 'Delicious' apple fruits was retained in higher values as compared to control.

An increase in respiration by the action of ethylene triggers the degradation of chlorophyll and biosynthesis of carotenoids. In addition, Ding and Syakirah [37] stated that the low chlorophyll content in bagged fruits is due to low light radiation. In our study, bagged fruits exhibited brighter skin colour compared to non-bagged fruits. In a previous study [38], brown paper bags improved the yellow skin in the 'Keitt' mango when compared to the non-bagged fruits. Similar findings were also found in bagged litchi and peach fruits [39,40].

SSC is considered as a sweetness index, mainly determined by the total concentration of sugar compounds. Sugar metabolism is a complex mechanism that is regulated by enzymatic reactions, and genetic and environmental factors [41]. Ni et al. [42] reported that due to lower activities of sucrose degradation enzyme (acid invertase) and higher activities of sucrose synthesis enzymes (sucrose synthase and sucrose-phosphate synthase), bagged fruits had more sugar accumulation than non-bagged fruits. In the present study, significantly higher SSC contents were observed in bagged fruits as compared to non-bagged fruits which might be due to the changes in enzymatic activities. Moreover, it was observed that SSC contents were increased under ambient conditions followed by cold storage. Likewise, an increase in SSC at the shelf followed by cold storage was observed in mango fruit [17,43,44].

During fruit ripening, a decrease in TA is linked to the conversion of organic acids into sugars [45]. In our study, low TA and high juice pH values were observed in bagged fruits as compared to non-bagged fruits. Previously, Sarker et al. [46] and Islam, Akter, Rahman,

Uddin, Bari, Islam and Rahman [17] reported a decrease in TA in bagged mango fruits. In addition, similar changes have been also reported in other fruit crops due to bagging treatment [42,47].

Vitamin C (*L*-ascorbic acid) is a water-soluble vitamin that is abundant in mango fruits and is considered a major non-enzymatic antioxidant [48]. In plants, L-ascorbic acid metabolism is correlated with oxidative stress defence, while ascorbic acid accumulation in plant tissues and organs is altered by physiological phenomena such as senescence, cell expansion development and, various biotic and abiotic stimulants [49,50]. In our study, vitamin C content was decreased over time during cold storage and ambient storage, however, less reduction was observed in bagged fruits as compared to non-bagged fruits. The reduction in vitamin C content during prolonged cold storage could be due to an imbalance of ascorbic acid-redox homeostasis in fruits [50]. Moreover, it is suggested that fruits in brown paper bags were not directly exposed to sunlight, resulting in increased xanthophyll content, and hence bagged fruits preserved more vitamin C than non-bagged fruits. Previous reports [44,51] have also reported enhanced vitamin C content in bagged mango fruits than in non-bagged.

Total phenols and flavonoids create an active oxygen scavenging defence mechanism that ensures plant cell homeostasis [52]. Phenolic substances in fruits are considered vital molecules since they contribute to the antioxidant protection of the plant cell [53]. According to our study, the amount of TPC was high in bagged fruits as compared to non-bagged fruits during cold storage and ambient storage. Hudina and Stampar [54] observed the higher content of phenolic compounds (epicatechin and caffeic acid) in bagged pear fruits which ultimately resulted in improved TPC. The higher content of phenolic content in bagged fruits could be attributed to the decreased degradation rate of the phenolic molecules, due to the reduced rate of oxidation and polymerization reactions provoked by light exposure [55].

Fruit bagging changes the microclimate around the developing fruits. According to Zhang et al. [56], bagged fruits are exposed to higher average temperatures and humidity which can stimulate the production of reactive oxygen species. Zhu et al. [57] quantified significantly more H_2O_2 and O^{2-} content in bagged peach fruits than in non-bagged fruits. Taken together, these results support our finding that possibly because of the increase in the reactive oxygen species, SOD and CAT activity were significantly more activated in bagged fruits. During ambient storage, the SOD activity also remained higher in bagged fruits as compared to non-bagged fruits. Similarly, Nagamani et al. [58] reported the increased SOD activity in the pulp tissue of the mango 'Alphonso' during fruit ripening. Moreover, SOD, POD, CAT, and ascorbate peroxidase activity were greater in bagged apple fruit than in non-bagged fruit [59]. According to Razzaq et al. [60], cold storage treatment reduced the rate of enzyme activity in mango 'Samar Bahisht Chaunsa'. However, in the present study, only POD activity was significantly reduced during cold storage. Overall, the irregular behaviour of enzyme activities is considered to be caused by variations in CAT activity [60,61].

5. Conclusions

In conclusion, 0 d and 10 d cold-stored mangoes, regardless of bagging treatments, had a 7 d and 5 d shelf life under ambient conditions, respectively. Although bagged and non-bagged mango fruits did not differ significantly in terms of their storage life, bagged fruits had more fruit firmness, soluble solids, vitamin C, total phenolic contents and higher antioxidative enzyme activities, as compared to non-bagged fruits. On the other hand, when 20 d cold-stored fruits were kept under ambient conditions, fruits immediately lost their quality and exhibited poor shelf life. Overall, on-tree fruit bagging plus cold storage (up to 10 d) is a potential practice to preserve the high-quality status of postharvest mangoes.

Author Contributions: Conceptualization, S.B.H. and M.A.; methodology, S.M.A.-E., S.N. and A.M.M.; software, A.N., A.E.-D.K.O., S.J. and A.A.; formal analysis, A.N., S.J. and A.A.; investigation, A.N., S.J. and A.A.; resources, S.U. and K.R.; writing—original draft preparation, A.N.; writing—review and editing, S.B.H., V.Z. and Z.F.R.A.; supervision, S.B.H., I.A.R. and M.A.; funding acquisition, Z.F.R.A. All authors have read and agreed to the published version of the manuscript.

Funding: This research was partially funded by UAE University, grant number G00003332.

Institutional Review Board Statement: Not applicable.

Informed Consent Statement: Not applicable.

Acknowledgments: The authors are sincerely thankful to Roomi Foods, Kabir Wala, District Khanewal (Pakistan) for providing the cold storage facility. The gratitude is extended to the College of Agriculture and Veterinary Medicine, United Arab Emirates University, for partly funding this project.

Conflicts of Interest: The authors declare no conflict of interest.

References

1. Singh, Z.; Singh, R.K.; Sane, V.A.; Nath, P. Mango-postharvest biology and biotechnology. *Crit. Rev. Plant. Sci.* **2013**, *32*, 217–236. [CrossRef]
2. FAO. Food and Agricultural Organization of United Nations. Available online: http://faostat.fao.org (accessed on 27 August 2019).
3. MNFSR. *Fruit, Vegetable, and Condiments Statistics of Pakistan (2018–2019)*; Ministry of National Food Security & Research Economic Wing Islamabad: Islamabad, Pakistan, 2019.
4. Rajwana, I.A.; Malik, A.U.; Khan, A.S.; Saleem, B.A.; Malik, S.A. A new mango hybrid shows better shelf life and fruit quality. *Pak. J. Bot.* **2010**, *42*, 2503–2512.
5. Jabbar, A.; Malik, A.U.; Saeed, M.; Malik, O.H.; Amin, M.; Khan, A.S.; Rajwana, I.A.; Saleem, B.A.; Hameed, R.; Mazhar, M.S. Performance of hot water phytosanitary treated mangoes for intended export from Pakistan to Iran and China. *Int. J. Agric. Biol.* **2011**, *13*, 645–651.
6. Hafeez, O.; Malik, A.U.; Khan, A.S.; Rehman, A.; Javaid, Q.A. Impact of different packaging types and low temperature shipping durations on fruit quality and marketability of Pakistani mangoes. *Int. J. Agric. Biol.* **2012**, *14*, 47–54.
7. Léchaudel, M.; Joas, J. An overview of preharvest factors influencing mango fruit growth, quality and postharvest behaviour. *Braz. J. Plant Physiol.* **2007**, *19*, 287–298. [CrossRef]
8. Qian, M.; Zhang, D.; Yue, X.; Wang, S.; Li, X.; Teng, Y. Analysis of different pigmentation patterns in 'Mantianhong' (*Pyrus pyrifolia* Nakai) and 'Cascade' (*Pyrus communis* L.) under bagging treatment and postharvest UV-B/visible irradiation conditions. *Sci. Hortic.* **2013**, *151*, 75–82. [CrossRef]
9. Bai, S.; Tuan, P.A.; Saito, T.; Honda, C.; Hatsuyama, Y.; Ito, A.; Moriguchi, T. Epigenetic regulation of MdMYB1 is associated with paper bagging-induced red pigmentation of apples. *Planta* **2016**, *244*, 573–586. [CrossRef]
10. Zheng, Y.; Li, J.; Xin, H.; Wang, N.; Guan, L.; Wu, B.-H.; Li, S. Anthocyanin profile and gene expression in berry skin of two red Vitis vinifera grape cultivars that are sunlight dependent versus sunlight independent. *Aust. J. Grape Wine Res.* **2013**, *19*, 238–248. [CrossRef]
11. Karanjalker, G.; Ravishankar, K.; Shivashankara, K.; Dinesh, M. Influence of bagging on color, anthocyanin and anthocyanin biosynthetic genes in peel of red colored mango cv. 'Lily'. *Erwerbs-Obstbau* **2018**, *60*, 281–287. [CrossRef]
12. Kanzaki, S.; Ichihi, A.; Tanaka, Y.; Fujishige, S.; Koeda, S.; Shimizu, K. The R2R3-MYB transcription factor MiMYB1 regulates light dependent red coloration of 'Irwin'mango fruit skin. *Sci. Hortic.* **2020**, *272*, 109567. [CrossRef]
13. Ali, M.M.; Anwar, R.; Yousef, A.F.; Li, B.; Luvisi, A.; Bellis, L.D.; Aprile, A.; Chen, F. Influence of Bagging on the Development and Quality of Fruits. *Plants* **2021**, *10*, 358. [CrossRef] [PubMed]
14. Nagaharshitha, D.; Haldankar, P. Effect of Preharvest Fruit Bagging on quality of Mango (*Mangifera indica*) Cv. Alphonso. *Int. J. Agric. Sci.* **2017**, *7*, 531–536.
15. Devalla, N.; Haldankar, P.; Khopkar, R. Effect of bagging on chemical properties of mango (*Mangifera indica* L.) cv. Alphonso. *Int. J. Hortic. Sci. Crop Sci. Res.* **2016**, *6*, 1–8.
16. Ahmed, T.; Hasan, M.; Hassan, K.; Ahmed, J.D.; Ahmed, K.S.D.; Azam, A.; Mondal, M. Fruit bagging of custard apple (*Annona reticulata*) as an eco-friendly protection approach against mealybug (*Phenacoccus solenopsis*) infestation in the north-eastern Bangladesh. *Int. J. Trop. Insect Sci.* **2022**, *42*, 723–732. [CrossRef]
17. Islam, M.; Akter, M.; Rahman, M.; Uddin, M.; Bari, M.; Islam, M.; Rahman, M. Effect of bagging on quality and shelf life of mango (*Mangifera indica* L.) cv. BARI Mango-4. *Asian J. Agric. Hortic. Res.* **2020**, *6*, 37–45. [CrossRef]
18. Hossain, M.; Sarkar, B.; Hossain, M.; Mian, M.; Rajotte, E.; Muniappan, R.; O'Rourke, M. Comparison of biorational management approaches against mango fruit fly (*Bactrocera dorsalis* Hendel) in Bangladesh. *J. Crop Prot.* **2020**, *135*, 104807. [CrossRef]
19. Akter, M.; Islam, M.; Akter, N.; Amin, M.; Bari, M.; Uddin, M. Pre-harvest fruit bagging enhanced quality and shelf-life of mango (*Mangifera indica* L.) cv. Amrapali. *Asian J. Agric. Hortic. Res.* **2020**, *5*, 45–54. [CrossRef]

20. Sharma, R.R.; Reddy, S.; Jhalegar, M. Pre-harvest fruit bagging: A useful approach for plant protection and improved post-harvest fruit quality–a review. *Int. J. Hortic. Sci.* **2014**, *89*, 101–113. [CrossRef]
21. Ayele, L.; Bayleyegn, A. Effect of low temperature storage on ripening and quality of Mango (*Mangifera indica* L.). *J. Agric. Sci.* **2017**, *5*, 570–576.
22. Purwanto, Y.A.; Budiastra, I.W.; Darmawati, E. Effect of low temperature storage on the ripestage eating period of 'gedong gincu'mango fruits. *J. Appl. Eng. Sci.* **2016**, *11*, 10365–10367.
23. Ahmed, Z.F.; Kaur, N.; Maqsood, S.; Schmeda-Hirschmann, G. Preharvest Applications of Chitosan, Salicylic Acid, and Calcium Chloride Have a Synergistic Effect on Quality and Storability of Date Palm Fruit (Phoenix dactylifera L.). *HortScience* **2022**, *57*, 422–430. [CrossRef]
24. Xylia, P.; Chrysargyris, A.; Ahmed, Z.F.; Tzortzakis, N. Application of rosemary and eucalyptus essential oils and their main component on the preservation of apple and pear fruits. *Horticulturae* **2021**, *7*, 479. [CrossRef]
25. Ruck, J. *Chemical Methods for Analysis of Fruits and Vegetables*; Canada Department of Agriculture: Ottawa, ON, Canada, 1963.
26. Saleem, M.S.; Anjum, M.A.; Naz, S.; Ali, S.; Hussain, S.; Azam, M.; Sardar, H.; Khaliq, G.; Canan, İ.; Ejaz, S. Incorporation of ascorbic acid in chitosan-based edible coating improves postharvest quality and storability of strawberry fruits. *Int. J. Biol. Macromol.* **2021**, *189*, 160–169. [CrossRef] [PubMed]
27. Khalil, H.A.; El-Ansary, D.O.; Ahmed, Z.F. Mitigation of Salinity Stress on Pomegranate (*Punica granatum* L. cv. Wonderful) Plant Using Salicylic Acid Foliar Spray. *Horticulturae* **2022**, *8*, 375. [CrossRef]
28. Bambalele, N.L.; Mditshwa, A.; Magwaza, L.S.; Tesfay, S.Z. Recent Advances on Postharvest Technologies of Mango Fruit: A Review. *Int. J. Fruit Sci.* **2021**, *21*, 565–586. [CrossRef]
29. Montalvo, E.; García, H.S.; Tovar, B.; Mata, M. Application of exogenous ethylene on postharvest ripening of refrigerated 'Ataulfo' mangoes. *J. Food Sci. Technol.* **2007**, *40*, 1466–1472. [CrossRef]
30. Amarante, C.; Banks, N.H.; Max, S. Effect of preharvest bagging on fruit quality and postharvest physiology of pears (*Pyrus communis*). *N. Z. J. Crop Hortic. Sci.* **2002**, *30*, 99–107. [CrossRef]
31. Sharma, R.R.; Pal, R.K.; Asrey, R.; Sagar, V.R.; Dhiman, M.R.; Rana, M.R. Pre-harvest fruit bagging influences fruit color and quality of apple cv. Delicious. *J. Agric. Sci.* **2013**, *4*, 443. [CrossRef]
32. Narayana, C.; Pal, R.; Roy, S. Effect of pre-storage treatments and temperature regimes on shelf-life and respiratory behaviour of ripe Baneshan mango. *J. Food Sci. Technol.* **1996**, *33*, 79–82.
33. Karki, D. Effect of harvesting states on the quality of tomato (*Lycopersicon esculentum* Mill CV. Avinash-2, hybrid). *J. Trib. Uni.* **2005**, *25*, 141–147. [CrossRef]
34. Ali, Z.M.; Chin, L.-H.; Lazan, H. A comparative study on wall degrading enzymes, pectin modifications and softening during ripening of selected tropical fruits. *J. Plant Sci.* **2004**, *167*, 317–327. [CrossRef]
35. Gonzalez-Aguilar, G.A.; Celis, J.; Sotelo-Mundo, R.R.; De La Rosa, L.A.; Rodrigo-Garcia, J.; Alvarez-Parrilla, E. Physiological and biochemical changes of different fresh-cut mango cultivars stored at 5 °C. *Int. J. Food Sci. Technol.* **2008**, *43*, 91–101. [CrossRef]
36. Robles-Sánchez, R.; Islas-Osuna, M.; Astiazarán-García, H.; Vazquez-Ortiz, F.; Martín-Belloso, O.; Gorinstein, S.; González-Aguilar, G. Quality index, consumer acceptability, bioactive compounds, and antioxidant activity of fresh-cut "Ataulfo" mangoes (*Mangifera indica* L.) as affected by low-temperature storage. *J. Food Sci. Technol.* **2009**, *74*, S126–S134.
37. Ding, P.; Syakirah, M. Influence of Fruit Bagging on Postharvest Quality of 'harumanis' mango (*Mangifera Indica* L.). *Acta Hortic.* **2010**, *877*, 169–174. [CrossRef]
38. Hofman, P. Mango fruit quality at harvest is affected by production condition. In *Proceedings of the Mango 2000 Marketing Seminar and Production Workshop*; DPI: Brisbane, Australia, 1995; pp. 199–207.
39. Wang, H.-C.; Huang, X.-M.; Hu, G.-B.; Yang, Z.-y.; Huang, H.-B. A comparative study of chlorophyll loss and its related mechanism during fruit maturation in the pericarp of fast-and slow-degreening litchi pericarp. *J. Sci. Hortic.* **2005**, *106*, 247–257. [CrossRef]
40. Jia, H.-J.; Araki, A.; Okamoto, G. Influence of fruit bagging on aroma volatiles and skin coloration of 'Hakuho' peach (*Prunus persica* Batsch). *Postharvest Biol. Technol.* **2005**, *35*, 61–68. [CrossRef]
41. Beauvoit, B.; Belouah, I.; Bertin, N.; Cakpo, C.B.; Colombié, S.; Dai, Z.; Gautier, H.; Génard, M.; Moing, A.; Roch, L.; et al. Putting primary metabolism into perspective to obtain better fruits. *Ann. Bot.* **2018**, *122*, 1–21. [CrossRef]
42. Ni, Z.; Zhang, Z.; Gao, Z.; Gu, L.; Huang, L. Effects of bagging on sugar metabolism and the activity of sugar metabolism related enzymes during fruit development of Qingzhong loquat. *Afr. J. Biotechnol.* **2011**, *10*, 4212–4216.
43. Hasan, M.U.; Malik, A.U.; Khan, A.S.; Anwar, R.; Latif, M.; Amjad, A.; Shah, M.S.; Amin, M. Impact of postharvest hot water treatment on two commercial mango cultivars of Pakistan under simulated air freight conditions for China. *J. Agric. Sci.* **2020**, *57*, 1381–1391.
44. Wu, H.; Wang, S.; Shi, S.; Ma, W.; Zhou, Y.; Zhan, R. Effects of bagging on fruit quality in Zill mango. *J. Plant Sci.* **2009**, *26*, 644–648.
45. Anthon, G.E.; LeStrange, M.; Barrett, D.M. Changes in pH, acids, sugars and other quality parameters during extended vine holding of ripe processing tomatoes. *J. Sci. Food Agric.* **2011**, *91*, 1175–1181. [CrossRef] [PubMed]
46. Sarker, D.; Rahman, M.; Barman, J. Efficacy of different bagging materials for the control of mango fruit fly. *Bangladesh J. Agric. Res.* **2009**, *34*, 165–168. [CrossRef]
47. Zhou, J.; Zhong, G.; Lin, Z.; Xu, H.-l. The effects of bagging on fresh fruit quality of *Canarium album*. *J. Plant Stud.* **2012**, *10*, 505–508.

48. Ribeiro, S.M.R.; Queiroz, J.H.; de Queiroz, M.E.L.R.; Campos, F.M.; Sant'Ana, H.M.P. Antioxidant in mango (*Mangifera indica* L.) pulp. *Plant Foods Hum. Nutr.* **2007**, *62*, 13–17. [CrossRef] [PubMed]
49. Davey, M.W.; Kenis, K.; Keulemans, J. Genetic control of fruit vitamin C contents. *Plant Physiol.* **2006**, *142*, 343–351. [CrossRef] [PubMed]
50. Tsaniklidis, G.; Delis, C.; Nikoloudakis, N.; Katinakis, P.; Aivalakis, G. Low temperature storage affects the ascorbic acid metabolism of cherry tomato fruits. *Plant Physiol. Biochem.* **2014**, *84*, 149–157. [CrossRef] [PubMed]
51. Islam, M.; Rahman, M.; Shamsuzzoha, M.; Alom, R. Influence of pre-harvest bagging on fruit quality of Mango (*Mangifera indica* L.) cv. Mishribhog. *Int. J. Biosci.* **2017**, *11*, 59–68.
52. Ma, X.; Wu, H.; Liu, L.; Yao, Q.; Wang, S.; Zhan, R.; Xing, S.; Zhou, Y. Polyphenolic compounds and antioxidant properties in mango fruits. *Sci. Hortic.* **2011**, *129*, 102–107. [CrossRef]
53. Ahmed, Z.F.; Al Shaibani, F.Y.; Kaur, N.; Maqsood, S.; Schmeda-Hirschmann, G. Improving fruit quality, bioactive compounds, and storage life of date palm (*Phoenix dactylifera* L., cv. Barhi) using natural elicitors. *Horticulturae* **2021**, *7*, 293. [CrossRef]
54. Hudina, M.; Stampar, F. Bagging of 'Concorde' pears (*Pyrus communis* L.) influences fruit quality. In Proceedings of the XI International Pear Symposium 909, General Roca, Argentine, 23–26 November 2010; pp. 625–630.
55. Cunha, M.C.D.; Silva, J.S.; Elias, H.H.D.S.; Carvalho, E.E.N.; Vilas Boas, E.V.D.B. Effects of processing and packaging on bioactive compounds of curriola jelly [*Pouteria ramiflora* (Mart.) Radlk.] during storage. *Food Sci. Technol.* **2020**, *41*, 96–104. [CrossRef]
56. Zhang, B.B.; Guo, J.Y.; Ma, R.J.; Cai, Z.X.; Yan, J.; Zhang, C.H. Relationship between the bagging microenvironment and fruit quality in 'Guibao' peach [*Prunus persica* (L.) Batsch]. *J. Hortic. Sci. Biotechnol.* **2015**, *90*, 303–310. [CrossRef]
57. Zhu, M.; Fang, W.; Chen, C.; Wang, L.; Cao, K. Effects of Shading by Bagging on Carotenoid Accumulation in Peach Fruit Flesh. *J. Plant Growth Regul.* **2021**, *40*, 1912–1921. [CrossRef]
58. Nagamani, J.; Shivashankara, K.; Roy, T. Role of oxidative stress and the activity of ethylene biosynthetic enzymes on the formation of spongy tissue in 'Alphonso' mango. *J. Food Sci. Technol.* **2010**, *47*, 295–299. [CrossRef] [PubMed]
59. Wang, J.; Chen, H.; Zhou, Q.; Zhang, X. Effects of bagging on the fruit quality in *Litchi chinensis* fruit and pesticide residues in it. *J. Appl. Ecol.* **2003**, *14*, 710–712.
60. Razzaq, K.; Khan, A.S.; Malik, A.U.; Shahid, M. Ripening period influences fruit softening and antioxidative system of 'Samar Bahisht Chaunsa' mango. *Sci. Hortic.* **2013**, *160*, 108–114. [CrossRef]
61. Ahmed, Z.F.R.; Palta, J.P. Hormone-like effect of a natural lipid, lysophosphatidylethanolamine, can calcium deficiency injury in potato shoot cultures. *HortScience* **2011**, *46*, S196.

horticulturae

MDPI

Article

Preharvest Foliar Application of Si–Ca-Based Biostimulant Affects Postharvest Quality and Shelf-Life of Clementine Mandarin (*Citrus clementina* Hort. Ex Tan)

Vasileios Ziogas [1,*], Nikos Bravos [2] and Syed Bilal Hussain [3]

1 Institute of Olive Tree, Subtropical Plants and Viticulture, ELGO—DIMITRA, 73134 Chania, Greece
2 Ministry of Rural Development and Food, 10432 Athens, Greece
3 Citrus Research and Education Center, University of Florida, 700 Experiment Station Rd., Lake Alfred, FL 33850, USA
* Correspondence: ziogas@elgo.gr

Abstract: Citriculture and the postharvest industry are in the quest for biostimulants that favour fruit quality and extend shelf-life. Recently, Si has emerged as a biostimulant and its impact on fruit quality and postharvest shelf-life needs to be elucidated. The experiment is conducted for two consecutive years (2019 and 2020) in a commercial citrus orchard. In the present study, a Si–Ca-based product (Gravital® Force SC, AGROLOGY SA, Sindos, Greece) is foliar sprayed upon clementine mandarin (*Citrus clementina* Hort. Ex Tan cv. SRA 63) trees from August to November, while unsprayed trees are kept as controls. At commercial maturity, both sprayed and unsprayed fruits are harvested and stored for thirty (30) days at 5 °C with 90–95% relative humidity. Afterwards, they are kept at shelf temperature (20 °C) for six (6) days (shelf-life). At different intervals [at harvest, after cold storage (30 d at 5 °C), at the third day of shelf-life (30 d at 5 °C plus 3 d at 20 °C) and sixth day of shelf-life (30 d at 5 °C plus 6 d at 20 °C)], fruits are sampled and analysed for their qualitative characteristics. According to the results, the preharvest foliar application of the Si–Ca-based product delayed fruit maturation, increased peel firmness, total soluble content, total acidity, ascorbic acid, total phenols and antioxidant capacity, and reduced fruit decay during shelf storage. Results suggest that the preharvest foliar spray of Si–Ca products is able to maintain the postharvest quality of mid-ripening mandarin fruit.

Keywords: citrus; biostimulants; postharvest; fruit quality; phytochemicals

Citation: Ziogas, V.; Bravos, N.; Hussain, S.B. Preharvest Foliar Application of Si–Ca-Based Biostimulant Affects Postharvest Quality and Shelf-Life of Clementine Mandarin (*Citrus clementina* Hort. Ex Tan). *Horticulturae* **2022**, *8*, 996. https://doi.org/10.3390/horticulturae8110996

Academic Editors: Yang Bi, Yongcai Li and Di Gong

Received: 5 October 2022
Accepted: 24 October 2022
Published: 26 October 2022

1. Introduction

Citrus is cultivated in more than 140 countries worldwide for its nutritional properties, flavour and by-products, reaching an annual production of more than 146 million tonnes [1]. Mandarins (*Citrus reticulata* Blanco) are among the most important citrus fruits since, in 2020, 30.86 million tonnes were produced globally [2]. In Greece, mandarins are the second most important citrus fruit, with an annual production of over 171,870 tonnes in 2020/2021 [2]. Among the most cultivated mandarin cultivars in Greece, Clementine mandarin cv. SRA 63 (*C. clementina* Hort. Ex Tan) is one of them due to its seedless characteristic and superior taste. In Greece, SRA 63 fruit ripe from mid-December to mid-January, while it is a common practice to store fruits at cold storage facilities in order to extend their market availability and commercial value.

Citriculture and postharvest distribution facilities are searching for compounds that can limit the use of hazardous chemical substances. Furthermore, consumers' awareness towards food safety and nutritional value has increased the demand for excellent-quality fruit [3]. During postharvest storage, it is important to sustain the quality attributes and reduce the use of hazardous chemicals, which could leave harmful residues and negatively affect public health and environmental security [3]. When citrus fruits are subjected to postharvest storage conditions, several biotic or abiotic stress conditions occur, which modify metabolic cascades, leading to the loss of fruit quality, nutrient deprivation and decay syndromes [4].

It is well known that many factors influence the quality of citrus fruit, such as cultivar and harvest day [5], fruit canopy position [6] and mineral nutrition [7]. Nowadays, there is a growing interest in the potential of foliar application of biostimulants towards the enchantment of tree characteristics [8] and to achieve better crop management [9]. Silicon (Si) is the second most abundant element on the earth, after oxygen. Even though Si is not an essential element for plant growth, in its classical term, according to Epstein [10], it can act as a biostimulant since it can modulate the growth, development and stress response of plants [11]. Plants uptake some amount of Si from the soil in the form of silicic acid [$Si(OH)_4$] and distribute it to plant tissues [12–14]. Furthermore, Si participates in the modulation of various metabolic cascades, or the change of gene expression that alleviates plant tissues or fruits from deleterious abiotic stress conditions [15].

Additionally, calcium (Ca) is an important macro-element in plant nutrition, which acts as a signalling molecule in various metabolic cascades and is a critical component of the cell wall structure. It participates in multiple biochemical processes in plants and fruits, including respiration, development of chlorophyll, senescence, ethylene production, fruit firmness and prevention of rot incidents [16,17]. It has been reported that preharvest Ca spray upon fruits minimises postharvest decay syndromes and minimises the appearance of physiological disorders [18,19]. In Verna lemons, Ca application increased fruit firmness and prevented colour break for 21 days of storage at 15 °C [20]. Moreover, in the work of Cid-López et al. [21], the application of calcium oxide (CaO) nanoparticles to polymeric coatings, increased the shelf life of cucumbers by up to 24 days and improved quality attributes such as appearance, colour and antioxidant activity.

Concerning the impact of Si–Ca on plant physiology, previous studies mainly focused on its role as a biotic or abiotic stress tolerance [22–24]. In the work of Wutscher et al. [25], Si application (66 mg/L) as a nutrient solution enhanced the growth of young orange trees [*C. sinensis* (L.) Osbeck cv. Valencia]. Moreover, in grapefruit trees, supplementation of monosilicic acid in the irrigation water increased root weight by 40% [26]. So far, the combined action of Si and Ca is not yet elucidated, particularly for citrus. The scope of the current work was to investigate the impact of the preharvest foliar application of Si–Ca biostimulant upon the postharvest quality and shelf-life behaviour of mandarin fruit.

2. Materials and Methods

2.1. Plant Material and Treatments

The experiment was performed on thirty-five years-old citrus trees (*C. clementina* Hort. Ex Tan, cv. SRA 63), planted at 5 m × 5 m spacing between rows and along the row, grafted upon sour orange (*C. aurantium* L.). The soil type of the citrus grove was deep, well-drained and sandy loam. At the end of January (2019 and 2020), 1.5 kg of 20-20-20 NPK fertilizer was applied to the roots of each tree. Moreover, at the end of August, 0.5 kg of potassium nitrate fertilizer was also applied to the roots of each tree. Experimental trees were selected from a commercial citrus grove located in Corinth, Greece. For the experimental needs, for each treatment (Control and Si–Ca-Based compound) five (5) trees were used for each replication (three replications), counting a total number of 30 trees. Additional 10 trees were used as border trees for each replication of each treatment, in order to isolate each replication since all trees were within the same citrus grove (60 trees in total). During the years 2019 and 2020, a novel Si–Ca-based compound, (Gravital® Force SC, 35% *w*/*v* SiO_2, 35% *w*/*v* CaO and 30% *w*/*v* inactive compounds, originated from pulverized rock, creating a particle film of Si–Ca, being certified for organic farming use and commercially available by AGROLOGY SA, Sindos, Greece), was sprayed on fifteen trees at the rate of 10 g/L water, while another set of fifteen trees was unsprayed and served as control (five trees per replication). Overall, three sprays were performed each year as follows: at the beginning of August, mid-September and late November.

When the fruit reached commercial maturation (12 January 2020 and 22 December 2020, experimental years 2019 and 2020, respectively), one hundred and twenty (120) healthy and uniform size fruits were collected from each treatment and divided into three lots, forty (40) fruits in each. After harvest, the fruits were transferred into a cold storage facility (5 °C temperature and 90–95% relative humidity) for thirty days. After cold storage, the fruits were moved out and transported to the lab where they remained for 6 days at 20 °C, to stimulate shelf-life environment conditions. In total, thirty (30) fruits (ten in each replication) of each treatment were sampled at four (4) time points i.e., at harvest day, day 0 (exit from cold storage), day 3 of shelf-life and day 6 of shelf-life and used for the determination of qualitative physicochemical parameters.

2.2. Fruit Characteristics (Weight, Juice Content, Peel Thickness and Peel Colour)

Fruit weight (g) and juice content (%) was measured on an electronic precision balance (Kern 440-47 N, Kern, Germany). The peel thickness was measured with an electronic digital slide gauge (Parkside, Germany) with 0.01 mm accuracy. Peel colour was measured on two opposite portions of the equatorial area of each fruit via the use of a Minolta CR-300 chroma-meter. Values L*, a* and b* were converted to the Citrus Color Index (CCI) by the use of the formula CCI = (a* 1000)/(b*L*) [27].

2.3. Fruit Firmness

Fruit firmness was determined by a stable GY-4 Digital Fruit sclerometer (Beijing Channel Scientific Instruments Co., Ltd., Beijing, China), equipped with a flat compression plate. For each fruit, two opposite areas were penetrated and the resistance to compression of 10 mm was recorded and expressed as kg [28].

2.4. Juice Chemical Parameters (Juice pH, Total Soluble Solids, Total Acidity, Ascorbic Acid Content)

The juice pH was determined by using a potentiometer (Thermo Scientific ORION STAR A211, Waltham, MA, USA). Total soluble content (TSS) and total acidity (TA) were measured at 25 °C with a refractometer PAL-BX/ACID 1 Master Kit (Atago CO. Ltd., Tokyo, Japan), and expressed as °Brix and % citric acid, respectively, according to the manual of the instrument.

For the determination of ascorbic acid, total phenolics and antioxidant activity, the juice sample was centrifuged at 1650× *g* for 10 min at 4 °C. The quantification of ascorbic acid was performed according to the AOAC method [29]. Briefly, a 1 mL centrifuged juice sample was mixed with 1 mL of extraction solution (HPO_3-acetic acid). The mixture was used for the determination of the ascorbic acid content by titration with 2,6-dichlorophenolindophenol sodium salt hydrate (2,6-DCPI). The results were expressed as mg ascorbic acid/100 mL juice.

2.5. Total Phenolic Content

Total phenolic content was determined according to Singleton et al. [30]. In detail, in order to extract the juice phenolic content, 1 mL juice sample was mixed with 9 mL 80% (*v*/*v*) methanol and the mixture was incubated for 24 in 4 °C. Afterwards, the mixture was centrifuged at 4 °C for 15 min at 5000 rpm and an aliquot of the supernatant was used for the determination of total phenolics via the Folin-Ciocalteu method. Total phenolics were measured at 760 nm. The results were expressed as Gallic Acid Equivalent (GAE) (mg GAE/mL juice).

2.6. Total Antioxidant Capacity (FRAP Assay)

The antioxidant capacity of the juice was estimated according to Benzie and Strain [31], with some modifications. For the Ferric Reducing Antioxidant Power (FRAP) assay, fresh FRAP reagent solution (300 mM sodium acetate buffer, pH 3.6, 10 mM Fe (II)-TPTZ prepared in 40 mM HCl, 20 mM $FeCl_3 \cdot H_2O$ (10:1:1)) was freshly prepared prior to analysis. An aliquot of the supernatant was mixed with the FRAP reagent and the absorbance was measured at 593 nm after 30 min incubation at 37 °C. The ascorbic acid solution was used to express the antioxidant capacity as μmol Ascorbic acid/mL juice.

2.7. Fruit Decay during Storage

For the determination of fruit decay during shelf storage, forty-five fruits were randomly selected at the harvest date from each treatment. All collected fruits were divided into 3 replications, each replication was comprised of fifteen fruits. Fruit decay was counted after the 30 days of cold storage, at the 3rd and 6th day of shelf storage at 20 °C. The number of fruits that showed any sign of surface rot by mycelia development (*Penicillium* sp.) was counted [32]. Fruit decay was expressed as a percentage (%).

2.8. Statistical Analysis

The two-year data were analysed by using year and treatment as fixed effects in ANOVA models. In all tables, each year's data are presented along with pooled data since a similar trend was observed between the two years. Since for all measurements there was a similar trend between the years (2019 and 2020), in the results section and during the discussion section, the mean values of each measurement are presented and analysed. The data reported are the mean of replicates and expressed as mean ± standard deviation. Statistical analysis was performed via the use of the SPSS v.27 package (SPSS Inc., Chicago, IL, USA), with the use of a one-way analysis of variance (One-way ANOVA), with six replicates for each treatment. When there was a significant difference ($p < 0.05$), means were separated using Duncan's test.

A principal component analysis [33] was applied via the use of the SPSS v.27 package (SPSS Inc., Chicago, IL, USA). Principal component analysis (PCA) is used as a protocol for the determination of variables or as a tool to pinpoint factors that provide a better explanation of the correlation or covariance matrix of several attributes.

3. Results

3.1. Fruit Parameters and Fruit Decay

The weight of the citrus fruit is one of the basic quality attributes since it affects the overall tree productivity measurement. In our work, the mean fruit weight of the mandarin fruits was not affected by the preharvest foliar application of Si–Ca at all four postharvest time points (Table 1). However, irrespective of foliar treatments, a significant reduction in fruit weight was observed after one month of cold storage.

The CIE L*, a*, b* colour scale was used for the evaluation of peel colour during harvest, cold storage and ambient room storage. In this study, it was observed that Si–Ca spray did not significantly affect the fruit skin colour (CCI) at the harvest date and after one-month cold storage (30 d at 5 °C). During shelf storage, at day 3 (30 d at 5 °C plus 3 d at 20 °C), Si–Ca treated mandarin fruit exhibited a 14.37% higher CCI value than control fruit, whereas, at day 6 (30 d at 5 °C plus 6 d at 20 °C), no difference of CCI value was observed between treatments (Table 1).

Skin firmness is also an important attribute for the determination of citrus fruit quality and is tightly related to their ability to travel long distances. Si–Ca treated mandarin fruit exhibited a significant increase in fruit firmness values by 14.2% at harvest date, by 19.3% after one-month cold storage, by 16.28% on the third day (30 d at 5 °C plus 3 d at 20 °C) of shelf storage and by 10.95% on the sixth day of shelf storage (30 d at 5 °C plus 6 d at 20 °C), as compared to their respective control fruits (Table 1).

Peel thickness is a significant quality attribute of citrus fruit. In the current work, the foliar spray of Si–Ca significantly increased the peel thickness by 11.44% on the harvest day. After one-month cold storage (30 d at 5 °C) and at both shelf storage intervals (30 d at 5 °C plus 3 d at 20 °C and 30 d at 5 °C plus 6 d at 20 °C), Si–Ca treated mandarin fruits had no significant changes in peel thickness over their respective control (Table 1).

Table 1. Effect of preharvest Si–Ca foliar application on physical attributes of mandarin fruit. Means ± S.D. different letters indicate that there are statistically significant differences ($p < 0.05$) between the values of each treatment at each time point.

	Postharvest Time Point	2019		2020		Mean	
		Control	Si–Ca	Control	Si–Ca	Control	Si–Ca
Fruit weight (g)	Harvest day	95.13 ± 12.93 b	90.46 ± 17.45 b	75.67± 2.96 b	77.45 ± 2.00 b	85.4 ± 5.08 c	83.95 ±7.77 c
	30 d at 5 °C	53.10 ± 2.53 a	61.41 ± 5.43 a	67.48 ± 1.7 a	67.56 ± 1.05 a	60.29 ± 0.71 ab	64.49 ± 2.66 b
	30 d at 5 °C plus 3 d at 20 °C	50.01 ± 2.76 a	60.91 ± 0.31 a	66.32 ± 1.77 a	67.43 ± 1.36 a	58.17 ± 0.66 ab	64.17 ± 0.71 b
	30 d at 5 °C plus 6 d at 20 °C	48.52 ± 3.05 a	61.27 ± 0.22 a	65.17 ± 1.89 a	63.71 ± 1.61 a	56.84 ± 0.81 a	62.49 ± 0.77 ab
CCI	Harvest day	11.32 ± 1.40 a	10.76 ±0.86 a	12.39 ±0.92 a	14.04 ± 0.28 b	11.85 ± 0.26 a	12.40 ± 0.57 a
	30 d at 5 °C	10.56 ± 1.33 a	12.07 ± 1.56 a	13.31 ± 0.62 ab	12.33 ± 0.54 a	11.93 ± 0.36 a	12.20 ± 0.54 a
	30 d at 5 °C plus 3 d at 20 °C	11.24 ± 3.82 a	15.53 ± 1.33 b	13.11 ± 0.37 ab	12.32 ± 0.57 a	12.17 ± 1.74 a	13.92 ± 0.88 b
	30 d at 5 °C plus 6 d at 20 °C	10.55 ± 1.42 a	13.56 ± 0.45 ab	12.74 ± 0.57 a	12.34 ± 0.26 a	11.64 ± 0.71 a	12.96 ± 0.23 ab
Skin Firmness (kg)	Harvest day	2.78 ± 0.07 ab	3.36 ± 0.34 de	3.04 ± 0.12 bc	3.30 ± 0.18 cd	2.91 ± 0.09 ab	3.33 ± 0.26 cd
	30 d at 5 °C	2.81 ±0.12 abc	3.09 ± 5.00 cde	2.71 ± 0.14 a	3.50 ± 0.10 c	2.76 ± 0.1 a	3.3 ± 0.05 cd
	30 d at 5 °C plus 3 d at 20 °C	2.74 ± 0.12 a	3.20 ± 0.09 de	2.83 ± 0.08 ab	3.27 ± 0.14 cd	2.78 ± 0.10 a	3.24 ± 0.11 cd
	30 d at 5 °C plus 6 d at 20 °C	3.07 ± 3.38 bcd	3.38 ± 0.18 e	3.08 ± 0.08 bc	3.44 ± 0.23 c	3.07 ± 0.10 bc	3.41 ± 0.20 d
Rind Thickness (mm)	Harvest day	3.00 ± 0.38 b	3.52 ± 0.14 c	3.40 ± 0.10 c	3.61 ± 0.10 d	3.20 ± 0.24 b	3.57 ± 0.12 c
	30 d at 5 °C	2.29 ± 0.12 a	2.31 ± 0.15 a	2.47 ± 0.13 ab	2.56 ± 0.15 b	2.38 ± 0.12 a	2.44 ± 0.15 a
	30 d at 5 °C plus 3 d at 20 °C	2.10 ± 0.2 a	2.24 ± 0.08 a	2.32 ± 0.13 ab	2.37 ± 0.07 ab	2.21 ± 0.16 a	2.31 ± 0.07 a
	30 d at 5 °C plus 6 d at 20 °C	2.13 ± 0.06 a	2.41 ± 0.35 a	2.26 ± 0.07 a	2.54 ± 0.34 ab	2.20 ± 0.06 a	2.48 ± 0.35 a
Juice Content (%)	Harvest day	31.89 ± 1.64 a	32.60 ± 2.95 a	34.23 ± 2.00 a	35.08 ± 2.40 a	33.06 ± 1.81 a	33.84 ± 2.61 a
	30 d at 5 °C	37.01 ± 7.04 a	35.64 ± 3.69 a	40.76 ± 6.08 a	37.88 ± 3.26 a	38.88 ± 6.53 a	36.76 ± 3.35 a
	30 d at 5 °C plus 3 d at 20 °C	34.36 ± 3.16 a	33.03 ± 3.27 a	35.49 ± 3.85 a	35.98 ± 2.08 a	34.92 ± 3.49 a	34.51 ± 2.67 a
	30 d at 5 °C plus 6 d at 20 °C	36.00 ± 2.69 a	37.40 ± 1.98 a	37.50 ± 3.88 a	39.46 ± 0.81 a	36.75 ± 3.27 a	38.43 ± 1.24 a
Fruit Decay (%)	Harvest day	-	-	-	-	-	-
	30 d at 5 °C	33.49 ± 1.41 a	32.94 ± 2.68 a	34.66 ± 2.69 a	34.57 ± 2.26 a	34.07 ± 1.89 a	33.75 ± 2.47 a
	30 d at 5 °C plus 3 d at 20 °C	40.70 ± 1.46 b	34.80 ± 1.90 a	42.82 ± 1.62 cd	37.70 ± 1.69 ab	41.76 ± 1.54 cd	36.25 ± 1.79 ab
	30 d at 5 °C plus 6 d at 20 °C	43.90 ± 1.71 c	38.14 ± 0.65 b	46.21 ± 2.97 d	39.36 ± 0.93 bc	45.05 ± 2.32 d	38.75 ± 0.78 bc

In addition, the juice content did not show any significant change between the examined treatments at all four time points (Table 1).

Postharvest fruit decay incidence during shelf storage was significantly decreased by the preharvest foliar spray of Si–Ca (Table 1). After one month of cold storage (30 d at 5 °C), no significant difference in fruit decay was observed between treatments. During shelf storage (30 d at 5 °C plus 3 d at 20 °C and 30 d at 5 °C plus 6 d at 20 °C), preharvest foliar sprayed Si–Ca mandarin fruit exhibited a significant lower fruit decay by 13%, as compared to the control (Table 1).

3.2. *Juice Parameters (Total Soluble Solids Content, Total Acidity, TSS/TA Ratio and pH)*

TSS and acidity are very important quality indexes since they determine the taste of mandarin fruit. In our study, TSS and TA were determined at harvest day, one month after cold storage (30 d at 5 °C) and during shelf storage (30 d at 5 °C plus 3 d at 20 °C and 30 d at 5 °C plus 6 d at 20 °C). In the present study, it was observed that preharvest foliar Si–Ca spray significantly increased the fruit TSS content by 19.70% at harvest date, by 35.47% after one month of cold storage (30 d at 5 °C), by 40.18% at the third day of shelf storage (30 d at 5 °C plus 3 d at 20 °C) and by 32.10% at the sixth day of shelf storage (30 d at 5 °C plus 6 d at 20 °C), with respect to their control fruits (Table 2).

Moreover, preharvest foliar Si–Ca application significantly increased fruit TA by 40.74% at harvest date, by 81.65% after one-month cold storage (30 d at 5 °C), by 75% at the third day of shelf storage (30 d at 5 °C plus 3 d at 20 °C) and by 76.34% at the sixth day of shelf storage (30 d at 5 °C plus 6 d at 20 °C), with respect to their control fruits (Table 2).

Furthermore, the ratio of TSS/TA is used in order to determine the taste of citrus fruits, and mandarin fruits as an indicator of the fruit's maturity stage. In the current work, Si–Ca treated fruit exhibited a significantly lower ratio of TSS/TA by 14.6% at the harvest date, by 25.40% after one month of cold storage (30 d at 5 °C), by 19.66% at the third day of shelf storage (30 d at 5 °C plus 3 d at 20 °C) and by 24.81% at the sixth day of shelf storage (30 d at 5 °C plus 6 d at 20 °C), with respect to their control fruits (Table 2).

Juice pH is a significant quality attribute of citrus fruit. In the current work, Si–Ca treated fruit exhibited a significantly lower ratio of juice pH by 10.99% at the harvest date, by 10.17 % after one month of cold storage (30 d at 5 °C), by 11.06% on the third day of shelf storage (30 d at 5 °C plus 3 d at 20 °C) and by 14.57% on the sixth day of shelf storage (30 d at 5 °C plus 6 d at 20 °C), with respect to their control fruits (Table 2).

3.3. *Juice Ascorbic Acid Content, Phenolics and Antioxidant Capacity*

Ascorbic acid (AA) content and total phenolic content, along with the determination of the total antioxidant capacity, are crucial factors that link the consumption of fruit for health-related benefits. In our study, the preharvest foliar spray of Si–Ca significantly affected the AA content of mandarin fruit at the harvest date. After one-month cold storage (30 d at 5 °C) of Si–Ca treated fruit, a significant change of AA content by 69.63% was observed over control ones. During shelf storage, at day 3 (30 d at 5 °C plus 3 d at 20 °C), Si–Ca treated fruit exhibited 55.76% higher AA content than control fruit, whereas, at day 6 (30 d at 5 °C plus 6 d at 20 °C), a significant increase of 51.76% AA content was observed in Si–Ca treated fruit (Table 3).

Moreover, in our study, preharvest Si–Ca foliar spray on mandarin trees significantly affected the total phenolic (TP) content at harvest date by 21.90% with respect to control fruits. After one-month cold storage (30 d at 5 °C) upon Si–Ca treated fruits, a significant change of TP content by 39.45 % was observed over control ones. During shelf storage, at day 3 (30 d at 5 °C plus 3 d at 20 °C), Si–Ca treated mandarin fruit exhibited 30.53% higher TP content than control fruit, whereas, at day 6 (30 d at 5 °C plus 6 d at 20 °C), 32.17% more TP content was observed in Si–Ca treated mandarin fruit (Table 3).

Table 2. Effect of preharvest Si–Ca foliar application on quality attributes of mandarin fruit. Means ± S.D. different letters indicate that there are statistically significant differences ($p < 0.05$) between the values of each treatment at each time point.

	Postharvest Time Point	2019		2020		Mean	
		Control	Si–Ca	Control	Si–Ca	Control	Si–Ca
Total Soluble Solids (°Brix)	Harvest day	10.80 ± 0.56 a	13.43 ± 0.81 b	11.53 ± 0.21 a	13.30 ± 0.17 bc	11.17 ± 0.35 a	13.37 ± 0.49 b
	30 d at 5 °C	11.05 ± 2.15 a	15.95 ± 0.94 c	11.89 ± 1.14 ab	15.12 ± 0.89 d	11.47 ± 0.50 a	15.53 ± 0.12 d
	30 d at 5 °C plus 3 d at 20 °C	10.54 ± 0.27 a	16.13 ± 2.19 c	11.86 ± 0.85 ab	15.27 ± 1.21 d	11.20 ± 0.53 a	15.70 ± 0.52 d
	30 d at 5 °C plus 6 d at 20 °C	10.10 ± 0.56 a	15.01 ± 1.82 b	11.50 ± 0.56 a	13.52 ± 0.81 c	10.80 ± 0.44 a	14.27 ± 0.55 c
Acidity (% citric acid)	Harvest day	0.62 ± 0.02 b	0.93 ± 0.17 d	0.53 ± 0.02 b	0.69 ± 0.02 c	0.58 ± 0 b	0.81 ± 0.08 de
	30 d at 5 °C	0.39 ± 0.01 a	0.86 ± 0.12 cd	0.54 ± 0.03 b	0.83 ± 0.04 d	0.47 ± 0.02 a	0.84 ± 0.06 e
	30 d at 5 °C plus 3 d at 20 °C	0.40 ± 0.04 a	0.78 ± 0.16 bcd	0.46 ± 0.02 a	0.71 ± 0.01 c	0.43 ± 0.03 a	0.75 ± 0.09 cd
	30 d at 5 °C plus 6 d at 20 °C	0.35 ± 0.02 a	0.72 ± 0.14 bc	0.45 ± 0.02 a	0.68 ± 0.06 c	0.40 ± 0 a	0.70 ± 0.04 c
TSS/Acidity	Harvest day	17.45 ± 0.32 a	14.71 ± 1.97 a	21.65 ± 1.10 b	19.20 ± 0.70 ab	19.39 ± 0.61 ab	16.56 ± 1.08 a
	30 d at 5 °C	28.40 ± 6.02 cd	18.75 ± 1.54 a	21.98 ± 0.92 b	18.18 ± 1.56 a	24.71 ± 2.09 c	18.44 ± 1.13 ab
	30 d at 5 °C plus 3 d at 20 °C	26.79 ± 3.30 bcd	20.80 ± 1.92 ab	26.02 ± 2.36 c	21.53 ± 2.00 b	26.33 ± 2.22 c	21.15 ± 1.62 b
	30 d at 5 °C plus 6 d at 20 °C	29.23 ± 3.03 d	21.75 ± 7.05 abc	25.74 ± 0.40 c	20.05 ± 2.82 ab	27.22 ± 1.10 c	20.47 ± 2.00 b
Juice pH	Harvest day	3.62 ± 0.05 cd	3.30± 0.18 a	3.72 ± 0.14 c	3.23 ± 0.06 a	3.67 ± 0.05 d	3.27 ± 0.07 a
	30 d at 5 °C	3.79 ± 0.12 d	3.33 ± 0.05 ab	3.75 ± 0.06 c	3.44 ± 0.07 b	3.77 ± 0.03 e	3.39 ± 0.04 b
	30 d at 5 °C plus 3 d at 20 °C	4.02 ± 0.08 e	3.53 ± 0.12 bc	3.94 ± 0.06 d	3.55 ± 0.12 b	3.98 ± 0.02 f	3.54 ± 0.08 c
	30 d at 5 °C plus 6 d at 20 °C	3.99 ± 0.07 e	3.39 ± 0.14 ab	3.97 ± 0.12 d	3.41 ± 0.05 b	3.98 ± 0.03 f	3.40 ± 0.07 b

Table 3. Effect of preharvest Si–Ca foliar application on phytochemical attributes of mandarin fruit. Means ± S.D. different letters indicate that there are statistically significant differences ($p < 0.05$) between the values of each treatment at each time point.

	Postharvest Time Point	2019		2020		Mean	
		Control	Si–Ca	Control	Si–Ca	Control	Si–Ca
Ascorbic Acid (mg/100 mL)	Harvest day	40.57 ± 9.45 a	58.33 ± 6.56 b	73.90 ± 4.51 b	86.06 ± 3.40 c	57.24 ± 4.91 b	72.20 ± 3.90 c
	30 d at 5 °C	36.58 ± 4.68 a	85.71 ± 17.19 c	74.15 ± 3.38 b	102.11 ± 2.98 d	55.36 ± 1.33 ab	93.91 ± 8.56 d
	30 d at 5 °C plus 3 d at 20 °C	30.61 ± 7.21 a	59.23 ± 4.71 b	65.04 ± 2.98 a	89.76 ± 7.04 c	47.83 ± 3.01 a	74.49 ± 5.52 c
	30 d at 5 °C plus 6 d at 20 °C	36.11 ± 5.21 a	61.76 ± 3.41 b	62.44 ± 1.95 a	87.80 ± 5.16 c	49.28 ± 1.81 ab	74.78 ± 1.74 c
Total Phenols (mg GAE/ mL)	Harvest day	0.83 ± 0.03 abc	0.96 ± 0.08 bcd	0.48 ± 0.04 a	0.63± 0.01 c	0.65 ± 0.03 a	0.80 ± 0.04 b
	30 d at 5 °C	0.75 ± 0.09 a	1.20 ± 0.11 e	0.58 ± 0.02 b	0.64 ± 0.01 c	0.66 ± 0.04 a	0.92± 0.05 d
	30 d at 5 °C plus 3 d at 20 °C	0.82 ± 0.08 abc	1.10 ± 0.24 de	0.58 ± 0.01 b	0.73 ± 0.03 d	0.70 ± 0.04 a	0.92 ± 0.11 cd
	30 d at 5 °C plus 6 d at 20 °C	0.77 ± 0.05 ab	1.02 ± 0.07 cde	0.48 ± 0.04 a	0.63 ± 0.01 c	0.63 ± 0.01 a	0.83 ± 0.04 bc
FRAP (μmole Asc.Acid/mL)	Harvest day	5.09 ± 0.70 ab	5.94 ± 0.48 ab	5.14 ± 0.08 a	6.53 ± 0.50 b	5.11 ± 0.31 a	6.23 ± 0.37 bc
	30 d at 5 °C	5.57 ± 0.60 ab	7.73 ± 0.90 c	5.22 ± 0.08 a	7.32 ± 0.26 c	5.40 ± 0.30 ab	7.53 ± 0.43 d
	30 d at 5 °C plus 3 d at 20 °C	5.46 ± 0.75 ab	6.35 ± 1.62 bc	5.24 ± 0.10 a	7.64 ± 0.49 c	5.35 ± 0.42 ab	6.99 ± 1.05 cd
	30 d at 5 °C plus 6 d at 20 °C	4.60 ± 0.59 a	6.27 ± 0.64 bc	5.10 ± 0.10 a	6.57 ± 0.25 b	4.85 ± 0.26 a	6.42 ± 0.44 c

Furthermore, in our study, a significant improvement in the overall antioxidant capacity was observed in the Si–Ca-treated mandarin fruit over untreated ones. In detail, Si–Ca spray on fruit significantly increased the antioxidant capacity (FRAP value) of the juice at harvest date by 21.90% with respect to control fruits. After one-month cold storage (30 d at 5 °C) of Si–Ca treated mandarin fruits, a significant increase of FRAP value by 39.45%, was observed over control ones. During shelf storage, at day 3 (30 d at 5 °C plus 3 d at 20 °C), Si–Ca treated mandarin fruit exhibited 30.72% higher FRAP values than control fruit, whereas, at day 6 (30 d at 5 °C plus 6 d at 20 °C), 32.18% more TP content was observed in Si–Ca treated mandarin fruit (Table 3).

In the end, principal components analysis (PCA) was used to detect possible patterns, groupings and differences in the Si–Ca treated mandarin fruit, at harvest and post-harvest period. A PCA analysis was conducted using 14 variables related to mandarin physiology and quality (Figure 1). The variance of data that was explained by the PCA model was 77.2%, where PC1 explained 49.4% and PC2 27.8% of the total variance. The Kaiser Meyer Olkin measure of sampling adequacy (KMO) on the data was 0.657. A clear separation between the Si–Ca treated mandarins (positive values) and control mandarins (negative values) was observed. Moreover, PC1 construction was closely related to total phenols content, FRAP, TSS, TA, AA content and peel firmness that increased in Si–Ca treated mandarins while it was decreased in the pH and TSS/TA. PC2 was more closely linked to stage separation between harvest and post-harvest storage. PC2 construction is related to fruit decay, fruit weight and rind thickness where these variables are mainly involved in the separation of harvest from post-harvest storage (Figure 1).

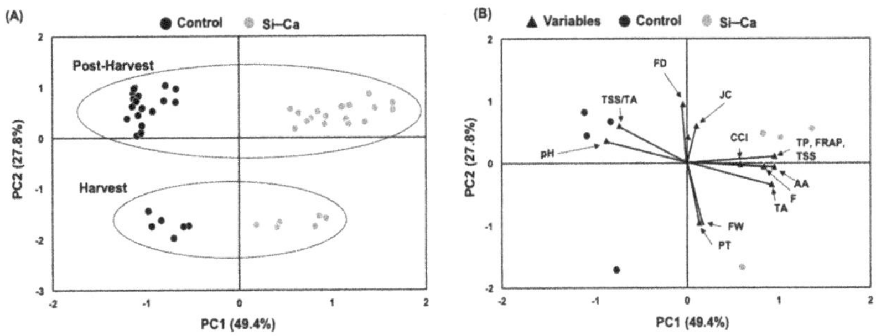

Figure 1. Principal component analysis (PCA) of physiological and quality traits is illustrated in mandarins treated with Si–Ca at harvest and post-harvest, using a score plot (**A**) and a biplot (**B**). Control: black circle, Si–Ca: grey circle. Variables are represented with a black triangle: fruit weight (FW); peel thickness (RT); colour index (CCI); flesh firmness (F), juice content (JC); total soluble solids (TSS); titratable acidity (TA); fruit decay (FD); ascorbic acid content (AA); total phenols content (TP), antioxidant activity (FRAP).

4. Discussion

Nowadays, there is a shift in agriculture towards the implementation of agricultural practises that preserve fruit quality attributes [34]. Fruit sprayed with biostimulants can sustain the quality attributes for an extensive period of shelf storage [34]. However, limited data are currently available regarding the effect of applied biostimulants towards the quality attributes of perennial tree crops [11].

In the present study, fruit weight was not significantly affected by the preharvest foliar application of Si–Ca biostimulant (Table 1). This result is aligned with that of Omar and El-Enin [35], who demonstrated that the foliar application of Ca did not significantly affect citrus fruit weight. Likewise, the application of Si to hydroponically cultivated strawberries, under Fe deficiency, did not alter fruit weight [36]. However, when lemon fruits were dipped in Si solution, a significant reduction in weight was recorded due to the proposed reduction of membrane permeability and increased membrane stability and integrity [37].

The difference in the effect of Si–Ca upon fruit commodities could be attributed to the method of application (fruit coating, fertigation, foliar spray) and time of application.

The colour of the mandarin fruit is a crucial quality attribute since it determines consumer choice [38]. In the current work, no significant differences in mandarin skin CCI were recorded at the various studied time intervals (Table 1). This result is in accordance with Bang et al. [39] and Peris-Felipo et al. [36], who demonstrated that the application of CaO or Si, respectively, did not influence the peel colour of the examined fruit. In the work of D' Aguino et al. [40], the application of the commercial compound surrounding WP (a Kaolin-Si-based compound) delayed cactus pear fruit colouration at harvest. Furthermore, in the work of Karagiannis et al. [14], foliar spray of a commercial Si-based biostimulant favoured the development of red colour on the peel of apples at harvest. These results demonstrate that the impact of Si on the development of colour peel during fruit ripening is contradictory and rather limited, as stated by several research groups [36,41,42].

In the current study, it was demonstrated that the preharvest foliar spray of the Si–Ca-based compound improved the postharvest fruit firmness (Table 1). This result is in line with the data from Weerahewa and David [43], who witnessed a significant increase in fruit firmness when Si was applied during flowering. Moreover, in the work of Yavad and Varu [44], the addition of Ca to papaya fruits resulted in the maintenance of fruit firmness during storage. The beneficial effect of foliar Si–Ca application upon tomato fruit firmness was demonstrated by Jing et al. [45]. The current beneficial effect of preharvest foliar Si–Ca spray could be linked with the ability of Ca to protect membranes from disorganization [16] and the ability of Si to stabilize the cell wall, via its protection from degradative enzymes [14], and the stimulation of the deposition of cellulose and hemicellulose [46].

In citrus fruit, it has been proposed that foliar application of Ca along with biostimulants and bioregulators is the most optimum method to alleviate fruit cracking and affect fruit rind thickness [47]. In the current study, preharvest spray of a Si–Ca-based compound resulted in increased rind thickness, only at harvest time, while no significant differences were recorded during postharvest cold storage and shelf storage (Table 1). This finding is in line with that of Hoda et al. [48], who demonstrated that the preharvest foliar application of Ca- or Si-based compounds can increase the mandarin peel thickness at harvest time. In another study by Hoda et al. [49], preharvest foliar application of diatoms (a source of Si) on Valencia orange fruit increased citrus peel thickness. It can be proposed that the preharvest foliar spray of Si–Ca compound provides the proper time interval for the plant leaves to absorb Si and Ca, translocate them into the fruit peel tissue, thus increase their tissue concentration and participation in the build-up of the mesocarp and epicarp of the fruit peel [48]. These findings support the role of Ca and Si in the citrus rind's development and rigidity when Ca and Si are applied at the pre-harvest stage [50].

It has been reported that the use of Si and Ca is beneficial towards the control of citrus fruit decay caused by *Penicillium spp.* [32,51]. In our work, mandarin fruit sprayed with a Si–Ca-based compound exhibited lower fruit decay than the control (Table 1). These data are in accordance with those observed by Moscoso-Ramírez and Palou [32], who found significant preventive (treatment before fungal inoculation) and curative (treatment after inoculation) activities of Si against *Penicillium spp.* after 6 days at 20 °C. Moreover, in the work of Weerahewa and Somapala [52], it is well documented that the application of Si can enhance the biotic resistance of tropical fruits. This beneficial role of Si was also supported by Mvondo-She et al. [53], who provided evidence that Si treatment can mitigate biotic stress syndromes in citrus fruits. It has been proposed that Si–Ca application in citrus fruit decreases fruit spoilage due to the ability of Ca to maintain cell wall structure and cell membrane integrity [16] and the ability of Si to act as a physical barrier against pathogen penetration or as a stimulant of defence responses [54]. It is known that Si's ability to initiate defence responses in plant pathosystems is linked with the ability of Si to induce physical resistance via reduced penetration and/or increased hardness and abrasiveness of the fruit tissue due to Si deposition when applied at preharvest time points [52]. Si can also induce chemical resistance via the enhanced production of defensive enzymes and the production of antifungal compounds such as phenolic compounds [52], a fact that was witnessed in the current work (Table 3).

It is well documented that in citrus fruit, sugars and especially organic acids are detrimental compounds for the determination of fruit taste quality [55]. It is stated that a high concentration of organic acids or a rather low pH value is an indicator of delayed senescence procedures [56], while the ratio between TSS and TA (which refers to maturity index) determines fruit sweetness [21]. In our work, the preharvest foliar spray of a Si–Ca-based compound increased TSS, TA and juice pH during postharvest treatments (Table 2). The positive impact of Si or Ca application upon TSS was also demonstrated by Mounika et al. [16]. In the work of Matichenkov and Calvert [26], the supply of Si in orange trees significantly increased fruit sugar content. Moreover, in the work of Sharma et al. [57], the application of kaolin (a Si-containing compound) improved or had no adverse effect on the TSS of pomegranate fruit during postharvest storage. Similar positive results, upon the TSS content, were also accounted via the application of Ca to cucumber fruit during postharvest storage [21]. In our work, the increased amount of TA and decreased pH values, along with lower TSS/TA values, at all-time points of the analysis, indicate that the preharvest foliar application of the Si–Ca compound delayed the maturation of the mandarin fruit at harvest time and during postharvest storage (Table 2).

The basic phytochemicals of citrus juice are ascorbic acid (AA) and phenolic compounds, as it has been well documented in various citrus species such as blood oranges [58], pomelo [59] and lemons [60]. It is stated that the application of Si or Ca-based fertilizers exerts a positive impact upon the concentration of AA and fruit sugars [61]. In our work, the foliar application of a Si–Ca-based compound increased the concentration of AA content when compared to the control (Table 3). The positive impact is in line with the work of Bang, Kim and Min [39], who demonstrated that the application of Ca (CaO) increased the concertation of AA in mandarin fruit. This Ca effect could be attributed to the movement of Ca within the mandarin tissue, which modulated the oxidation of AA, thus preventing its loss [62]. Furthermore, in the work of Ibrahim and Al-Wasfy [63], the preharvest foliar spray of potassium silicate upon Valencia orange trees resulted in enhanced levels of AA, TSS and total sugars. Additionally, the preharvest application of nano-silicon to soybean seedlings was able to increase the AA levels [64]. The increased levels of AA could be regulated by the increased levels of sugars since D-glucose is a precursor in the pathway of AA biosynthesis [65].

It has been proposed that Si and Ca induce the production of phenolics and enhance the antioxidant capacity of citrus juice [39,66]. In the current work, it was demonstrated that preharvest foliar application of Si–Ca-based compounds increased the total phenolic concentration of the juice along with the antioxidant capacity (FRAP values) of mandarin fruits examined at postharvest time intervals (Table 3). These results are in accordance with the work of Bang et al. [39], in which the application of Ca halted the loss of total phenolics and antioxidant capacity in mandarin fruit stored at cold temperatures or when exposed to shelf storage conditions. The beneficial impact of Ca on the preservation of phenolics was attributed to the ability of Ca to penetrate mandarin tissues and diminish the potent contact of polyphenol oxidase with its substrate [67]. Furthermore, the increased antioxidant capacity of mandarin fruits, under the effect of Ca, was linked also to the ability of Ca to penetrate the citrus tissue, resulting in a decreased amount of free radicals in the fruit tissue [68], thus preserving the antioxidant capacity at higher levels. In parallel, the ability of Si to exert a positive action towards the induction of total phenolic production and high antioxidant capacity in citrus fruit (lemons) was reported by Mditshwa et al. [37]. Furthermore, potassium silicate application to avocado fruit increased total phenolic concertation and the activity of antioxidant enzymes such as polyphenol oxidase and catalase [34]. Our results are in accordance with those of Mditshwa et al. [37], who proposed that Si can induce the production of phenolics and flavonoid content in the peel of lemon fruits. Additionally, Karagiannis et al. [14] demonstrated that the preharvest foliar application of a similar Si-based commercial compound (AGROLOGY SA) can induce the accumulation of total phenolics and total anthocyanin compounds in various tissues of apple fruit. The observed data in the current work can be attributed to the ability of Si to increase the photosynthetic active radiation of the fruit [69,70], which increases the enzymatic activity of phenyl ammonia lyase (PAL), and thus the concentration of the total phenolics within the fruit tissues [57].

5. Conclusions

From the current study, the extracted data revealed that the preharvest foliar spray of Si–Ca can positively change the postharvest quality traits of mandarin clementine cv. SRA 63 fruits. At harvest, TSS and TA content were increased, and the fruit maturation of this mid-ripening variety was delayed while ascorbic acid, total phenolics and antioxidant capacity were increased. This increase in quality attributes was preserved even after cold storage for 30 days (30 d at 5 °C) and after the exposure of the fruit at room temperature (30 d at 5 °C plus 3 d at 20 °C and 30 d at 5 °C plus 6 d at 20 °C). The beneficial impact of the Si–Ca-based compound (Gravital® Force SC, AGROLOGY SA, Sindos, Greece) application was also exerted via the decreased fruit decay syndromes during fruit shelf storage. Our results add valuable information regarding the effect of a Si–Ca-based biostimulant on fruit quality and postharvest manipulation of citrus fruit. Overall, this work provides valuable data for understanding metabolic cascades, which are triggered via the application of Si and Ca.

Author Contributions: V.Z. and N.B. conceived and designed the project and its components. V.Z. and N.B. collected the samples and performed the laboratory analysis. V.Z. analysed the data and wrote the initial draft of the manuscript. S.B.H. revised the manuscript and provided technical support. All authors have read and agreed to the published version of the manuscript.

Funding: This research was funded by AGROLOGY SA, grant number 22.1533.242 and the APC was funded by AGROLOGY SA.

Acknowledgments: The authors thank AGROLOGY SA for the kind offer of the Si–Ca compound.

Conflicts of Interest: The authors declare no conflict of interest. The funders had no role in the design of the study; in the collection, analyses, or interpretation of data; in the writing of the manuscript; or in the decision to publish the results.

References

1. Chen, Y.; Barzee, T.J.; Zhang, R.; Pan, Z. Citrus. In *Integrated Processing Technologies for Food and Agricultural By-Products*, 1st ed.; Elsevier: Amsterdam, The Netherlands, 2019; pp. 217–242, ISBN 978-0-12-814138-0.
2. FAO. *Citrus Fruit Statistical Compendium 2020*; FAO: Rome, Italy, 2021.
3. Papoutsis, K.; Mathioudakis, M.M.; Hasperué, J.H.; Ziogas, V. Non-chemical treatments for preventing the postharvest fungal rotting of citrus caused by *Penicillium digitatum* (green mold) and *Penicillium italicum* (blue mold). *Trends Food Sci. Technol.* **2019**, *86*, 479–491. [CrossRef]
4. Sheng, L.; Shen, D.; Luo, Y.; Sun, X.; Wang, J.; Luo, T.; Zeng, Y.; Xu, J.; Deng, X.; Cheng, Y. Exogenous γ-aminobutyric acid treatment affects citrate and amino acid accumulation to improve fruit quality and storage performance of postharvest citrus fruit. *Food Chem.* **2017**, *216*, 138–145. [CrossRef] [PubMed]
5. Giuffrè, A.M. Bergamot (*Citrus bergamia*, Risso): The Effects of Cultivar and Harvest Date on Functional Properties of Juice and Cloudy Juice. *Antioxidants* **2019**, *8*, 221. [CrossRef] [PubMed]
6. Adhikari, P.; Singh, Z.; Tokala, V.Y.; Kyaw, P.N.; Walsh, B. Fruit canopy position and harvest date influence on color and quality of Imperial mandarin (*Citrus reticulata* Blanco). *Aust. J. Crop Sci.* **2020**, *14*, 660–666. [CrossRef]
7. Aular, J.; Casares, M.; Natale, W. Factors affecting fruit quality: Emphasis on mineral nutrition. *Cientifica* **2017**, *45*, 64–72. [CrossRef]
8. Tanou, G.; Ziogas, V.; Molassiotis, A. Foliar nutrition, biostimulants and prime-like dynamics in fruit tree physiology: New insights on an old topic. *Front. Plant Sci.* **2017**, *8*, 75. [CrossRef] [PubMed]
9. Gomes, T.M.; Mazon, L.F.; Panceri, C.P.; Machado, B.D.; Brighenti, A.; Burin, V.M.; Bordignon-Luiz, M.T. Changes in vineyard productive attributes and phytochemical composition of sauvignon blanc grape and wine induced by the application of silicon and calcium. *J. Sci. Food Agric.* **2020**, *100*, 1547–1557. [CrossRef] [PubMed]
10. Epstein, E. The anomaly of silicon in plant biology. *Proc. Natl. Acad. Sci. USA* **1994**, *91*, 11–17. [CrossRef] [PubMed]
11. Savvas, D.; Ntatsi, G. Biostimulant activity of silicon in horticulture. *Sci. Hortic.* **2015**, *196*, 66–81. [CrossRef]

12. Hernandez-Apaolaza, L. Can silicon partially alleviate micronutrient deficiency in plants? A review. *Planta* **2014**, *240*, 447–458. [CrossRef] [PubMed]
13. Mandlik, R.; Thakral, V.; Raturi, G.; Shinde, S.; Nikolić, M.; Tripathi, D.K.; Sonah, H.; Deshmukh, R. Significance of silicon uptake, transport, and deposition in plants. *J. Exp. Bot.* **2020**, *71*, 6703–6718. [CrossRef]
14. Karagiannis, E.; Michailidis, M.; Skodra, C.; Molassiotis, A.; Tanou, G. Silicon influenced ripening metabolism and improved fruit quality traits in apples. *Plant Physiol. Biochem.* **2021**, *166*, 270–277. [CrossRef] [PubMed]
15. Carrasco-Gil, S.; Rodríguez-Menéndez, S.M.; Fernández, B.; Pereiro, R.; de la Fuente, V.; Hernandez-Apaolaza, L. Silicon induced Fe deficiency affects Fe, Mn, Cu and Zn distribution in rice (*Oryza sativa* L.) growth in calcareous conditions. *Plant Physiol. Biochem.* **2018**, *125*, 153–163. [CrossRef] [PubMed]
16. Mounika, M.; Kumar, T.S.; Kumar, A.K.; Joshi, V.; Sunil, N. Studies on the effect of foliar application of calcium, potassium and silicon on quality and shelf life of sweet orange (*Citrus sinensis* L.) cv. Sathgudi. *J. Pharmacogn. Phytochem.* **2021**, *10*, 1711–1713.
17. Poovaiah, B. Role of calcium in prolonging storage life of fruits and vegetables. *Food Technol.* **1986**, *40*, 86–89.
18. Madani, B.; Mohamed, M.T.M.; Watkins, C.B.; Kadir, J.; Awang, Y.; Shojaei, T.R. Preharvest calcium chloride sprays affect ripening of Eksotika II'papaya fruits during cold storage. *Sci. Hortic.* **2014**, *171*, 6–13. [CrossRef]
19. Kinay, P.; Yildiz, F.; Sen, F.; Yildiz, M.; Karacali, I. Integration of pre- and postharvest treatments to minimize Penicillium decay of Satsuma mandarins. *Postharvest Biol. Technol.* **2005**, *37*, 31–36. [CrossRef]
20. Martínez-Romero, D.; Valero, D.; Serrano, M.; Martínez-Sánchez, F.; Riquelme, F. Effects of post-harvest putrescine and calcium treatments on reducing mechanical damage and polyamines and abscisic acid levels during lemon storage. *J. Sci. Food Agric.* **1999**, *79*, 1589–1595. [CrossRef]
21. Cid-López, M.L.; Soriano-Melgar, L.D.A.A.; García-González, A.; Cortéz-Mazatán, G.; Mendoza-Mendoza, E.; Rivera-Cabrera, F.; Peralta-Rodríguez, R.D. The benefits of adding calcium oxide nanoparticles to biocompatible polymeric coatings during cucumber fruits postharvest storage. *Sci. Hortic.* **2021**, *287*, 110285. [CrossRef]
22. Alamri, S.; Hu, Y.; Mukherjee, S.; Aftab, T.; Fahad, S.; Raza, A.; Ahmad, M.; Siddiqui, M.H. Silicon-induced postponement of leaf senescence is accompanied by modulation of antioxidative defense and ion homeostasis in mustard (*Brassica juncea*) seedlings exposed to salinity and drought stress. *Plant Physiol. Biochem.* **2020**, *157*, 47–59. [CrossRef] [PubMed]
23. Bhat, J.A.; Shivaraj, S.M.; Singh, P.; Navadagi, D.B.; Tripathi, D.K.; Dash, P.K.; Solanke, A.U.; Sonah, H.; Deshmukh, R. Role of Silicon in Mitigation of Heavy Metal Stresses in Crop Plants. *Plants* **2019**, *8*, 71. [CrossRef] [PubMed]
24. Zargar, S.M.; Mahajan, R.; Bhat, J.A.; Nazir, M.; Deshmukh, R. Role of silicon in plant stress tolerance: Opportunities to achieve a sustainable cropping system. *3 Biotech* **2019**, *9*, 73. [CrossRef]
25. Wutscher, H.K. Growth and mineral nutrition of young orange trees grown with high levels of silicon. *Hortscience* **1989**, *24*, 275–277. [CrossRef]
26. Matichenkov, V.; Calvert, D. Silicon fertilizers for citrus in Florida. *Proc. Fla. State Hort. Soc.* **1999**, *112*, 5–8.
27. Singh, K.K.; Reddy, B.S. Post-harvest physico-mechanical properties of orange peel and fruit. *J. Food Eng.* **2006**, *73*, 112–120. [CrossRef]
28. Sogvar, O.B.; Koushesh Saba, M.; Emamifar, A. Aloe vera and ascorbic acid coatings maintain postharvest quality and reduce microbial load of strawberry fruit. *Postharvest Biol. Technol.* **2016**, *114*, 29–35. [CrossRef]
29. Hortwitz, W. *Official Methods of AOAC International*, 17th ed.; Association of Official Analytic Chemists: Gaithersburg, MD, USA, 2000.
30. Singleton, V.L.; Orthofer, R.; Lamuela-Raventós, R.M. Analysis of total phenols and other oxidation substrates and antioxidants by means of folin-ciocalteu reagent. *Methods Enzymol.* **1999**, *299*, 152–178. [CrossRef]
31. Benzie, I.F.; Strain, J.J. The ferric reducing ability of plasma (FRAP) as a measure of "antioxidant power": The FRAP assay. *Anal. Biochem.* **1996**, *239*, 70–76. [CrossRef] [PubMed]
32. Moscoso-Ramírez, P.A.; Palou, L. Preventive and curative activity of postharvest potassium silicate treatments to control green and blue molds on orange fruit. *Eur. J. Plant Pathol.* **2014**, *138*, 721–732. [CrossRef]
33. Ihaka, R.; Gentleman, R.R. A language for data analysis and graphics. *J. Comput. Graph. Stat.* **1996**, *5*, 299–314.
34. Tesfay, S.Z.; Bertling, I.; Bower, J.P. Effects of postharvest potassium silicate application on phenolics and other anti-oxidant systems aligned to avocado fruit quality. *Postharvest Biol. Technol.* **2011**, *60*, 92–99. [CrossRef]
35. Omar, A.E.-D.K.; El-Enin, M.S.A. Foliar spray with different agrochemicals on fruit quality and exportability of 'Washington' navel orange fruit (*Citrus sinensis* L.). *Int. J. Fruit Sci.* **2017**, *17*, 280–295. [CrossRef]
36. Peris-Felipo, F.J.; Benavent-Gil, Y.; Hernández-Apaolaza, L. Silicon beneficial effects on yield, fruit quality and shelf-life of strawberries grown in different culture substrates under different iron status. *Plant Physiol. Biochem.* **2020**, *152*, 23–31. [CrossRef]
37. Mditshwa, A.; Bower, J.P.; Bertling, I.; Mathaba, N.; Tesfay, S.Z. The potential of postharvest silicon dips to regulate phenolics in citrus peel as a method to mitigate chilling injury in lemons. *Afr. J. Biotechnol.* **2013**, *12*, 1482–1489.
38. Won, M.Y.; Lee, S.J.; Min, S.C. Mandarin preservation by microwave-powered cold plasma treatment. *Innov. Food Sci. Emerg. Technol.* **2017**, *39*, 25–32. [CrossRef]
39. Bang, I.H.; Kim, Y.E.; Min, S.C. Preservation of mandarins using a microbial decontamination system integrating calcium oxide solution washing, modified atmosphere packaging, and dielectric barrier discharge cold plasma treatment. *Food Packag. Shelf Life* **2021**, *29*, 100682. [CrossRef]

40. D'Aquino, S.; Palma, A.; Chessa, I.; Satta, D.; De Pau, L.; Inglese, P.; Ochoa, J.M.; Glenn, D.M. Effect of Surround WP (a Kaolin-based particle film) on Ceratitis capitata infestation, quality and postharvest behaviour of cactus pear fruit cv Gialla. *Sci. Hortic.* **2021**, *289*, 110484. [CrossRef]
41. Jayawardana, H.A.R.K.; Weerahewa, H.L.D.; Saparamadu, M.D.J.S. Effect of root or foliar application of soluble silicon on plant growth, fruit quality and anthracnose development of capsicum. *Trop. Agric. Res.* **2015**, *26*, 74. [CrossRef]
42. González-Terán, G.E.; Gómez-Merino, F.C.; Trejo-Téllez, L.I. Effects of silicon and calcium application on growth, yield and fruit quality parameters of cucumber established in a sodic soil. *Acta Sci. Pol. Hortorum Cultus* **2020**, *19*, 149–158. [CrossRef]
43. Weerahewa, D.; David, D. Effect of silicon and potassium on tomato anthracnose and on the postharvest quality of tomato fruit (*Lycopersicon esculentum* Mill.). *J. Natl. Sci. Found. Sri Lanka* **2015**, *43*, 273–280. [CrossRef]
44. Lokesh, Y.; Varu, D.K. Effect of pre harvest spray and post harvest dipping of fruit on shelf life and quality of papaya. *Asian J. Hortic.* **2013**, *8*, 581–587.
45. Jing, S.U.; Lan-chun, N.I.; Ying-bin, Q.; Miao-miao, W. Effect of foliar application of silicon and calcium on the firmness and related physiological metabolism of tomato fruits. *Acta Hortic. Sin.* **2016**, *43*, 789.
46. Głazowska, S.; Baldwin, L.; Mravec, J.; Bukh, C.; Hansen, T.H.; Jensen, M.M.; Fangel, J.U.; Willats, W.G.; Glasius, M.; Felby, C. The impact of silicon on cell wall composition and enzymatic saccharification of *Brachypodium distachyon*. *Biotechnol. Biofuels* **2018**, *11*, 171. [CrossRef]
47. García-Luis, A.; Duarte, A.; Kanduser, M.; Guardiola, J.L. The anatomy of the fruit in relation to the propensity of citrus species to split. *Sci. Hortic.* **2001**, *87*, 33–52. [CrossRef]
48. Mohamed, H.M.; Omran, M.A.A.; Mohamed, S.M. Effect of foliar spraying of some materials on protecting Murcott mandarin fruits from sunburn injuries. *Mid. East J. Agric. Res.* **2019**, *8*, 514–524.
49. Hoda, M.M.; Faten, A.A.-K.; Azza, A.M.A.-E. Effect of Magnetite and Some Biofertilizer Application on Growth and Yield of Valencia Orange Trees under El-Bustan Condition. *Nat. Sci.* **2013**, *11*, 46–61.
50. Treeby, M.T.; Storey, R. Calcium-spray treatments for ameliorating albedo breakdown in navel oranges. *Aust. J. Exp. Agric.* **2002**, *42*, 495–502. [CrossRef]
51. Liu, J.; Zong, Y.; Qin, G.; Li, B.; Tian, S. Plasma membrane damage contributes to antifungal activity of silicon against *Penicillium digitatum*. *Curr. Microbiol.* **2010**, *61*, 274–279. [CrossRef]
52. Weerahewa, D.; Somapala, K. Role of silicon on enhancing resistance in tropical fruits and vegetables: A review. *OUSL J.* **2016**, *11*, 135–162. [CrossRef]
53. Mvondo-She, M.A.; Gatabazi, A.; Laing, M.D.; Ndhlala, A.R. A review on the role of silicon treatment in biotic stress mitigation and citrus production. *Agronomy* **2021**, *11*, 2198. [CrossRef]
54. Shen, G.H.; Xue, Q.H.; Tang, M.; Chen, Q.; Wang, L.N.; Duan, C.M.; Xue, L.; Zhao, J. Inhibitory effects of potassium silicate on five soil-borne phytopathogenic fungi in vitro. *J. Plant Dis. Prot.* **2010**, *117*, 180–184. [CrossRef]
55. Hussain, S.B.; Shi, C.-Y.; Guo, L.-X.; Kamran, H.M.; Sadka, A.; Liu, Y.-Z. Recent advances in the regulation of citric acid metabolism in citrus fruit. *Crit. Rev. Plant Sci.* **2017**, *36*, 241–256. [CrossRef]
56. Angioni, A.; Schirra, M. Long-term frozen storage impact on the antioxidant capacity and chemical composition of Sardinianmyrtle (*Myrtus communis* L.) berries. *J. Agric. Sci. Technol. B* **2011**, *1*, 1168–1175.
57. Sharma, R.R.; Datta, S.C.; Varghese, E. Effect of Surround WP®, a kaolin-based particle film on sunburn, fruit cracking and postharvest quality of 'Kandhari' pomegranates. *Crop Prot.* **2018**, *114*, 18–22. [CrossRef]
58. Giuffrè, A.M.; Zappia, C.; Capocasale, M. Physicochemical stability of blood orange juice during frozen storage. *Int. J. Food Prop.* **2017**, *20*, 1930–1943. [CrossRef]
59. Cheong, M.W.; Liu, S.Q.; Zhou, W.; Curran, P.; Yu, B. Chemical composition and sensory profile of pomelo (*Citrus grandis* (L.) Osbeck) juice. *Food Chem.* **2012**, *135*, 2505–2513. [CrossRef] [PubMed]
60. Aguilar-Hernández, M.G.; Núñez-Gómez, D.; Forner-Giner, M.; Hernández, F.; Pastor-Pérez, J.J.; Legua, P. Quality parameters of Spanish lemons with commercial interest. *Foods* **2021**, *10*, 62. [CrossRef]
61. Poodeh, S.D.; Ghobadi, C.; Baninasab, B.; Gheysari, M.; Bidabadi, S.S. Effects of potassium silicate and nanosilica on quantitative and qualitative characteristics of a commercial strawberry (*Fragaria* × *ananassa* cv. 'camarosa'). *J. Plant Nutr.* **2016**, *39*, 502–507. [CrossRef]
62. Ayón-Reyna, L.E.; Tamayo-Limón, R.; Cárdenas-Torres, F.I.; López-López, M.E.; López-Angulo, G.; López-Moreno, H.S.; López-Cervantes, J.; López-Valenzuela, J.A.; Vega-García, M.O. Effectiveness of hydrothermal-calcium chloride treatment and chitosan on quality retention and microbial growth during storage of fresh-cut papaya. *J. Food Sci.* **2015**, *80*, C594–C601. [CrossRef] [PubMed]
63. Ibrahim, H.I.M.; Al-Wasfy, M.M. The promotive impact of using silicon and selenium with potassium and boron on fruiting of Valencia orange tree grown under Minia region conditions. *World Rural Obsrv.* **2014**, *6*, 28–36.
64. Farhangi-Abriz, S.; Torabian, S. Nano-silicon alters antioxidant activities of soybean seedlings under salt toxicity. *Protoplasma* **2018**, *255*, 953–962. [CrossRef] [PubMed]
65. Fenech, M.; Amaya, I.; Valpuesta, V.; Botella, M.A. Vitamin C content in fruits: Biosynthesis and regulation. *Front. Plant Sci.* **2019**, *9*, 2006. [CrossRef] [PubMed]
66. Liang, Y.; Sun, W.; Zhu, Y.G.; Christie, P. Mechanisms of silicon-mediated alleviation of abiotic stresses in higher plants: A review. *Environ. Pollut.* **2007**, *147*, 422–428. [CrossRef] [PubMed]

67. Shahkoomahally, S.; Ramezanian, A. Hot water combined with calcium treatment improves physical and physicochemical attributes of kiwifruit (*Actinidia deliciosa* cv. Hayward) during storage. *HortScience* **2015**, *50*, 412–415. [CrossRef]
68. Supapvanich, S.; Arkajak, R.; Yalai, K. Maintenance of postharvest quality and bioactive compounds of fresh-cut sweet leaf bush (*Sauropus androgynus* L. Merr.) through hot $CaCl_2$ dips. *Int. J. Food Sci.* **2012**, *47*, 2662–2670. [CrossRef]
69. Glenn, D.M.; Puterka, G.J.; Vanderzwet, T.; Byers, R.E.; Feldhake, C. Hydrophobic particle films: A new paradigm for suppression of arthropod pests and plant diseases. *J. Econ. Entomol.* **1999**, *92*, 759–771. [CrossRef]
70. Rosati, A.; Metcalf, S.G.; Buchner, R.P.; Fulton, A.E.; Lampinen, B.D. Effects of kaolin application on light absorption and distribution, radiation use efficiency and photosynthesis of almond and walnut canopies. *Ann. Bot.* **2007**, *99*, 255–263. [CrossRef] [PubMed]

horticulturae

MDPI

Review

Preharvest Elicitors Spray Improves Antioxidant Activity, Alleviates Chilling Injury, and Maintains Quality in Harvested Fruit

Di Gong [1,2], Yang Bi [1,*], Yongcai Li [1], Yi Wang [1], Dov Prusky [3] and Noam Alkan [3]

1 College of Food Science and Engineering, Gansu Agricultural University, No. 1 Yingmencun, Anning District, Lanzhou 730070, China
2 Department of Food Sciences, Institute of Postharvest and Food Sciences, Agricultural Research Organization, VolcaniCenter, RishonLeZion 7528809, Israel
3 Department of Postharvest, Institute of Postharvest and Food Sciences, Agricultural Research Organization, VolcaniCenter, RishonLeZion 7528809, Israel
* Correspondence: biyang@gsau.edu.cn; Tel.: +86-931-7631113; Fax: +86-931-7631201

Abstract: Antioxidant activity is an important feature for evaluating fruit quality and tolerance to biotic or abiotic stresses. Moreover, antioxidant activity is involved in chilling injury (CI) response and postharvest quality in fruit. Chemical elicitors can induce systemic acquired resistance in fruit against pathogens, which could partially replace synthetic fungicides. Recently, researchers have found that preharvest sprays with chemical elicitors can improve antioxidant activity, reduce CI, and maintain quality in harvested fruit. In this review, we summarize that preharvest elicitors spray improve antioxidant activity in harvested fruit by promoting antioxidant components biosynthesis as well as antioxidant ability in vitro. Moreover, preharvest elicitors spray alleviates CI in fruit by regulation of membrane lipid metabolism and reactive oxygen species metabolism. In addition, preharvest elicitors spray maintains fruit quality by modulation of respiration and ethylene release. Finally, this review points out the issues existing and proposes an outlook on preharvest elicitors spray to maintain postharvest fruit quality.

Keywords: fruit; preharvest elicitor spray; antioxidant activity; chilling injury; postharvest quality

Citation: Gong, D.; Bi, Y.; Li, Y.; Wang, Y.; Prusky, D.; Alkan, N. Preharvest Elicitors Spray Improves Antioxidant Activity, Alleviates Chilling Injury, and Maintains Quality in Harvested Fruit. *Horticulturae* **2022**, *8*, 1208. https://doi.org/10.3390/horticulturae8121208

Academic Editor: Isabel Lara

Received: 27 October 2022
Accepted: 12 December 2022
Published: 16 December 2022

1. Introduction

Fresh fruits have aesthetic appearance and unique flavor as well as nutrition. However, postharvest loss in fruit is huge. The global average of postharvest loss in fresh fruit is estimated to be 40% [1]. Fungal infection is the main reason for postharvest loss of fruit, but over-ripening and senescence, and chilling injury (CI) caused by inappropriate low temperature are also considered to be important causes of the loss [2,3]. Various measures can effectively control quality deterioration and CI in fruit. However, most of them focus on postharvest treatments, which are inconvenient and inevitably increase postharvest procedures and costs [2,4,5].

Chemical elicitors are synthetic chemicals, mainly including salicylic acid (SA), jasmonic acid (JA) and chitosan and their analogs or derivatives. These elicitors can induce fruit resistance against pathogens by eliciting systemic acquired resistance (SAR), which could partially replace synthetic fungicides [6]. Recently, in addition to inhibiting fruit diseases by inducing resistance, chemical elicitor shave been found to contribute to enhancing antioxidant activity, reducing CI, and maintaining postharvest quality in fruit [7–9].

The periods of fruit development are critical for the formation of postharvest resistance and quality in fruit. During the development, continuous cell division and expansion lead to fruit enlargement and the formation of barriers that can contribute to fruit resistant, including epidermal wax, cut in, cell wall, and cell membrane [10]. Meanwhile, products such as sugars, organic acids, aromas, ascorbic acid, carotenoids, and phenolic compounds are continuously accumulated in cells, which can affect fruit quality and tolerance biotic or abiotic stresses [11,12]. This review focuses on the effect of preharvest elicitors spray on improving antioxidant activity, alleviating CI, and maintaining quality in harvested fruit, and analyzes the possible reasons.

2. Improvement of Antioxidant Capacity

Phenolics, ascorbic acid, and carotenoids are important antioxidant compounds in fruit, which are known to be beneficial to human health [13].

2.1. Enhancing Phenolics, Ascorbic Acid and Carotenoids Contents

Preharvest elicitors spray effectively enhances the accumulation of phenolic compounds in fruit. Similarly, 0.1 mM methyl jasmonate (MeJA) or 0.5 mM SA that was sprayed 4 times on 'Fino' lemon before harvest increased the contents of total phenolics and two main flavonoids (hesperidin and eriocitrin) [14]. 'Xiaobai' Apricot was sprayed twice with 0.05% chitosan oligochitosan or 1 mmol L^{-1} SA at 7 and 2 days before harvest, resulting in higher contents of total phenolics, total flavonoids and three main phenolic compounds (5-O-caffeoylquinic acid, 3-O-caffeoylquinic acid, quercetin-3-rutinoside) at harvest and during cold storage [7]. Preharvest with phenylalanine increased the total flavonoids and anthocyanins in 'Kent' and 'Shelly' mango fruits and 'Pink Lady' apples [15]. Spraying twice with 0.1%, 0.2%, or 0.4% prohydrojasmon on 4 and 2 weeks before harvest increased the contents of total anthocyanin and total flavonoid and the accumulation of anthocyanin monomers and flavonoids in 'Kent' mango fruit during cold storage, of which 0.4% treatment showed the best effect. On the 21st day of cold storage, the contents of total anthocyanin and total flavonoid in 0.4%-sprayed fruit showed 4 folds and 67% higher than those of the control. Moreover, the contents of anthocyanin monomers, including cyanidin, methyl-cyanidin, cyanidin-3-galactoside, 7-O-methylcyanidin-3-O-β-D-galactopyranoside, as well as quercetin and kaempferol content, were significantly higher than those of the control [16]. In addition, at harvest and during cold storage, contents of total phenolics, total anthocyanins, and anthocyanin monomers enhanced in 'Mollar de Elche' pomegranate sprayed by 1, 5, or 10 mmol L^{-1} MeJA at 94, 64, 34, and 4 days before harvest, of which 5 mmol L^{-1} MeJA spay showed the best effect [17].

Preharvest elicitors spray can significantly enhance the accumulation of ascorbic acid and carotenoids in fruit. At harvest, a higher content of ascorbic acid was observed in 'El-Bayadi' table grapefruit sprayed with 4.0 mM SA 4 times from the pea stage to the version stage [18]. A higher total carotenoid content was shown in 'Black Splendor' plum fruit that was sprayed 3 times with 0.5 mmol L^{-1} SA, 1 mmol L^{-1} acetyl salicylic acid (ASA), or 0.5 mmol L^{-1} Methyl salicylate (MeSA) at 61, 76, and 94 days after flowering. At harvest, the total carotenoid content in the three treated plum fruit increased by 17%, 25%, and 25% compared with the control, which was 18%, 55%, and 36% higher on the 40th day of cold storage, respectively [19]. Similarly, three sprays of 1 mM oxalic acid at 63, 77 and 98 after flowering increased total carotenoid content in 'Black Splendor' plum fruit, which increased by 20% compared with the control on the 35th of cold storage [20]. In addition, the total carotenoid content in 'Neelum', 'Bangalore', 'Banganapalli', and 'Alphonso' mango fruit sprayed twice with 2% hexanal at 30 and 15 days before harvest was 6.7%, 61%, 27%, and 32% higher than that of the control on the 9th day of cold storage [21]. Three sprays of 250 mmol L^{-1} MeJA at flowering, 24 days of turning green, and 7 days of turning red enhanced ascorbic acid content in 'Camarosa' strawberry, which increased by 20% and 63% compared with the control at harvest and 2 days of storage [22]. Further, two sprays of 1.2 mM hexanal4 and 2 weeks before harvest increased ascorbic acid content in 'Allahabad Safeda' guava fruit at harvest and during cold storage [23]. Preharvest elicitors spray on improving phenolics, ascorbic acid and carotenoids accumulation in harvested fruit are shown in Table 1.

Phenolics and flavonoids in fruit are mainly synthesized from phenylpropanoid metabolism pathway. L-phenylalanine is the substrate of phenylpropanoids metabolism and it can generate phenolics and flavonoids under the continuous activity of phenylalanine ammonialyase (PAL), cinnamate 4-hydroxylase (C4H), and 4-coumarate: coenzyme A ligase (4CL), and cinnamyl alcohol dehydrogenase (CAD) [24,25]. Preharvest phenylalanine spray improved the contents of total flavonoids and total anthocyanins in mango fruit [15]. Similarly, preharvest phenylalanine spray triggered PAL, C4H, 4CL, and CAD activity, increasing the accumulation of total phenolic and flavonoids in muskmelon fruit [26]. Preharvest melatonin up-regulated the levels of *SlPAL, SlC4H,* and *Sl4CL* expression in cherry tomato fruit during storage [27]. Moreover, preharvest prohydrojasmon spray up-regulated the expression of a key transcription factor (*MYB114*) that regulated key genes for anthocyanin biosynthesis (*PAL, CHS, CHI, F3H, ANS, UFGT, FLS,* and *LAR*) in pear fruit at harvest [28]. In addition, preharvest MeJA improved PAL activity, increasing flavonoid accumulation in raspberries at harvest [29]. Preharvest harpin spray increased PAL activity in muskmelon fruit during storage, improving total phenolics and flavonoid content [30]. Preharvest chitosan spray enhanced the activity of PAL, 4CL, CAD, and C4H, promoting the biosynthesis of phenolics and flavonoids in muskmelon fruit during storage [31,32]. Although the effects of preharvest elicitors spray on the biosynthesis of phenolics in harvested fruit have been studied at the biochemical level, the relevant molecular mechanisms remain to be further unlocked.

Ascorbic acid is mainly synthesized through the L-galactose pathway in fruit [33]. However, there is no report on the regulation of the L-galactose pathway in fruit by preharvest elicitors spray. Carotenoids are mainly synthesized through 2-c-methyl-d-erythritol 4-phosphate and/or mevalonate pathway in fruit [34]. The transcriptome results showed that preharvest of benzothiadiazole (BTH) and chitosan spray up-regulated the expression of important transcription factors as WRKYs and NACs, involved in regulating the metabolism of carotenoid compounds in fruit, thereby promoting the synthesis of carotenoids in harvested 'Alba' strawberry fruit [35]. There are few reports on the effect of preharvest elicitors spray on carotenoid biosynthesis in harvested fruit, and the related mechanism needs to be further studied.

Table 1. Preharvest elicitors spray improves antioxidant compounds and freeradicalscavengingactivity in fruit.

Elicitor	Manufacturer	Fruit	Cultivar	Spraying Concentration	Spraying Times	Spraying Stage	Antioxidant Compounds					Free Radical Scavenging Activity			References
							Phenols	Flavonoids	Anthocyanins	VC	Carotenoids	DPPH	ABTS⁺	FRAP	
SA	Sigma-Aldrich	Plum	Black Splendor	0.5 mmol L^{-1}	3	61, 76 and 94 days after full blossom	√ [a]	— [b]	√	—	√	—	—	—	[19,36]
	Oxford Laboratory Reagents	Apricot	Canino	2 or 4 mM	2	30 and 15 days before harvest	—	—	—	—	√	—	—	—	[37]
	—		Xiaobai	1 mmol L^{-1}	2	7 and 2 days before harvest	√	√	—	√	—	√	√	√	[7]
	—	Palm	Khesab	3%	3	5 and 15 weeks from pollination, and two weeks before harvest	√	√	—	—	—	—	—	—	[38]
ASA	Sigma-Aldrich	Plum	Black Splendor	1 mmol L^{-1}	3	61,76 and 94days after full blossom	√	—	√	—	√	—	—	—	[19,36]
MeSA	Sigma-Aldrich	Plum	Black Splendor	0.5 mmol L^{-1}	3	61,76 and 94days after full blossom	√	—	√	—	√	—	—	—	[19,36]
BTH	Novartis Crop Protection	Muskmelon	Yindi	100 mg L^{-1}	3	flowering and 14, 28, 42 days after flowering	√	√	—	—	—	—	—	—	[39]
Prohydrojasmon	—	Pear	Nanhong	50 or 100 mg L^{-1}	2	100 and 103 days after blooming	√	√	√	—	—	—	—	—	[28]
	—	Mango	Kent, Shelly and Maya	0.1, 0.2 or 0.4%	2	4 and 2 weeks before harvest	—	√	√	—	—	—	—	—	[16]
Chitosan	—	Strawberry	Seascape	2, 4, or 6 g L^{-1}	1	just turning red	—	—	√	—	—	—	—	—	[40]
	—	Grape	Jingxiu	1 g L^{-1}	1	10 days before harvest	√	—	—	—	—	—	—	—	[40]
	—		Yaghouti	2% or 3%	3	fruit set, 25 and 50 days after fruit set	—	—	√	√	—	√	—	—	[41]
	WN Group of Publishers	Muskmelon	Manao	0.001	4	14, 21, 28, 40 days after flowering	√	√	—	—	—	—	—	—	[31]
	Cornell Lab	Apricot	Canino	1.5% or 2.5%	2	30 and 15 days before harvest	—	—	—	—	√	—	—	—	[37]
	—	Palm	Khesab	1%	3	5 and 15 weeks from pollination and two weeks before harvest	√	√	—	—	—	—	—	—	[38]
	Huarun Bioengineering	Kiwifruit	guichang	28.6%	3	budding phase, fruit setting phase and expanding final phase	√	√	—	√	—	—	—	—	[42]

Table 1. *Cont.*

Elicitor	Manufacturer	Fruit	Cultivar	Spraying Concentration	Spraying Times	Spraying Stage	Antioxidant Compounds					Free Radical Scavenging Activity			References
							Phenols	Flavonoids	Anthocyanins	VC	Carotenoids	DPPH	ABTS⁺	FRAP	
Chitosan oligosaccharide	Dalian GlycoBio	strawberry	qingxiang	50 mg·L^{-1}	4	seedling stage, before flowering, fruit coloring and full bloom	√	√	√	√	—	√	—	—	[43]
Chitosan oligochitosan	—	Apricot	Xiaobai	0.05%	2	7 and 2 days before harvest	√	√	—	√	—	√	√	√	[7]
Oligochitosan	Jinan Haidebei Marine Bio-engineering	Navel orange	Osbeck	15 g L^{-1}	4	30, 60, and 90days after physiological fruit drop and 10 days before harvest	√	—	—	—	—	—	—	—	[44]
β-aminobutyric acid	Sigma-Aldrich	Blueberry	Bluecrop	20 mM	1	7 days before harvest	—	—	√	—	—	—	—	—	[45]
Oxalic acid	—	Peach	Anjirymaleki	1, 3 or 5 mmol L^{-1}	1	15 days before harvest	√	√	—	—	—	√	—	√	[46]
	—	Kiwifruit	Bruno	5 mM	3	130days after full blossom and 2 times at 7 days intervals	√	√	—	√	—	—	—	—	[47]
	Sigma-Aldrich	Sweet cherry	Sweet Heart and Sweet Late	0.5, 1.0 or 2.0 mM	3	98, 112, and 126 days after full blossom	√	—	√	—	—	—	—	—	[48]
	—	Plum	Black Splendor	1 mM	3	63,77 and 98 days after fullblossom	√	—	—	—	√	—	√	—	[20]
	—	Kiwifruit	Bruno	5 mmol L^{-1}	3	130, 137 and 144 days after the flowering	—	—	—	√	—	—	—	—	[49]
	Sigma-Aldrich	Pomegranate	Mollar deElche	1, 5 or 10 mM	3	80, 110, 140, and 170 days after full blossom	√	—	√	—	—	—	—	—	[50]
	—	Apricot	Red Flesh	0.5, 1 or 2 mM	1	fruit set stage	√	—	—	√	—	—	—	—	[51]
	Sigma-Aldrich	Lemon	Fino	0.1, 0.5 or 1.0 mM	5	from physiological fruit dropto 3 days before harvest	√	—	—	—	—	—	—	—	[52]
Harpin	Eden Bioscience	Muskmelon	Huanghemi	50 mg L^{-1}	3	flowering and 14, 28, 42 days after flowering	√	—	—	√	—	—	—	—	[53]

Table 1. *Cont.*

Elicitor	Manufacturer	Fruit	Cultivar	Spraying Concentration	Spraying Times	Spraying Stage	Antioxidant Compounds					Free Radical Scavenging Activity			References
							Phenols	Flavonoids	Anthocyanins	VC	Carotenoids	DPPH	ABTS+	FRAP	
Hexanal	Sigma-Aldrich	Mango	Neelum, Bangalora, Banganapalli and Alphonso	0.02%	1; 1; 2	15 days before harvest; 30 days before harvest; 15 and 30 days before harvest	—	—	—	√	√	—	—	—	[21]
	—	Strawberry	Jewel, Kent, Mira and St. Pierre	0.01	3	once per week before harvest	√	√	√	—	—	—	—	—	[54]
	—	Guava	Allahabad Safeda	0.8, 1.2 or 1.6 mM	2	4 and 2 weeks before harvest	√	—	—	√	—	—	—	—	[23]
SNP	Sigma-Aldrich	Muskmelon	Manao	0.5 mM	4	14, 21, 28, 40 days after flowering	√	√	—	—	—	—	—	—	[55]
	Sigma-Aldrich		Manao	0.5 mM	4	young fruit stage, early stage of en largement, late stage of enlargement and mature stage	√	√	—	—	—	—	—	—	[56]
	Sigma-Aldrich	Peach	G.H. Hill	0, 25, 50 or 100 mol L^{-1}	1	14 days before harvest	—	—	—	√	—	—	—	—	[57]
Putrescine and spermidine	—	Grape	Olhoghi and Rishbaba	1 or 2 mM/1 or 2 mM	2	40 and 20 days before harvest	√	—	√	—	—	—	—	—	[58]
Arginine	—	Pomegranate	MalaseSaveh	0, 0.5, 1 or 2 mM	3	20 days interval before commercial harvest	√	√	—	√	—	√	—	—	[59]
L-phenylalanine	Hunan-Shaofeng	Muskmelon	Manao	8 mM	4	young fruit stage, early expansion stage, late expansion stage and one week before harvest	√	√	—	—	—	—	—	—	[26]
Melatonin	—	Apricot	Colorado and Mikado	0.1 mM	3	pit hardening, final fruit growth, and 4 days before harvest	√	—	—	—	—	—	—	—	[60]
	Sigma-Aldrich	Sweet cherry	Ferrovia	0.5 mM	2	2 and 1 weeks prior to harvest	√	—	—	—	—	—	—	—	[61]
	—	Pear	Nanhong	50 or 200 μmol L^{-1}	2	dripping and 3 days after	√	√	—	—	—	—	—	—	[62]

[a] √ indicates detected. [b] — indicates no assessment.

2.2. Promoting the Antioxidant Activity

Preharvest elicitors spray also improves in vitro antioxidant ability in fruit (Table 1). For example, spraying 5 mmol L^{-1} oxalic acids 15 days before harvest had no significant effect on the v̀alues of DPPH and FRAP in 'Anjiry Maleki' peach fruit at harvest. However, the spray increased the values of DPPH and FRAP in the fruit by 25% and 20% compared with the control on the 4th of storage [46]. Another example is that four sprays of 4.0 mM SA at the 'pea stage' to the version stage increased the values of DPPH and $ABTS^+$ in 'El-Bayadi' table grapes, which was 63% and 25% higher than that of the control at harvest [18].

DPPH, $ABTS^+$, and FRAP are important indicators to evaluate antioxidant ability. DPPH mainly reflects the antioxidant ability of phenols, flavonoids, and terpenoids in fruits [63]. $ABTS^+$ and FRAP reflect the antioxidant ability of ascorbic acid and carotenoids in fruit, respectively [64]. Due to the increase of the biosynthesis of phenolics and flavonoids as well as ascorbic acid and carotenoids, preharvest elicitors spray improves antioxidant ability in fruit. The current studies are mainly focus on the antioxidant ability in vitro of harvested fruit, while further research is needed on the antioxidant ability of human cells or in vivo.

3. Preharvest Elicitors Spray Alleviates CI in Fruit

3.1. Reducing the Occurrence of CI

Many fruits are sensitive to low temperatures, which makes them prone to CI during storage at suboptimal temperatures, resulting in surface pitting, water-soaked spots, epidermal or internal browning, abnormal ripening, and other CI symptoms. Cold-sensitive fruits mainly grow in tropical or subtropical regions, including citrus, bananas, avocados, mangos, pineapples, peaches, apricots, plums, papayas, pomegranates, melons, etc. [65]. Preharvest elicitors spray can effectively alleviate CI in fruit. For example, 'Lane Late' sweet orange fruit sprayed with 2, 4, 6, or 8 mM SA 10 days before harvest reduced CI index of fruit stored for 93 days at 5 °C. Compared with the control, the CI index decreased by 76.15% in the fruit sprayed with 8 mM SA on the 93rd day of cold storage, before harvest [8]. Another example is that preharvest spray at 7 and 2 days with 0.05% chitosan, oligochitosan, or 1 mmol L^{-1} SA reduce CI in 'Xiaobai' apricot that was stored for 14 days at 2 °C [7]. In addition, 0.02% hexanal spray at 15 and 10 days before harvest decreased the CI index in 'Fantasia' nectarine fruit by 55% compared to the control on the 45th day of storage at 2 °C [66]. Preharvest elicitors spray on CI alleviation in fruit during cold storage are shown in Table 2.

Table 2. Preharvest elicitors spray alleviates chilling injury in fruit.

Elicitor	Manufacturer	Fruit	Cultivar	Spraying Concentration	Spraying Times	Spraying Stage	Storage Temperature (°C)/Days	References
SA	—	Navel orange	Lane Late	2, 4, 6 or 8 mM	1	10 days before harvest	3/93	[67]
	—	Sweet orange	Lane Late and Valencia Late	2, 3, 4, 6, 8 or 9 mM	1	10 days before harvest	3/93	[8]
	Sigma-Aldrich	Grapefruit	Ray Ruby	6, 8 or 12 mM	— [a]	20 days intervals before harvest	8/90	[68]
	—	Peach	Flordaking	1, 2 or 3 mM	3	the cell division, cell enlargement and pit-hardening stages	1/42	[69]
	—	Apricots	Xiaobai	1 mmol L^{-1}	2	7 and 2 days before harvest	2/70	[7]
	—	Pineapple	Comte de Paris	2.0 mM	4	15-day intervals before harvest	10/20	[70]
	—	Jujube	Peyuan	150 mg L^{-1}	1	3 weeks before harvest	—/35	[71]
MeSA	Aladdin	Apricots	Kate	0.05, 0.1 or 0.2 mmol L^{-1}	2	72 d and 74 d after full blossom	2/32	[72]
MeJA	Sigma-Aldrich	Grapefruit	Ray Ruby	3, 4 or 5 mM	—	20 days intervals before harvest	8/90	[68]
	Sigma-Aldrich	Pomegranates	Malas	1 or 2 mM	1	15 days before harvest	4/28	[73]
	Sigma-Aldrich	Pomegranate	Mollar de Elche	5 mM	5	80, 110, 140 and 170 days after full blossom, and 4 d before harvest	2/90	[74]
Chitosan oligochitosan	—	Apricots	Xiaobai	0.05%	2	7 and 2 days before harvest	2/70	[7]
β-aminobutyric acid	—	Apple	Honeycrisp	40 mM	2	2 and 4 weeks before harvest; 1 and 2 weeks before harvest	0.5 or 3/4 months; 0.5/5 months	[75]
Arginine	—	Pomegranate	MalaseSaveh	0, 0.5, 1 or 2 mM	3	20 days interval before commercial harvest	4/120	[59]
SNP	Sigma-Aldrich	Peach	G.H. Hill	0, 25, 50 or 100 mol L^{-1}	1	14 days before harvest	4/28	[57]
Melatonin	—	Apricot	Colorado and Mikado	0.1 mM	3	pit hardening, final fruit growth, and 4 days before harvest	1 and 8 /21 and 28	[60]
Hexanal	—	Nectarine	Fantasia	0.02%	2	15 and 10days before harvest	2/45	[66]
Putrescine	—	Pear	Spadona	0.5, 1 or 2 mM	3	—	0/21	[76]

[a] — indicates unspecified.

3.2. Improving ROS Scavenging Ability

Oxygen burst caused by CI leads to excessive H_2O_2 accumulation that accelerates peroxidation of the cell membrane, leading to the production of malondialdehyde (MDA), which also destroys membrane integrity [5]. Preharvest elicitors sprayed as SA can increase the activity of antioxidant enzymes as well as the concentration of antioxidant compounds, which contribute to reducing H_2O_2 damage in fruit and vegetables [14]. Preharvest MeJA and SA spray improved the activities of catalase (CAT), ascorbate peroxidase (APX), and peroxidase (POD) as well as the biosynthesis of phenols and flavonoids in 'Fino' and 'Verna' lemon fruit during cold storage [14]. Moreover, during cold storage, higher activities of superoxide dismutase (SOD), CAT, APX, and POD and accumulations of total phenols, total anthocyanins, total carotenoids, and total flavonoids were found in 'Black Splendor' and 'Royal Rosa' plum fruit that was preharvest sprayed with SA, acetylsalicylic acid (ASA), or MeSA [19,36]. Preharvest MeJA spray increased total phenols, total anthocyanins, total flavonoids, and vitamin C contents, thereby improving ROS scavenging and antioxidant activity in fruit [17]. Moreover, preharvest spray with SA or chitosan oligochitosan in the 'Xiaobai' apricot orchard decreased MDA content and increased contents of total phenols, total flavonoids, and vitamin C. Additionally, preharvest chitosan oligochitosan and SA spray enhanced free radical-scavenging capacity, including DPPH free radical scavenging capacity, ABTS free radical scavenging capacity, cupric ion reducing antioxidant activity, and ferric reducing antioxidant activity, elevating scavenging ability of free radicals in fruit [7]. Although the effects of preharvest elicitors spray on activity of antioxidant enzymes and biosynthesis of antioxidants in harvested fruit have been studied at the biochemical level, the relevant molecular mechanisms remain to be further revealed.

3.3. Decreasing Membrane Lipid Metabolism

Destruction of the cell membrane is considered to bean important characteristic of CI. The content of unsaturated fatty acids (USFAs) in the cell membrane is closely related to CI in fruit [52]. A high content of USFAs contributes to the fluidity of cell membrane and membrane integrity and increases the cold tolerance of fruit [65,77]. Preharvest MeJA spray maintained USFAs content in 'Mollar de Elche' pomegranate fruit during cold storage and reduced the ratio of unsaturated/saturated fatty acids, contributing to maintaining the integrity of cell membrane and alleviating CI in fruit [78]. Low temperature leads to degradation and peroxidation of membrane lipids that cause the loss of membrane integrity and functionality, resulting in the appearance of CI symptoms in fruit [79,80]. Phospholipase D (PLD) is a key enzyme in membrane phospholipid metabolism, which can directly hydrolyze ester bonds of phosphoglyceride and catalyze the decomposition of phospholipid into phospholipid acid (PA) and diacylglycerol [81]. The occurrence of CI increases PLD activity in fruit that induces PA accumulation, while excessive production of PAleads to H_2O_2 accumulation, which destroys the integrity of the cell membrane [82]. Preharvest hexanal spray inhibited membrane phospholipid metabolism in 'Fantasia' nectarine fruit by down-regulating *PLD* expression, thereby maintaining the integrity of the cell membrane [66]. Although the effects of preharvest elicitors spray on membrane lipid metabolism in harvested fruit have been studied at the biochemical level, the relevant molecular mechanisms remain to be further proclaimed.

3.4. Reducing Pulp Browning

Epidermal or internal browning is a typical symptom of fruit CI. Preharvest SA and chitosan oligochitosan spray decreased polyphenol oxidase (PPO) activity in 'Xiaobai' apricot fruit, alleviating pulp browning in fruit during cold storage [7]. Preharvest sodium nitroprusside (SNP) spray alleviated CI-induced browning of pulp tissue of 'G.H. Hill' peach fruit by decreasing PPO activity [57]. Cell membrane degradation caused by CI increases the possibility of interaction between polyphenols and PPO, which oxidizes polyphenols to quinones, and further polymerizes and forms dark brown products [70].

However, the specific regulation of enzymatic browning in fruit by preharvest elicitors spray remains to be further explained.

4. Preharvest Elicitors Spray Delays Ripening and Maintains Postharvest Quality of Fruit

Fruit ripening caused by ethylene is an important cause of quality deterioration in harvested fruit. Many reports have elucidated that preharvest elicitors spray can effectively retard fruit ripening by inhibiting ethylene production, thereby maintaining postharvest quality and prolonging shelf life of fruit [83].

4.1. Retarding Ripening

Ethylene release and increased respiration are typical physical features of climacteric fruit, which play a critical role in fruit maturation and ripening [84]. The peak of ethylene in 'Black Splendor' plum fruit retarded after spraying with 0.5 mmol L^{-1} SA, 1 mmol L^{-1} ASA, or 0.5 mmol L^{-1} MeSA 3 times at 61, 76, and 94 days after full blossom, respectively [36]. Moreover, the respiration rate in 'Dashehari' mango fruit retarded by 7 days during cool storage after spraying with 1200, 1600, or 2000 μM hexanal before harvest [85]. Preharvest elicitors spray can retard the peak of ethylene and respiration in climacteric fruit as well as decrease ethylene release and respiratory rate. In 'Zill' mango fruit, sprayed with 150 μM SA before harvest delayed the peak of ethylene by 2 days and decreased the ethylene accumulation by 82% compared with the control during cool storage [83]. These elicitors spray also can inhibit ethylene release and respiratory rate in non-climacteric fruit. 'Fino' lemon fruit that was sprayed with 0.1 mM MeJA or 0.5 mM SA 4 times before harvest had a decrease in ethylene production and respiration rate [14]. Similarly, compared with the control, the respiratory rate in 'Mollar de Elche' pomegranate fruit decreased by 33.33% on the 60th of storage after 5 mmol L^{-1} MeJA sprayed 4 times at 94, 64, 34 and 4 days before harvest [17]. Preharvest elicitors spray on inhibiting ethylene production and respiration rate in fruit after harvest are shown in Table 3. Although the effect of preharvest elicitors spray on respiratory rate in fruit has been studied, how the spray affects respiratory metabolism needs to be further revealed.

Ethylene plays a dual role in fruit as it participates both in the fruit in the ripening and defense response of climacteric fruits [86]. 1-Aminocyclopropane-1-carboxylic acid synthase (ACS) and 1-aminocyclopropane-1-carboxylic acid oxidase (ACO) are two key enzymes involved in ethylene biosynthesis. Preharvest elicitors spray reduces ethylene release in climacteric and non-climacteric fruit during storage. The gene expression of *MiACS* and *MiACO* in 'Zill' mango fruit preharvest sprayed with SA was reduced by 90% and 96% on the 9th and 7th day of storage, respectively [83]. Preharvest chitosan spray decreased the gene expression of *AdACS2* and *AdACO2* by 30% and 50% in 'Garmrok' kiwifruit compared with the control on the 60th day of storage [86]. Similarly, preharvest chitosan oligosaccharides spray reduced *FaACS* and *FaACO* gene expression in 'Qingxiang' strawberry fruit by 92% and 66% compared with the control at harvest [43]. Ethylene receptors negatively regulate ethylene signal transduction, while ethylene response factors (ERFs) are positively regulated by ethylene [87]. Transcriptome analysis showed that preharvest BTH and chitosan spray decreased *ERF* expression in 'Alba' strawberries after harvest, thereby inhibiting ethylene signal transduction [35]. Moreover, at harvest, *ETR2* was down-regulated in 'Stark Red Gold' peach fruit that was preharvest sprayed with MeJA [88]. Although the effects of preharvest elicitors spray on biosynthesis and signal transduction of ethylene in harvested fruit have been studied, the relevant regulation mechanisms remain to be further announced.

Table 3. Preharvest elicitors spray maintains postharvest quality in fruit.

Elicitor	Manufacturer	Fruit	Cultivar	Spraying Concentration	Spraying Times	Spraying Stage	Storage Temperature (°C)/Days	Postharvest Quality				References
								TSS	TA	TSS/TA	Firmness	
SA	—	Navel orange	Lane Late	2, 4, 6 or 8 mM	1	10 days before harvest	5/93	√ [a]	√	— [c]	√	[67]
	Sigma-Aldrich	Lemon	Fino	0.5 mmol L^{-1}	4	21 days intervals untile 3 days before harvest	8/35	√	√	—	√	[14]
	Sigma-Aldrich	Mango	Zill	150 µM	1	100 days after anthesis	25/11	× [b]	√	—	√	[83]
	Sigma-Aldrich	Peach	Cresthaven	1 or 2 mM	2	23 and 15 days before harvest	20/2	×	√	—	√	[89]
	—		Flordaking	1, 2 or 3 mM	3	cell division, pit hardening or lag phase and cell enlargement stage	1/42	×	√	×	√	[69]
	—	Apricot	Xiaobai	1 mmol L^{-1}	2	7 and 2 days before harvest	2/70	√	√	—	√	[7]
	Cornell Lab		Canino	2 or 4 mM	2	30 and 15 days before harvest	0/28	×	√	×	√	[37]
	—	Grape	Thompson Seedless	100 mg L^{-1}	2	pea and veraison stage	20/7	—	—	—	√	[90]
	Sigma-Aldrich		Flame Seedless	1.0, 1.5 or 2.0 mM	2	pea and veraision stage	3–4/75	√	√	√	√	[91]
	Sigma-Aldrich		Magenta and Crimson	1 mM	3	40% berries, veraison stage and 3 days before harvest	2/45	√	×	—	—	[50]
	—		Superior Seedless	1, 2, or 4 mM	3	fruit set, berry variation and 14 days before harvest	28/44	√	√	√	√	[92]
	Sigma-Aldrich	Plum	Black Splendor	0.5 mmol L^{-1}	3	61,76 and 94 days after full blossom	2/50	√	×	√	√	[19]
	—	Strawberry	Festival	2 or 4 mmol	3	full flowering, green fruits and pink stage	4/12	√	—	—	√	[93]
	—	Sweet cherry	Sweet Heart, Sweet Late and Lapins	0.5 mmol L^{-1}	3	98, 112 and 126 days after full blossom (SE/SL)/66, 75 and 81 days after full blossom (L)	2/28	√	√	—	√	[94]
	Sigma-Aldrich	Jujube	Dongzao	2 mM	4	30, 60, 90, and 110 days after full blossom	0/60	—	—	—	—	[95]
	—	Wax apple	Taaptipjaan	0.5 or 1.0 mM	1	24 h before harvest	13/9	—	—	—	√	[96]
	—	Palm	Khesab	3%	3	5 and 15 weeks from pollination, and two weeks before harvest	2/60	×	—	—	—	[38]
ASA	Sigma-Aldrich	Grape	Magenta and Crimson	1 mM	3	40% berries, veraison stage and 3 days before harvest	2/45	√	×	—	—	[17]
	—	Strawberry	Festival	0.25 or 0.50 mmoL	3	full flowering, green fruits and pink stage	4/12	√	—	—	√	[93]
	—	Sweet cherry	Sweet Heart, Sweet Late and Lapins	1.0 mmol L^{-1}	3	98, 112 and 126 days after full blossom	2/28	√	√	—	√	[94]

Table 3. *Cont.*

Elicitor	Manufacturer	Fruit	Cultivar	Spraying Concentration	Spraying Times	Spraying Stage	Storage Temperature (°C)/Days	Postharvest Quality				References
								TSS	TA	TSS/TA	Firmness	
MeSA	Sigma-Aldrich	Plum	Black Splendor	1 mmol L^{-1}	3	61,76 and 94 days after full blossom	2/50	√	×	√	√	[19,36]
	—	Apricot	Kate	0.05, 0.1 or 0.2 mmol L^{-1}	2	72 d and 74 days after full blossom	2/32	√	√	—	√	[72]
	Sigma-Aldrich	Grape	Magenta and Crimson	1 mM	3	40% berries, veraison stage and 3 days before harvest	2/45	√	×	—	—	[17]
MeJA	Sigma-Aldrich	Plum	Black Splendor	0.5 mmol L^{-1}	3	61,76 and 94days after full blossom	2/50	√	×	√	√	[19]
	Sigma-Aldrich	Lemon	Fino	0.1 mmol L^{-1}	4	21 days intervals untile 3 days before harvest	8/35	√	√	—	√	[14]
	Sigma-Aldrich	Mango	Mahachanok	20, 40, 80 or 120 μL mL^{-1}	1	90 days after anthesis	15/24	—	—	√	√	[97]
	—	Plum	Black Splendor and Royal Rosa	0.5 or 1.0 mM	3	-	20/9; 2/50	×	√	—	√	[98]
	—	Strawberry	Fortune	1120 or 2240 mg L^{-1}	1	115 days after full blossom	0/28	×	—	—	√	[99]
	—		Chilean	0.25 mM	3	80% flowering, turning fruit and full ripe fruit stage	22/3	—	—	√	√	[100]
	Sigma-Aldrich		Camarosa	250 mmol L^{-1}	1; 2; 3	100% red stage; large green andafter 7 days at 100% red receptacle stages; flowering, after 24 days at the large green, and after 7 days at 100% red receptacle stages	25/3	—	—	√	√	[22]
	Sigma-Aldrich	Pomegranate	Malas	1 or 2 mM	1	15 days before harvest	4/80	√	√	—	√	[73]
	Sigma-Aldrich		Mollar de Elche	1, 5 or 10 mmol L^{-1}	4	94, 64, 34 and 4 days before harvest	10/60	—	—	—	√	[78]
Prohydrojasmon	—	Mango	Kent, Shelly and Maya	0.1%, 0.2% or 0.4%	2	4 and 2 weeks before harvest	12/21	√	×	—	×	[26]
Chitosan	Oxford Laboratory Reagents	Apricot	Canino	1.5% or 2.5%	2	30 and 15 days before harvest	0/28	×	√	×	√	[37]
	—	Grape	Jingxiu	1 g L^{-1}	1	10 days before harvest	20/16;0/42	×	√	×	—	[40]
	—		Yaghouti	2% or 3%	3	fruit set, 25 and 50 days after fruit set	-/40	√	√	√	√	[41]
	—	Strawberry	Alba and Romina	0.5% or 1%	5	flowering and followed every 5 days	0.5/7	—	—	—	—	[101]

Table 3. *Cont.*

Elicitor	Manufacturer	Fruit	Cultivar	Spraying Concentration	Spraying Times	Spraying Stage	Storage Temperature (°C)/Days	Postharvest Quality				References
								TSS	TA	TSS/TA	Firmness	
	—		Chilean	1.5%	3	80% flowering, turning fruit stage and full ripe fruit stage	22/3	—	—	√	√	[100]
	Nova-Chem		Seascape	2, 4, or 6 g L^{-1}	1	just turning red	3/28;13/35		×	—	√	[102]
	Huarun Bioengineering	Kiwifruit	Guichang	28.6%	3	budding phase, fruit setting phase and expanding final phase	25/25	×	×	—	√	[42]
	Sigma-Aldrich		Garmrok	100 or 500 mg·L^{-1}	4	146, 154, 161 and 170 daysafter full blossom	0/90	×	×	—	√	[103]
	—	Palm	Khesab	1%	3	5 and 15 weeks from pollination and two weeks before harvest	2/60	×	—	—	—	[38]
Chitosan oligo-chi-tosan	—	Apricot	Xiaobai	0.05%	2	7 and 2 days before harvest	2/70	×	—	—	—	[7]
	Haidebei Marine Bioengineering	Jujube	Dongzao	0.7 g L^{-1} 2 kDa, 5 kDa or 10 kDa; 0.3, 0.7 or 1.0 g L^{-1} 10 kDa	4	30, 60, 90 and 110 days after full blossom	0/60	—	—	—	√	[104]
β-aminobutyric acid	—	Apple	Honeycrisp	40 mM	2	4, 2 and 1 weeks before harvest	0.5 or 3/4 months; 0.5/5 months	—	√	—	√	[75]
	Sigma-Aldrich	Blueberry	Bluecrop	20 mM	1	7 days before harvest	2/20	×	√	—	√	[45]
Oxalic acid	Sigma-Aldrich	Lemon	Fino	0.1, 0.5 or 1.0 mM	5	from physiological fruit dropto 3 days before harvest	10/35	√	√	—	√	[52]
	—	Apricot	Red Flesh	0.5, 1 or 2 mM	1	fruit set stage	25/5	×	√	×	√	[51]
	—	Peach	Anjirymaleki	1, 3 or 5 mmol L^{-1}	1	15 days before harvest	1/28	—	—	—	√	[46]
	—	Kiwifruit	Bruno	5 mM	3	130 days after full blossom and 2 times at 7 days intervals	20/15	×	×	—	√	[47]
	—		Bruno	5 mmol L^{-1}	3	130, 137 and 144 days after the flowering	20/13	×	×	—	—	[49]
	Sigma-Aldrich	Plum	Black Splendor	1 mM	3	63, 77 and 98 days after full blossom	2/35	×	√	×	√	[20]
Hexanal	—	Apple	Honeycrisp	0.02%	2	30 and 15 days before harvest	2.5/120	√	—	—	√	[105]
	—		Honeycrisp	0.02%	2	30 and 15 days before harvest	2.5/120	√	—	—	√	[106]
	—	Mango	Dashehari	800, 1200, 1600 or 2000 µM	2	15 and 30 days before harvest	12/35	√	√	—	—	[85]

Table 3. *Cont.*

Elicitor	Manufacturer	Fruit	Cultivar	Spraying Concentration	Spraying Times	Spraying Stage	Storage Temperature (°C)/Days	Postharvest Quality				References
								TSS	TA	TSS/TA	Firmness	
	Sigma-Aldrich		Neelum, Bangalora, Banganapalli and Alphonso	0.02%	1; 1; 2	15 days before harvest; 30 days before harvest; 15 and 30 days before harvest	25/-;13/-	×	—	—	√	[21]
	—		Alphonso and Banganapalli	2%	2	30 and 15 days before harvest	14/21; 28/21	√	√	√	—	[107]
	—	Nectarines	Fantasia	2%	2	15 and 10 days before harvest	2/45	—	—	—	√	[66]
	—	Strawberry	Jewel and Wendy	0.01% or 0.02%	2	7 and 3 days before harvest	4/9	—	×	—	√	[108]
	—	Guava	Allahabad Safeda	0.8, 1.2 or 1.6 mM	2	4 and 2 weeks before harvest	6–8/35	√	√	—	√	[23]
Putrescine and spermidine	Olhoghi and Rishbaba	Grape	Olhoghi and Rishbaba	1 or 2 mM/1 or 2 mM	2	40 and 20 days before harvest	1.5/55	×	√	—	—	[58]
Melatonin	—	Apricot	Colorado and Mikado	0.1 mM	3	pit hardening, final fruit growth and 4 days before harvest	1 or 8/28	√	×	—	√	[60]

[a] √ indicates an increase. [b] × indicates a decrease. [c] — indicates no assessment.

4.2. Maintaining Sugarand Organic Acid

The ratio of sugars and organic acids is an important characteristic of fruit quality. Generally, total soluble solid (TSS) and titratable acid (TA) content are used to indicate sugars and organic acids in fruit. Preharvest elicitors spray can delay the increase of TSS content and maintain TA content [69].

In climacteric fruit, TSS content increases with ripening and then decreases with senescence, while TA only decreases after harvest [109,110]. On the 6th week of storage, compared with the control, TSS and TSS/TA content were 12.3% and 31.56% lower, while TA content was 23.19% higher in 'Flordaking' peach fruit sprayed with 3 mM SA before harvest [69]. Moreover, 0.5 mmol L^{-1} SA, 1 mmol L^{-1} ASA, or 0.5 mmol L^{-1} MeSA sprayed 'Black Splendor' plum fruit 3 times at 61, 76, and 94 days after anthesis, which increased TA content in fruit by 44.09%, 39.48% and 35.48% on the 50th of storage compared with the control, respectively [19]. In non-climacteric fruit, the concentration of TSS and TA gradually declined after harvest. Higher TSS and TA content were found in 'Magenta' and 'Crimson' seedless grapefruit sprayed with 1 mM ASA before harvest [50]. TSS content in 'Fino' lemon was 11% higher than that of the control on the 35th of storage after spraying with 0.1 mM MeJA 4 times before harvest [14]. Preharvest elicitors spray o maintaining TSS and TA in harvested fruit are shown in Table 3.

In fruits, sugars are stored in the form of starch, but there are few reports on the effect of preharvest elicitors spray on the degradation or accumulation of starch. However, preharvest elicitors spray delays ethylene biosynthesis and fruit ripening. Meanwhile, the TSS accumulation in fruit is delayed, which is associated with the inhibition of sucrose conversion. Higher sucrose content was found in 'Kate' apricot preharvest sprayed with MeSA during 32 days of storage [72]. In addition, a higher accumulation of sucrose, glucose, and fructose was found in 'Fino' lemon fruit preharvest sprayed with SA during 35 days of storage [14]. During 60 days of storage, preharvest MeJA spray kept fructose and glucose content in 'Mollar de Elche' pomegranate [17]. With fruit ripening and, organic acids are consumed as substrates of the tricarboxylic acid cycle (TCA) cycle in fruit. Preharvest elicitors spray delays respiration and fruit ripening [69]. Indeed, during 32 days of storage, the concentration of citric acid, malic acid, and total organic acids was higher in 'Kate' apricot fruit preharvest sprayed with MeSA [72]. Similarly, preharvest MeJA spray maintained the concentration of malic acid, succinic acid, oxalic acid, and ascorbic acid in 'Mollar de Elche' pomegranate during 60 days of storage [17].

4.3. Maintaing Fruit Firmness

Softening is an important cause of quality deterioration and shortens shelf life in climacteric and non-climacteric fruit. Preharvest elicitors spray can maintain firmness in fruit during storage.1, 2 or 3 mM SA sprayed 'Flordaking' peach fruit 3 times before harvest maintained the firmness during storage. On the 6th week of storage, the firmness in peach fruit sprayed with 3 mM SA was 1.68 fold higher than that of the control [69]. Moreover, 0.05% chitosan oligochitosan or 1 mmol L^{-1} SA sprayed twice on 'Xiaobai' apricot fruit at 7 and 2 days before harvest increased the firmness in fruit by 24% and 35% compared to the control on the 14th day of storage, respectively [7]. Similarly, the firmness in the 'Mahachanok' mango sprayed with 20, 40, 80, or 120 μL mL^{-1} MeJA at 90 days after anthesis was higher than that of the control during storage. Among them, 80 μL mL^{-1} MeJA spray showed the best effect, which increased the firmness by 20.75% compared with the control on the 24th of storage [97]. Compared with the control, the firmness showed 2 folds higher in 'Flame Seedless' table grapes sprayed with 2.0 mM SA twice at the pea stage and veraison stage on the 75th day of storage [91]. Preharvest elicitors spray on maintaining firmness in harvested fruit is shown in Table 3. This delay in fruit softening that was observed in the preharvest spray of elicitors could be connected to the inhibition in the ethylene released and the respiration rate that delayed the fruit ripening.

Fruit respiration is associated with the degree of water transpiration and weight loss [111]. Preharvest elicitors spray inhibited the water transpiration in fruit after harvest [92]. This is correlated to the inhibition of respiration that was observed in fruit preharvest treated with elicitors. On the other hand, preharvest application of elicitors strengthened the structure of the epidermal tissue of fruit and enhanced the thickness of the cuticle and density of wax, thereby further inhibiting water transpiration in harvested fruit [39,112]. Additionally, some elicitors spray, such as chitosan, can form a protective film on the surface of the fruit, which could further hinder water transpiration of the fruit [7].

Fruit softening also relates to the activity of cell wall degradation enzymes (CWDEs) [111]. Pectin, hemicellulose, and cellulose are the main components of the cell wall in fruit, and they can be depolymerized and degraded under the action of CWDEs, which causes the collapse of the cell wall, leading to fruit softening. The CWDEs mainly include polygalacturonases (PG), pectin methyl esterases (PME), and cellulases [44]. Cellulose content increased by 20.08% in 'Qingxiang' strawberry fruit preharvest sprayed with chitosan oligosaccharides compared with the control [43]. On the 7th day of storage, the activities of PME, PG, and cellulose were 26%, 17.43%, and 79.16% lower in 'Agate' muskmelon fruit sprayed with ASA before harvest than that in the control, respectively [113]. On the 9th day of storage, preharvest hexanal spray reduced the activities of PME and PG in 'Neelum' mango fruit, which was 50% and 42% lower than that of control, respectively [21]. Moreover, on the 14th day of storage, the PME activity decreased and protopectin content increased in 'Allahabad Safeda' guava fruit preharvest sprayed with hexanal, which was 33.3% lower and 1.3 folds higher than that of the control, respectively [23]. Compared with the control, the PME activity was 25% lower in 'Flame Seedless' grapefruit preharvest sprayed with SA [91]. In addition, preharvest SA and chitosan spray decreased PG activity in 'Tupi' blackberry fruit during storage. At the 6 days of storage, the PG activity in SA and chitosan-sprayed fruit was 5.5% and 14.4% lower than that of the control, respectively [9]. On the 21stday of storage, higher protopectin content was found in 'Osbeck' navel orange preharvest sprayed with oligochitosan, which was 19.6% higher than that of the control. Moreover, the sprayed fruit showed lower PG and PME activity, which was 58.3% and 30.8% lower than that of the control, respectively [44]. Furthermore, preharvest hexanal spray down-regulated *PME* expression in 'Jewel' and 'Wendy' strawberry fruit during storage [108]. Thus, in all, preharvest treatments with elicitors inhibit the expression of CWDE in harvested fruit during storage, while fruit firmness also involves in water transpiration and epidermal structure. Further studies are required to elucidate the relevant molecular mechanisms modulated by preharvest elicitors spray.

5. Concluding Remarks and Perspectives

In addition to inducing SAR in fruit, preharvest elicitors spray can improve phenylpropanoids metabolism and carotenoids biosynthesis, whichin crease the accumulation of flavonoids, carotenoids, and ascorbic acid, thereby enhancing antioxidant activity in harvested fruit. Antioxidant activity is an important property of fruits, which is correlated to scavenging ability of ROS. The increase in antioxidant activity maintains the integrity of the cell membrane and reduces lipid peroxidation, which is probably related to the reduced ROS in fruits, thereby improving cold tolerance of fruit. Moreover, the spray also inhibits membrane phospholipid metabolism, and maintains the function and fluidity of cell membrane in fruits, which alleviate CI in harvested fruit during cold storage reduces CI, maintaining postharvest quality of fruits. In addition, preharvest elicitors spray retards fruit ripening by inhibiting ethylene production and respiration rate, which maintains the levels of sugars and organic acids as well as firmness in harvested fruit during storage (Figure 1).

A possible mode of action of how preharvest elicitors spray alleviates CI, maintains quality, and improves antioxidant capacity in harvested fruit is shown in Figure 1. Preharvest elicitors spray is a simple and effective strategy in controlling CI and maintaining quality in harvested fruit, which reduces the postharvest process and costs. However, preharvest elicitors spray still has challenges in practical agricultural applications. Firstly, most of the elicitors are registered for disease control, but only a few of them have been registered for improving antioxidant activity, controlling CI, maintaining quality, and in harvested fruit. Therefore, it is necessary to improve the register of the relevance elicitors. Secondly, there are tremendous differences in the effect of different elicitors on different species or cultivars of fruits. Therefore, it is important to screen the appropriate elicitor, concentration and application time according to the different fruits. Thirdly, fertilization, irrigation, pruning and other practices modulate the elicitors' action. Moreover, the spraying method, such as the density and intensity of the spray, and the equipment used also affect the effect of the application of preharvest elicitor. Hence, the time, concentration, and frequency of elicitors spraying need to be screened based on different cultivation techniques. Fourthly, climate factors, such as temperature, light and rainfall, affect preharvest application. Therefore, it is important to investigate the influence of these climatic factors and their changes on the effect of applying these measures. Fifthly, postharvest environmental factors, including temperature, relative humidity, and gas composition, also affect the application effects of the elicitor, the optimum combination of pre-and postharvest environmental factors needs to be screened. Finally, although some researchers have elucidated the effects of preharvest elicitors spray on postharvest properties of fruits and revealed the preliminary mechanisms of action at physiological and cytological levels, the in-depth mechanisms of action are still unclear. Hence, approaches of multi-omic, epigenetic and molecular biology need to be used to further reveal the mechanism of preharvest elicitors spray on controlling CI and maintaining the quality of fruit during storage.

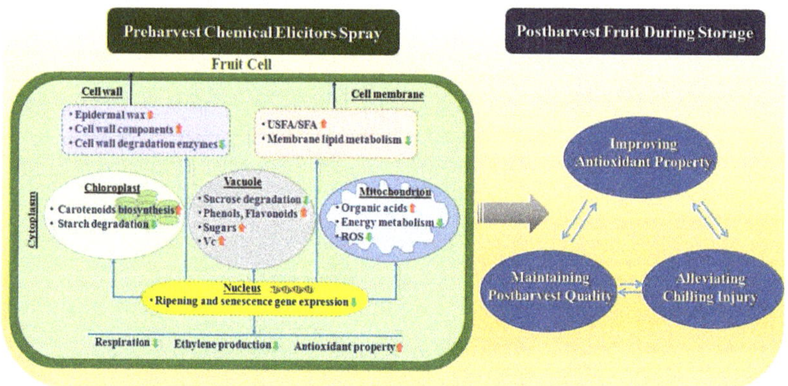

Figure 1. A possible mode of action of preharvest elicitors spray improves antioxidant property, alleviates chilling injury, and maintains quality in harvested fruit. ROS indicates reactive oxygen species; USFS/SFA indicates unsaturated/saturated fatty acids; Vc indicates vitamin C.

Due to the rapid cell division and significant cell expansion in fruit during the growth and development stage, the fruit is more sensitive to the stimuli of external elicitors. Additionally, preharvest fruit has a better active metabolism than the harvested fruit which is usually stored at cold temperatures. Therefore, in a way, the effects of preharvest elicitors spray on delaying ripening and senescence and improving resistance are better than that of postharvest treatment. Moreover, the elicitors are metabolized in the fruit and induce the biosynthesis of various natural metabolites that are safer than chemical fungicides, have a broader spectrum of activity, and will not be easily braked by a resistant isolate. Finally, more attention needs to be paid to strengthening the relationship between pre- and postharvest, expanding the breadth and depth of pre- and postharvest research.

Author Contributions: Writing—original draft preparation, D.G.; writing—review and editing, Y.B.; investigation, Y.L.; visualization, Y.W.; supervision, D.P. and N.A.; funding acquisition, Y.B. All authors have read and agreed to the published version of the manuscript.

Funding: This research was funded by the Israel–China Project of the National Natural Science Foundation of China (31861143046). And the APC was funded by 31861143046.

Acknowledgments: This work was supported by the National Natural Science Foundation of China-Israel International Cooperation Project (31861143046).

Conflicts of Interest: The authors declare no conflict of interest.

References

1. FAO. Fruit and vegetables-your dietary essentials. In *The International Year of Fruits and Vegetables, 2021, Background Paper*; FAO: Rome, Italy, 2020. [CrossRef]
2. Ziv, C.; Fallik, E. Postharvest storage techniques and quality evaluation of fruits and vegetables for reducing food loss. *Agronomy* **2021**, *11*, 1133. [CrossRef]
3. Thompson, A.K. Chaper 17-Postharvest chemical and physical deterioration of fruit and vegetables. In *Chemical Deterioration and Physical Instability of Food and Beverages*; Skibsted, L.H., Risbo, J., Andersen, M.L., Eds.; Woodhead Publishing: London, UK, 2010; pp. 483–518.
4. Aghdam, M.S.; Asghari, M.; Babalar, M.; Sarcheshmeh, M.A.A. Chapter 8-Impact of salicylic acid on postharvest physiology of fruits and vegetables. In *Eco-Friendly Technology for Postharvest Produce Quality*; Siddiqui, M.W., Ed.; Academic Press: Cambridge, MA, USA, 2016; pp. 243–268.
5. Sevillano, L.; Sanchez-Ballesta, M.T.; Romojaro, F.; Flores, F.B. Physiological, hormonal and molecular mechanisms regulating chilling injury in horticultural species. Postharvest technologies applied to reduce its impact. *J. Sci. Food Agric.* **2009**, *89*, 555–573. [CrossRef]
6. Romanazzi, G.; Sanzani, S.M.; Bi, Y.; Tian, S.; Guti´errez Martínez, P.; Alkan, N. Induced resistance to control postharvest decay of fruit and vegetables. *Postharvest Biol. Technol.* **2016**, *122*, 82–94. [CrossRef]
7. Cui, K.B.; Shu, C.; Zhao, H.D.; Fan, X.G.; Cao, J.K.; Jiang, W.B. Preharvest chitosan oligochitosan and salicylic acid treatments enhance phenol metabolism and maintain the postharvest quality of apricots (*Prunus armeniaca* L.). *Sci. Hortic.* **2020**, *267*, 109334. [CrossRef]
8. Ahmad, S.; Singh, Z.; Khan, A.S.; Iqbal, Z. Pre-harvest application of salicylic acid maintain the rind textural properties and reduce fruit rot and chilling injury of sweet orange during cold storage. *Pak. J Agric. Sci.* **2013**, *50*, 559–569.
9. Martínez-Camacho, J.E.; Guevara-González, R.G.; Rico-García, E.; Tovar-Pérez, E.G.; Torres-Pacheco, I. Delayed senescence and marketability index preservation of blackberry fruit by preharvest application of chitosan and salicylic acid. *Front. Plant Sci.* **2022**, *13*, 796393. [CrossRef]
10. Segado, P.; Domínguez, E.; Heredia, A. Ultrastructure of the epidermal cell wall and cuticle of tomato fruit (*Solanum lycopersicum* L.) during development. *Plant Physiol.* **2016**, *170*, 935–946. [CrossRef]
11. Seymour, G.B.; Østergaard, L.; Chapman, N.H.; Knapp, S.; Martin, C. Fruit development and ripening. *Annu. Rev. Plant Biol.* **2013**, *64*, 219. [CrossRef]
12. Bi, X.Y.; Liao, L.; Deng, L.J.; Jin, Z.H.; Huang, Z.H.; Sun, G.C.; Xiong, B.; Wang, Z.H. Combined transcriptome and metabolome analyses reveal candidate genes involved in Tangor (*Citrus reticulata* × *Citrus sinensis*) fruit development and quality formation. *Int. J. Mol. Sci.* **2022**, *23*, 5457. [CrossRef]
13. Harasym, J.; Oledzki, R. Effect of fruit and vegetable antioxidants on total antioxidant capacity of blood plasma. *Nutrition* **2014**, *30*, 511–517. [CrossRef]
14. Serna-Escolano, V.; Martínez-Romero, D.; Giménez, M.J.; Serrano, M.; García-Martínez, S.; Valero, D.; Valverde, J.M.; Zapata, P.J. Enhancing antioxidant systems by preharvest treatments with methyl jasmonate and salicylic acid leads to maintain lemon quality during cold storage. *Food Chem.* **2021**, *338*, 128044. [CrossRef] [PubMed]
15. Fanyuk, M.; Kumar Patel, M.; Ovadia, R.; Maurer, D.; Feygenberg, O.; Oren-Shamir, M.; Alkan, N. Preharvest application of phenylalanine induces red color in mango and apple fruit's skin. *Antioxidants* **2022**, *11*, 491. [CrossRef] [PubMed]
16. Sudheeran, P.K.; Love, C.; Feygenberg, O.; Maurer, D.; Ovadia, R.; Oren-Shamir, M.; Alkan, N. Induction of red skin and improvement of fruit quality in 'Kent', 'Shelly' and 'Maya' mangoes by preharvest spraying of prohydrojasmon at the orchard. *Postharvest Biol. Technol.* **2019**, *149*, 18–26. [CrossRef]
17. García-Pastor, M.E.; Serrano, M.; Guillén, F.; Giménez, M.J.; Martínez-Romero, D.; Valero, D.; Zapata, P.J. Preharvest application of methyl jasmonate increases crop yield, fruit quality and bioactive compounds in pomegranate 'Mollar de Elche' at harvest and during postharvest storage. *J. Sci. Food Agric.* **2020**, *100*, 145–153. [CrossRef] [PubMed]
18. Alrashdi, A.M.A.; Al-Qurashi, A.D.; Awad, M.A.; Mohamed, S.A.; Alrashdi, A.A. Quality, antioxidant compounds, antioxidant capacity and enzymes activity of 'El-Bayadi' table grapes at harvest as affected by preharvest salicylic acid and gibberellic acid spray. *Sci. Hortic.* **2017**, *220*, 243–249. [CrossRef]
19. Martínez-Esplá, A.; Serrano, M.; Valero, D.; Martínez-Romero, D.; Castillo, S.; Zapata, P.J. Enhancement of antioxidant systems and storability of two plum cultivars by preharvest treatments with salicylates. *Int. J. Mol. Sci.* **2017**, *18*, 1911. [CrossRef]
20. Serrano, M.; Marínez-Esplá, A.; Giménez, M.J.; Valero, D.; Zapata, P.J.; Guillén, F.; Castillo, S. Preharvest application of oxalic acid improves antioxidant systems in plums. *Acta Hortic.* **2016**, *19*, 1194. [CrossRef]

21. Preethi, P.; Soorianathasundaram, K.; Sadasakthi, A.; Subramanian, K.S.; Vijay Rakesh Reddy, S.; Paliyath, G.; Subramanian, J. Preharvest application of hexanal as a surface treatment improved the storage life and quality of mango fruits. *Coatings* **2021**, *11*, 1267. [CrossRef]
22. Zuñiga, P.E.; Castañeda, Y.; Arrey-Salas, O.; Fuentes, L.; Aburto, F.; Figueroa, C.R. Methyl jasmonate applications from flowering to ripe fruit stages of strawberry (*Fragaria* × *ananassa* 'Camarosa') reinforce the fruit antioxidant response at post-harvest. *Front. Plant Sci.* **2020**, *11*, 538. [CrossRef]
23. Gill, K.S.; Dhaliwal, H.S.; Mahajan, B.V.C.; Paliyath, G.; Boora, R.S. Enhancing postharvest shelf life and quality of guava (*Psidium guajava* L.) cv. Allahabad Safeda by pre-harvest application of hexanal containing aqueous formulation. *Postharvest Biol. Technol.* **2016**, *112*, 224–232. [CrossRef]
24. Ruiz-García, Y.; Gómez-Plaza, E. Elicitors: A tool for improving fruit phenolic content. *Agriculture* **2013**, *3*, 33–52. [CrossRef]
25. Shen, N.; Wang, T.F.; Gan, Q.; Liu, S.; Wang, L.; Jin, B. Plant flavonoids: Classification, distribution, biosynthesis, and antioxidant activity. *Food Chem.* **2022**, *383*, 132531. [CrossRef] [PubMed]
26. Xie, P.D.; Yang, Y.Y.; Gong, D.; Li, Z.C.; Wang, B.; Xue, S.L.; Oyom, W.; Prusky, D.; Bi, Y. Preharvest L-phenylalanine sprays accelerated wound healing of harvested muskmelons by eliciting phenylpropanoid metabolism and enzymatic browning. *Postharvest Biol. Technol.* **2022**, *193*, 112053. [CrossRef]
27. Li, S.E.; Cheng, Y.; Yan, R.; Liu, Y.; Huan, C.; Zheng, X.L. Preharvest spray with melatonin improves postharvest disease resistance in cherry tomato fruit. *Postharvest Biol. Technol.* **2022**, *193*, 112055. [CrossRef]
28. Wang, X.; Cao, X.; Shang, Y.; Bu, H.; Wang, T.; Lyu, D.; Du, G. Preharvest application of prohydrojasmon affects color development, phenolic metabolism, and pigment-related gene expression in red pear (*Pyrus ussuriensis*). *J. Sci. Food Agric.* **2020**, *100*, 4766–4775. [CrossRef]
29. Flores, G.; del Castillo, M.L.R. Influence of preharvest and postharvest methyl jasmonate treatments on flavonoid content and metabolomic enzymes in red raspberry. *Postharvest Biol. Technol.* **2014**, *97*, 77–82. [CrossRef]
30. Wang, J.; Bi, Y.; Zhang, Z.; Zhang, H.; Ge, Y.H. Reduction of latent infection and enhancement of disease resistance in muskmelon by preharvest application of harpin. *J. Agric. Food Chem.* **2011**, *59*, 12527. [CrossRef]
31. Li, Z.C.; Xu, X.Q.; Xue, S.L.; Gong, D.; Wang, B.; Zheng, X.Y.; Xie, P.D.; Bi, Y.; Prusky, D. Preharvest multiple sprays with chitosan promotes the synthesis and deposition of lignin at wounds of harvested muskmelons. *Int. J. Biol. Macromol.* **2022**, *206*, 167–174. [CrossRef]
32. Li, Z.C.; Xue, S.L.; Xu, X.Q.; Wang, B.; Zheng, X.Y.; Li, B.J.; Xie, P.D.; Bi, Y.; Prusky, D. Preharvest multiple sprays with chitosan accelerate the deposition of suberin poly phenolic at wound sites of harvested muskmelons. *Postharvest Biol. Technol.* **2021**, *179*, 111565. [CrossRef]
33. Fenech, M.; Amaya, I.; Valpuesta, V.; Botella, M.A. Vitamin C content in fruits: Biosynthesis and regulation. *Front. Plant Sci.* **2019**, *9*, 2006. [CrossRef]
34. Kato, M.; Ikoma, Y.; Matsumoto, H.; Sugiura, M.; Hyodo, H.; Yano, M. Accumulation of carotenoids and expression of carotenoid biosynthetic genes during maturation in citrus fruit. *Plant Physiol.* **2004**, *134*, 824–837. [CrossRef] [PubMed]
35. Landi, L.; De Miccolis Angelini, R.M.; Pollastro, S.; Feliziani, E.; Faretra, F.; Romanazzi, G. Global transcriptome analysis and identification of differentially expressed genes in strawberry after preharvest application of benzothiadiazole and chitosan. *Front. Plant Sci.* **2017**, *8*, 235. [CrossRef] [PubMed]
36. Martínez-Esplá, A.; Zapata, P.J.; Valero, D.; Martínez-Romero, D.; Díaz-Mula, H.M.; Serrano, M. Preharvest treatments with salicylates enhance nutrient and antioxidant compounds in plum at harvest and after storage. *J. Sci. Food Agric.* **2018**, *98*, 2742–2750. [CrossRef] [PubMed]
37. Elmenofy, H.M.; Okba, S.K.; Salama, A.M.; Alam-Eldein, S.M. Yield, fruit quality, and storability of 'Canino' apricot in response to aminoethoxyvinylglycine, salicylic acid, and chitosan. *Plants* **2021**, *10*, 1838. [CrossRef]
38. Ahmed, Z.F.R.; Alblooshi, S.S.N.A.; Kaur, N.; Maqsood, S.; Schmeda-Hirschmann, G. Synergistic effect of preharvest spray application of natural elicitors on storage life and bioactive compounds of date palm (*Phoenix dactylifera* L., cv. Khesab). *Horticulturae* **2021**, *7*, 145. [CrossRef]
39. Zhang, Z.K.; Bi, Y.; Ge, Y.H.; Wang, J.J.; Deng, J.J.; Xie, D.F.; Wang, Y. Multiple pre-harvest treatments with acibenzolar-S-methyl reduce latent infection and induce resistance in muskmelon fruit. *Sci. Hortic.* **2011**, *130*, 126–132. [CrossRef]
40. Meng, X.H.; Li, B.Q.; Liu, J.; Tian, S.P. Physiological responses and quality attributes of table grape fruit to chitosan preharvest spray and postharvest coating during storage. *Food Chem.* **2008**, *106*, 501–508. [CrossRef]
41. Nia, A.E.; Taghipour, S.; Siahmansour, S. Pre-harvest application of chitosan and postharvest Aloe vera gel coating enhances quality of table grape (*Vitis vinifera* L. cv. 'Yaghouti') during postharvest period. *Food Chem.* **2021**, *347*, 129012. [CrossRef]
42. Zhang, C.; Long, Y.H.; Wang, P.Q.; Li, J.H.; An, H.M.; Wu, X.M.; Li, M. The effect of preharvest 28.6% chitosan composite film sprays for controlling the soft rot on kiwifruit and its defence responses. *Hortic. Sci.* **2019**, *46*, 180–194. [CrossRef]
43. He, Y.; Bose, S.K.; Wang, W.; Jia, X.; Lu, H.; Yin, H. Pre-harvest treatment of chitosan oligosaccharides improved strawberry fruit quality. *Int. J. Mol. Sci.* **2018**, *19*, 2194. [CrossRef]
44. Deng, L.L.; Zhou, Y.H.; Zeng, K.F. Pre-harvest spray of oligochitosan induced the resistance of harvested navel oranges to anthracnose during ambient temperature storage. *Crop. Prot.* **2015**, *70*, 70–76. [CrossRef]
45. Chea, S.; Yu, D.J.; Park, J.; Oh, H.D.; Chung, S.W.; Lee, H.J. Preharvest β-aminobutyric acid treatment alleviates postharvest deterioration of 'Bluecrop' highbush blueberry fruit during refrigerated storage. *Sci. Hortic.* **2019**, *246*, 95–103. [CrossRef]

46. Razavi, F.; Hajilou, J. Enhancement of postharvest nutritional quality and antioxidant capacity of peach fruits by preharvest oxalic acid treatment. *Sci. Hortic.* **2016**, *200*, 95–101. [CrossRef]
47. Zhu, Y.Y.; Yu, J.; Brecht, J.K.; Jiang, T.J.; Zheng, X.L. Pre-harvest application of oxalic acid increases quality and resistance to *Penicillium expansum* in kiwifruit during postharvest storage. *Food Chem.* **2016**, *190*, 537–543. [CrossRef] [PubMed]
48. Martinez-Espla, A.; Zapata, P.J.; Valero, D.; Garcia-Viguera, C.; Castillo, S.; Serrano, M. Preharvest application of oxalic acid increased fruit size, bioactive compounds, and antioxidant capacity in sweet cherry cultivars (*Prunus avium* L.). *J. Agric. Food Chem.* **2014**, *62*, 3432–3437. [CrossRef] [PubMed]
49. Ali, M.; Liu, M.M.; Wang, Z.E.; Li, S.E.; Jiang, T.J.; Zheng, X.L. Pre-harvest spraying of oxalic acid improves postharvest quality associated with increase in ascorbic acid and regulation of ethanol fermentation in kiwifruit cv. Bruno during storage. *J. Integr. Agric.* **2019**, *18*, 2514–2520. [CrossRef]
50. García-Pastor, M.E.; Zapata, P.J.; Castillo, S.; Martínez-Romero, D.; Valero, D.; Serrano, M.; Guillén, F. Preharvest salicylate treatments enhance antioxidant compounds, color and crop yield in low pigmented-table grape cultivars and preserve quality traits during storage. *Antioxidants* **2020**, *9*, 832. [CrossRef]
51. Ahmed, M.; Ullah, S.; Razzaq, K.; Rajwana, I.A.; Akhtar, G.; Naz, A.; Amin, M.; Khalid, M.S.; Khalid, S. Pre-harvest oxalic acid application improves fruit size at harvest, physico-chemical and sensory attributes of 'Red Flesh' apricot during fruit ripening. *J. Hortic. Sci. Technol.* **2021**, *4*, 48–55. [CrossRef]
52. Serna-Escolano, V.; Giménez, M.J.; Castillo, S.; Valverde, J.M.; Martínez-Romero, D.; Guillén, F.; Serrano, M.; Valero, D.; Zapata, P.J. Preharvest Treatment with Oxalic Acid Improves Postharvest Storage of Lemon Fruit by Stimulation of the Antioxidant System and Phenolic Content. *Antioxidants* **2021**, *10*, 963. [CrossRef]
53. Wang, J.; Bi, Y.; Wang, Y.; Deng, J.; Zhang, H.; Zhang, Z. Multiple preharvest treatments with harpin reduce postharvest disease and maintain quality in muskmelon fruit (cv. Huanghemi). *Phytoparasitica* **2014**, *42*, 155–163. [CrossRef]
54. Misran, A.; Padmanabhan, P.; Sullivan, J.A.; Khanizadeh, S.; Paliyath, G. Composition of phenolics and volatiles in strawberry cultivars and influence of preharvest hexanal treatment on their profiles. *Can. J. Plant Sci.* **2015**, *95*, 115–126. [CrossRef]
55. Wang, B.; He, X.F.; Bi, Y. Preharvest sprays with sodium nitroprusside induce resistance in harvested muskmelon against the pink rot disease. *J. Food Process Preserv.* **2021**, *45*, e15339. [CrossRef]
56. Wang, B.; Jiang, H.; Bi, Y.; He, X.F.; Wang, Y.; Li, Y.C.; Zheng, X.Y.; Prusky, D. Preharvest multiple sprays with sodium nitroprusside promote wound healing of harvested muskmelons by activation of phenylpropanoid metabolism. *Postharvest Biol. Technol.* **2019**, *158*, 110988. [CrossRef]
57. Saba, M.K.; Moradi, S. Sodium nitroprusside (SNP) spray to maintain fruit quality and alleviate postharvest chilling injury of peach fruit. *Sci. Hortic.* **2017**, *216*, 193–199. [CrossRef]
58. Mirdehghan, S.H.; Rahimi, S. Pre-harvest application of polyamines enhances antioxidants and table grape (*Vitis vinifera* L.) quality during postharvest period. *Food Chem.* **2016**, *196*, 1040–1047. [CrossRef] [PubMed]
59. Babalar, M.; Pirzad, F.; Sarcheshmeh, M.A.A.; Talaei, A.; Lessani, H. Arginine treatment attenuates chilling injury of pomegranate fruit during cold storage by enhancing antioxidant system activity. *Postharvest Biol. Technol.* **2018**, *137*, 31–37. [CrossRef]
60. Medina-Santamarina, J.; Zapata, P.J.; Valverde, J.M.; Valero, D.; Serrano, M.; Guillén, F. Melatonin treatment of apricot trees leads to maintenance of fruit quality attributes during storage at chilling and non-chilling temperatures. *Agronomy* **2021**, *11*, 917. [CrossRef]
61. Michailidis, M.; Tanou, G.; Sarrou, E.; Karagiannis, E.; Ganopoulos, I.; Martens, S.; Molassiotis, A. Pre-and post-harvest melatonin application boosted phenolic compounds accumulation and altered respiratory characters in sweet cherry fruit. *Front. Nutr.* **2021**, *8*, 306. [CrossRef]
62. Sun, H.L.; Wang, X.Y.; Shang, Y.; Wang, X.Q.; Du, G.D.; Lv, D.G. Preharvest application of melatonin induces anthocyanin accumulation and related gene upregulation in red pear (*Pyrus ussuriensis*). *J. Integr. Agric.* **2021**, *20*, 2126–2137. [CrossRef]
63. Andrys, D.; Kulpa, D.; Grzeszczuk, M.; Bihun, M.; Dobrowolska, A. Antioxidant and antimicrobial activities of *Lavandulaangustifolia* Mill. field-grown and propagated in vitro. *Folia Hortic.* **2017**, *29*, 161–180. [CrossRef]
64. Gupta, M.; Karmakar, N.; Sasmal, S.; Chowdhury, S.; Biswas, S. Free radical scavenging activity of aqueous and alcoholic ex-tracts of *Glycyrrhizaglabra* Linn. measured by ferric reducing antioxidant power (FRAP), ABTS bleaching assay (αTEAC), DPPH assay and peroxyl radical antioxidant assay. *Int. J. Pharmacol. Toxicol.* **2016**, *4*, 235–240. [CrossRef]
65. Zhang, W.L.; Jiang, H.T.; Cao, J.K.; Jiang, W.B. Advances in biochemical mechanisms and control technologies to treat chilling injury in postharvest fruits and vegetables. *Trends Food Sci. Technol.* **2021**, *113*, 355–365. [CrossRef]
66. Kumar, S.K.; Kayal, W.E.; Sullivan, J.A.; Paliyath, G.; Jayasankar, S. Pre-harvest application of hexanal formulation enhances shelf life and quality of 'Fantasia' nectarines by regulating membrane and cell wall catabolism-associated genes. *Sci. Hortic.* **2018**, *229*, 117–124. [CrossRef]
67. Ahmad, S.; Singh, Z.; Iqbal, Z. Effect of preharvest sprays of salicylic acid on the shelf life and quality of 'Lane Late' sweet orange (*Citrus sinensis* L.) cold storage. *Acta Hortic.* **2013**, *1012*, 103–112. [CrossRef]
68. Ahmed, W.; Ahmed, S.; Ali, L. Effect of pre-harvest spray of salicylic (SA) and methyl jasmonate (MeJA) on the phyrtochemicals and physiological changes during the storage of grapefruit cv. Ray ruby. *Int. J. Biosci.* **2015**, *6*, 269–282.
69. Ali, I.; Wang, X.; Tareen, M.J.; Wattoo, F.M.; Qayyum, A.; Hassan, M.U.; Shafique, M.; Liaquat, M.; Asghar, S.; Hussain, T.; et al. Foliar application of salicylic acid at different phenological stages of peach fruit CV. 'Flordaking' improves harvest quality and reduces chilling injury during low temperature storage. *Plants* **2021**, *10*, 1981. [CrossRef]

70. Lu, X.H.; Sun, D.Q.; Li, Y.H.; Shi, W.Q.; Sun, G.M. Pre- and post-harvest salicylic acid treatments alleviate internal browning and maintain quality of winter pineapple fruit. *Sci. Hortic.* **2011**, *130*, 97–101. [CrossRef]
71. Al-Obeed, R.S. Jujube post-harvest fruit quality and storability in response to agro-chemicals preharvest application. *Afr. J. Agric. Res.* **2012**, *7*, 5099–5107.
72. Fan, X.G.; Du, Z.L.; Cui, X.Z.; Ji, W.J.; Ma, J.T.; Li, X.L.; Wang, X.M.; Zhao, H.D.; Liu, B.D.; Guo, F.J.; et al. Preharvest methyl salicylate treatment enhance the chilling tolerance and improve the postharvest quality of apricot during low temperature storage. *Postharvest Biol. Technol.* **2021**, *177*, 111535. [CrossRef]
73. Saba, K.M.; Zarei, L. Preharvest methyl jasmonate's impact on postharvest chilling sensitivity, antioxidant activity, and pomegranate fruit quality. *J. Food Biochem.* **2019**, *43*, e12763. [CrossRef]
74. García-Pastor, M.E.; Giménez, M.J.; Zapata, P.J.; Guillén, F.; Valverde, J.M.; Serrano, M.; Valero, D. Preharvest application of methyl salicylate, acetyl salicylic acid and salicylic acid alleviated disease caused by *Botrytis cinerea* through stimulation of antioxidant system in table grapes. *Int. J. Food Microbiol.* **2020**, *334*, 108807. [CrossRef] [PubMed]
75. Shoffe, Y.A.; Nock, J.F.; Zhang, Y.Y.; Watkins, C.B. Pre- and post-harvest γ-aminobutyric acid application in relation to fruit quality and physiological disorder development in 'Honeycrisp' apples. *Sci. Hortic.* **2021**, *289*, 110431. [CrossRef]
76. Hosseini, M.S.; Fakhar, Z.; Babalar, M.; Askari, M.A. Effect of pre-harvest putrescine treatment on quality and postharvest life of pear cv. Spadona. *Adv. Hortic. Sci.* **2017**, *31*, 11–17.
77. Zhang, L.; Cao, X.; Wang, Z.Q.; Zhang, Z.K.; Li, J.K.; Wang, Q.; Xu, X.B. Brassinolide alleviated chilling injury of banana fruit by regulating unsaturated fatty acids and phenolic compounds. *Sci. Hortic.* **2022**, *297*, 110922. [CrossRef]
78. García-Pastor, M.E.; Serrano, M.; Guillén, F.; Zapata, P.J.; Valero, D. Preharvest or a combination of preharvest and postharvest treatments with methyl jasmonate reduced chilling injury, by maintaining higher unsaturated fatty acids, and increased aril colour and phenolics content in pomegranate. *Postharvest Biol. Technol.* **2020**, *167*, 111226. [CrossRef]
79. Zhang, X.; Liu, T.; Zhu, S.; Wang, D.; Sun, S.; Xin, L. Short-term hypobaric treatment alleviates chilling injury by regulating membrane fatty acids metabolism in peach fruit. *J. Food Biochem.* **2022**, *64*, e14113. [CrossRef]
80. Chen, G.F.; Hou, Y.Y.; Zheng, Y.H.; Jin, P. 2,4-Epibrassinolide enhance chilling tolerance of loquat fruit by regulating cell wall and membrane fatty acid metabolism. *Sci. Hortic.* **2022**, *295*, 110813. [CrossRef]
81. Gong, D.; Bi, Y.; Zhang, X.M.; Han, Z.H.; Zong, Y.Y.; Li, Y.C.; Sionov, E.; Prusky, D. Benzothiadiazole treatment inhibits membrane lipid metabolism and straight-chain volatile compound release in *Penicillium expansum*-inoculated apple fruit. *Postharvest Biol. Technol.* **2021**, *181*, 111671. [CrossRef]
82. Song, C.B.; Wang, K.; Xiao, X.; Liu, Q.L.; Yang, M.J.; Li, X.; Feng, Y.B.; Li, S.S.; Shi, L.Y.; Chen, W.; et al. Membrane lipid metabolism influences chilling injury during cold storage of peach fruit. *Food Res. Int.* **2022**, *157*, 111249. [CrossRef]
83. Hong, K.; Gong, D.; Xu, H.; Wang, S.; Jia, Z.; Chen, J.; Zhang, L. Effects of salicylic acid and nitric oxide pretreatment on the expression of genes involved in the ethylene signaling pathway and the quality of postharvest mango fruit. *N. Z. J. Crop. Hortic. Sci.* **2014**, *42*, 205–216. [CrossRef]
84. Asghari, M.; Aghdam, M.S. Impact of salicylic acid on post-harvest physiology of horticultural crops. *Trends Food Sci. Technol.* **2010**, *21*, 502–509. [CrossRef]
85. Kaur, K.; Kaur, G.; Brar, J.S. Pre-harvest application of hexanal formulations for improving post-harvest life and quality of mango (*Mangifera indica* L.) cv. Dashehari. *J. Food Sci. Technol.* **2020**, *57*, 4257–4264. [CrossRef] [PubMed]
86. Alkan, N.; Fortes, A.M. Insights into molecular and metabolic events associated with fruit response to post-harvest fungal pathogens. *Front. Plant Sci.* **2015**, *6*, 889. [CrossRef] [PubMed]
87. Binder, B.M. Ethylene signaling in plants. *J. Biol. Chem.* **2020**, *295*, 7710–7725. [CrossRef]
88. Ruiz, K.B.; Trainotti, L.; Bonghi, C.; Ziosi, V.; Costa, G.; Torrigiani, P. Early methyl jasmonate application to peach delays fruit/seed development by altering the expression of multiple hormone-related genes. *J. Plant Growth Regul.* **2013**, *32*, 852–864. [CrossRef]
89. Erogul, D.; Özsoydan, İ. Effect of pre-harvest salicylic acid treatments on the quality and shelf life of the 'Cresthaven' peach cultivar. *Folia Hortic.* **2020**, *32*, 221–227. [CrossRef]
90. Marzouk, H.A.; Kassem, H.A. Improving yield, quality, and shelf life of Thompson seedless grapevine by preharvest foliar applications. *Sci. Hortic.* **2011**, *130*, 425–430. [CrossRef]
91. Champa, W.A.H.; Gill, M.I.S.; Mahajan, B.V.C.; Arora, N.K. Preharvest salicylic acid treatments to improve quality and postharvest life of table grapes (*Vitisvinifera* L.) cv. Flame Seedless. *J. Food Sci. Technol.* **2015**, *52*, 3607–3616. [CrossRef]
92. Lo'ay, A.A. Preharvest salicylic acid and delay ripening of 'superior seedless' grapes. *Egypt J. Basic Appl. Sci.* **2017**, *4*, 227–230. [CrossRef]
93. Darwish, O.S.; Ali, M.R.; Khojah, E.; Samra, B.N.; Ramadan, K.M.; El-Mogy, M.M. Pre-harvest application of salicylic acid, abscisic acid, and methyl jasmonate conserve bioactive compounds of strawberry fruits during refrigerated storage. *Horticulturae* **2021**, *7*, 568. [CrossRef]
94. Giménez, M.J.; Serrano, M.; Valverde, J.M.; Martínez-Romero, D.; Castillo, S.; Valero, D.; Guillén, F. Preharvest salicylic acid and acetylsalicylic acid treatments preserve quality and enhance antioxidant systems during postharvest storage of sweet cherry cultivars. *J. Sci. Food Agric.* **2017**, *97*, 1220–1228. [CrossRef] [PubMed]
95. Cao, J.K.; Yan, J.Q.; Zhao, Y.M.; Jiang, W.B. Effects of fourpre-harvest foliar sprays with β-aminobutyric acid or salicylic acid on the incidence of postharvest disease and induced defence responses in jujube (*Zizyphus jujuba* Mill.) fruit after storage. *J. Hortic. Sci. Biotechnol.* **2013**, *88*, 338–344. [CrossRef]

96. Supapvanich, S.; Mitsang, P.; Youryon, P. Preharvest salicylic acid application maintains physicochemical quality of 'Taaptimjaan' wax apple fruit (*Syzygium samarangenese*) during short-term storage. *Sci. Hortic.* **2017**, *215*, 178–183. [CrossRef]
97. Muengkaew, R.; Chaiprasart, P.; Warrington, I. Changing of physiochemical properties and color development of mango fruit sprayed methyl jasmonate. *Sci. Hortic.* **2016**, *198*, 70–77. [CrossRef]
98. Zapata, P.J.; Martínez-Esplá, A.; Guillén, F.; Díaz-Mula, H.M.; Martínez-Romero, D.; Serrano, M.; Valero, D. Preharvest application of methyl jasmonate (MeJA) in two plum cultivars. 2. Improvement of fruit quality and antioxidant systems during postharvest storage. *Postharvest Biol. Technol.* **2014**, *98*, 115–122. [CrossRef]
99. Karaman, S.; Ozturk, B.; Genc, N.; Celik, S.M. Effect of preharvest application of methyl jasmonate on fruit quality of plum (*Prunus Salicina L indell* cv."Fortune") at harvest and during cold storage. *J. Food Process. Preserv.* **2012**, *37*, 1049–1059. [CrossRef]
100. Saavedra, G.M.; Figueroa, N.E.; Poblete, L.A.; Cherian, S.; Figueroa, C.R. Effects of preharvest applications of methyl jasmonate and chitosan on postharvest decay, quality and chemical attributes of *Fragaria chiloensis* fruit. *Food Chem.* **2016**, *190*, 448–453. [CrossRef]
101. Feliziani, E.; Landi, L.; Romanazzi, G. Preharvest treatments with chitosan and other alternatives to conventional fungicides to control postharvest decay of strawberry. *Carbohydr. Polym.* **2015**, *5*, 111–117. [CrossRef]
102. Bhaskara-Reddy, M.V.; Belkacemi, K.; Corcuff, R.; Castaigne, F.; Arul, J. Effect of pre-harvest chitosan sprays on post-harvest infection by *Botrytis cinerea* and quality of strawberry fruit. *Postharvest Biol. Technol.* **2000**, *20*, 39–51. [CrossRef]
103. Kumarihami, H.M.P.C.; Kim, J.G.; Kim, Y.H.; Lee, M.; Lee, Y.S.; Kwack, Y.B.; Kim, J. Preharvest application of chitosan improves the postharvest life of 'Garmrok' kiwifruit through the modulation of genes related to ethylene biosynthesis, cell wall modification and lignin metabolism. *Foods* **2021**, *10*, 373. [CrossRef]
104. Yan, J.Q.; Cao, J.K.; Jiang, W.B.; Zhao, Y.M. Effects of preharvest oligochitosan sprays on postharvest fungal diseases, storage quality, and defense responses in jujube (*Zizyphus jujuba* Mill. cv. Dongzao) fruit. *Sci. Hortic.* **2012**, *142*, 196–204. [CrossRef]
105. DeBrouwer, E.J.; Sriskantharajah, K.; Kayal, W.E.; Sullivan, J.A.; Paliyath, G.; Subramanian, J. Pre-harvest hexanal spray reduces bitter pit and enhances post-harvest quality in 'Honeycrisp' apples (*Malus domestica* Borkh.). *Sci. Hortic.* **2020**, *273*, 109610. [CrossRef]
106. Sriskantharajah, K.; El Kayal, W.; Torkamaneh, D.; Ayyanath, M.M.; Saxena, P.K.; Sullivan, A.J.; Paliyath, G.; Subramanian, J. Transcriptomics of improved fruit retention by hexanal in 'Honeycrisp' reveals hormonal crosstalk and reduced cell wall degradation in the fruit abscission zone. *Int. J. Mol. Sci.* **2021**, *22*, 8830. [CrossRef] [PubMed]
107. Anusuya, P.; Nagaraj, R.; Janavi, G.J.; Subramanian, K.S.; Paliyath, G.; Subramanian, J. Pre-harvest sprays of hexanal formulation for extending retention and shelf-life of mango (*Mangifera indica* L.) fruits. *Sci. Hortic.* **2016**, *211*, 231–240. [CrossRef]
108. El Kayal, W.; El-Sharkawy, I.; Dowling, C.; Paliyath, G.; Sullivan, J.A.; Subramanian, J. Effect of preharvest application of hexanal and growth regulators in enhancing shelf life and regulation of membrane-associated genes in strawberry. *Can. J. Plant Sci.* **2017**, *97*, 1109–1120. [CrossRef]
109. Durán-Soria, S.; Pott, D.M.; Osorio, S.; Vallarino, J.G. Sugar signaling during fruit ripening. *Front. Plant Sci.* **2020**, *11*, 564917. [CrossRef]
110. Batista-Silva, W.; Nascimento, V.L.; Medeiros, D.B.; Nunes-Nesi, A.; Ribeiro, D.M.; Zsögön, A.; Araújo, W.L. Modifications in organic acid profiles during fruit development and ripening: Correlation or causation? *Front. Plant Sci.* **2018**, *9*, 1689. [CrossRef]
111. Liu, S.M.; Huang, H.; Huber, D.J.; Pan, Y.G.; Shi, X.Q.; Zhang, Z.K. Delay of ripening and softening in 'Guifei' mango fruit by postharvest application of melatonin. *Postharvest Biol. Technol.* **2020**, *163*, 111136. [CrossRef]
112. Fekry, W.M.E.; Rashad, Y.M.; Alaraidh, I.A.; Mehany, T. Exogenous application of melatonin and methyl jasmonate as a pre-harvest treatment enhances growth of Barhi date palm trees, prolongs storability, and maintains quality of their fruits under storage conditions. *Plants* **2022**, *11*, 96. [CrossRef]
113. Liu, Y.N.; Wang, Y.; Bi, Y.; Li, S.E.; Jiang, H.; Zhu, Y.; Wang, B. Effect of preharvest acetylsalicylic acid treatments on ripening and softening of harvested muskmelon fruit. *Sci. Agric. Sin.* **2017**, *50*, 1862–1872, (English abstract).

horticulturae

Article

Responses of High Carbon Dioxide Concentration on Postharvest Quality of Fresh Fig Fruit during Storage

Adem Dogan [1,2] and Mustafa Erkan [2,*]

1 Department of Plant and Animal Production, Elmalı Vocational School, Akdeniz University, Antalya 07700, Turkey
2 Department of Horticulture, Faculty of Agriculture, Akdeniz University, Antalya 07058, Turkey
* Correspondence: erkan@akdeniz.edu.tr; Tel.: +90-0242-310-24-28

Abstract: The aim of this study was to determine the effects of high CO_2 with the constant O_2 level on the postharvest quality of fig cv. Bursa Siyahi. For this purpose, the atmospheric compositions of 3% O_2 + 10% CO_2 (PA-1), 3% O_2 + 15% CO_2 (PA-2), 3% O_2 + 20% CO_2 (PA-3), and 21% O_2 + 0.03% CO_2 (RA) were tested under a palliflex controlled atmosphere (PA) storage system at 0°C for 28 days. At the end of the storage, weight loss increased during the storage period, but this increase slowed down in all tested PAs compared to RA. PA-1 and PA-2 delayed softening while PA-3 accelerated this process. There were no side effects in fruits stored under PAs for taste. The lowest total microorganism and decay rates were found in PA-2 and PA-3. The fig fruits stored under PAs had higher sugar and organic acid contents compared to the figs stored under the RA. Respiration rate decreased in all PAs compared to the RA. Ethylene productions increased with senescence in all atmospheres, but PA-3 inhibited this increase. Consequently, 15% CO_2 (PA-2) can be used to maintain postharvest quality of Bursa Siyahi fresh fig for 28 days at 0 °C.

Keywords: *Ficus carica* L.; postharvest losses; respiration rate; softening; storage condition; sugar content

Citation: Dogan, A.; Erkan, M. Responses of High Carbon Dioxide Concentration on Postharvest Quality of Fresh Fig Fruit during Storage. *Horticulturae* **2023**, *9*, 293. https://doi.org/10.3390/horticulturae9030293

Academic Editors: Yang Bi, Yongcai Li and Di Gong

Received: 30 December 2022
Revised: 11 February 2023
Accepted: 18 February 2023
Published: 22 February 2023

1. Introduction

Turkey provides approximately 24% of the world total fig production with 320,000 tons [1]. A large part of this production is considered dried fruit; however, the demand for fresh consumption has shown an increasing trend in the past decade. Bursa Siyahi is the dominant fresh consumption cultivar and fresh fig exportation of Turkey is based on this cultivar. This cultivar is highly attractive and demanded due to its delicious taste, flavor, and biochemical properties as well as attractive colors [2]. However, it is highly perishable due to its delicate fruit characteristics such as soft fruit tissue, thin peel, and ostiole-end opening. The expected postharvest life of fresh fig is limited to generally one or two weeks in ambient atmosphere at 0 °C [3].

The most common postharvest problems of fresh fig are water loss, softening, peel cracking, ostiole leakage and internal/external decays. Fig fruit is extremely sensitive to various pathogens, such as *Alternaria* spp., *Botrytis* spp., *Fusarium* spp., *Aspergillus* spp., *Penicillium* spp., and *Rhizopus* spp. [2,4–6], that cause economic losses especially in shelf-life at high temperatures. Karabulut et al. [7] reported that postharvest losses can reach up to 30–50% in shelf-life following 5–7 days of cold storage or transportation, especially when fruits are harvested during the rainy season.

Controlled atmosphere (CA) storage including low O_2 and high CO_2 have positive effects to maintain fruit quality and disease control in many horticultural commodities [8]. On the other hand, extremely high CO_2 levels may have an unfavorable effect on texture and promote off-flavor development [9]. Fresh fig fruit tolerates a high level of CO_2 in the storage atmosphere like other berries. Colelli et al. [10] tested the atmosphere enrichment with 15 and 20% CO_2 in the Mission fig variety at different storage temperatures for

28 days and highlighted that high CO_2 levels decreased the decay incidence and increased postharvest life of figs up to 2–3 weeks. Lowering O_2 levels can positively affect the postharvest quality of fresh fig as in high CO_2 [11]. Tsantili et al. [12] reported that low O_2 (2%) reduced O_2 uptake, ethylene production and maintained firmness of Mavra Markopoulou fig fruit stored for 29 days at −1 °C. Regulating atmospheric compositions, especially in berry fruit including fig, provides positive response to high CO_2 levels and can be used an alternative to a non-chemical approach to maintain postharvest quality of fresh fig during the storage, transportation, and marketing.

There were a few studies conducted with CA storage on different fresh fig cultivars. Türk et al. [13] tested CA storage of the Bursa Siyahi fig variety at 0 °C containing different CO_2 and O_2 levels (3% O_2 + 3% CO_2, 5% O_2 + 5% CO_2, 10% O_2 + 5% CO_2, 20% O_2 + 2% CO_2, and 0.03% O_2 + 21% CO_2). As a result of this study, CA storage containing 3% O_2 + 3% CO_2 and 5% O_2 + 5% CO_2 were recommended for the Bursa Siyahi fig cultivar for four weeks of storage due to their positive effects on overall external appearance and taste. Similarly, CA storage (6% O_2 + 17% CO_2) was also tested for the Brown Turkey, Kadota, and Mission fig varieties stored at 0 °C for 19 and 31 days. In this study, CA storage reduced and delayed the decay incidence for all cultivars [14]. In another study, Bahar and Lichter [15] studied different CA conditions (5, 10, 15% CO_2 with 5% O_2) on Ottomanit fresh fig. This study revealed that CA storage including 5% O_2 and 5% CO_2 maintained firmness, integrity, and decay control compared to the other atmospheric compositions and control.

Harvested fresh figs are generally stored under ambient atmosphere for up to two weeks. Storage duration can be prolonged up to four weeks under CA; however, this is not possible due to some challenges in practice. For example, loading the storage room and atmosphere adjustment take a long time due to the short harvest season and rapid fruit circulation after harvest. At this point, the palliflex storage system can be a solution to overcome these challenges in the storage of fresh fig. In this storage system, fruit are stored on a pallet basis, and the atmosphere of each pallet can be monitored and controlled independently. Higher CO_2 levels combined with a low O_2 level and the palliflex storage systems were not tested previously in this cultivar. For these reasons, this study was focused to investigate the effects of high CO_2 levels with low O_2 on the fruit quality of Bursa Siyahi fig cultivars stored under the palliflex storage system.

2. Materials and Methods

2.1. Fruit Material and Storage Experiment

Bursa Siyahi fresh fig fruit were harvested at fully colored (L^*: 30.60; C^*: 9.7; $h°$: 221.58) maturity stage from a commercial orchard in Kestel (Bursa, Turkey) (Figure 1). After harvest, the fruit were pre-cooled and then immediately transferred to the postharvest laboratory of Akdeniz University Antalya, Turkey, using a refrigerated truck at 0 °C. Then, the fruit without any visual defects were selected and used in the experiment.

(a) (b)

Figure 1. Skin (**a**) and flesh (**b**) color of fig fruit (cv. Bursa Siyahi) at harvest stage.

For establishment of different atmospheric compositions in the palliflex system, fruit were divided into four groups. The first group of fruit was stored in 3% O_2 + 10% CO_2 (PA-1), the second group was stored in 3% O_2 + 15% CO_2 (PA-2), the third group was stored 3% O_2 + 20% CO_2 (PA-3), and the last group was stored in 0.03% CO_2 + 21% O_2 (regular atmosphere (RA)). Then, all groups of fruit were stored at 0 °C with 90–95% RH for 28 days and sampled at 0, 7, 14, 21, and 28 days of storage. Palliflex storage system specially designed for laboratory scale was used for storage. For this purpose, a 12 m^2 cold room was designed to include 10 pallets with a computer-controlled system. The desired atmosphere conditions in the pallets were set up by a flow-through system, mixing N_2, CO_2, and air via pressure regulators. The atmosphere composition in the pallet was adjusted to the desired level within 15 min. The O_2 and CO_2 levels in each pallet were automatically controlled and monitored by a computer system (My Fruit Light, Van Amerongen CA Technology B.V., The Netherlands) for 1 h interval to ensure the proper composition.

2.2. Quality Analyses

The fruit were marked and weighed at the beginning of the experiment to determine an initial weight. They were then weighed at the end of each storage time, and the results were calculated as the percent loss from the initial weight. The firmness of fruit was determined by using a penetrometer (Chatillon DFI 10, Largo, FL, USA), measuring the peeled equatorial region on three different sides of the fruit. Fruit firmness was shown as average of these measurements and results were given as Newton (N). To determine the total soluble solids (TSS), fruit samples were homogenized for 5 min by using a hand blender and a 10 g sample was added to 40 mL distilled water and the mix was homogenized for 2 min. Extracts were centrifuged at 3000× g for 5 min at 20 °C (Sigma 2-16K, Osterodo am Harz, Germany). The amount of TSS content was measured by a digital refractometer (HI96801, Hanna, Rhode Island, USA) and the findings were represented as a percentage (%). The purees were diluted ten times with distilled water and used to measure titratable acidity (TA) according to the method described by Bahar and Lichter [15]. The TA was expressed as a percentage of citric acid. For taste evaluation, six panelists (three men and three women, 25–50 years old) tasted fruit, and the data were categorized using a five-point hedonic scale described by Selcuk and Erkan [16]. Decay incidence was recorded by counting the number of fruits with visible symptoms and expressed as a percentage of total fruits. The total microorganisms were examined according to the method described by Cantín et al. [5]. To measure the total microorganism, 1 cm^2 pieces were taken from the surface of the fruits and put into microtubes. Then, 1 mL of sterile water was added onto the samples, and they were vortexed for 30 s and 100 µL of the suspension was placed in a Petri dish containing potato dextrose agar (PDA) medium. Microorganisms were counted after incubation for 5 days at 25 °C. The respiration rate and ethylene production of fruit were determined by using a gas chromatography (GC) equipped with thermal conductivity and flame ionization detectors (Thermo Electron S.p.A., Strada Rivoltana, Milan, Italy). To determine respiration rate, 500 g fruit for each treatment was placed in 3 L gas tight jars that were specially designed with a cover seal for an hour at 20 °C. A 1 mL gas sample was taken out from head space of jars with gas-tight syringe and injected into thermal conductivity detector of the system. The calibration was done by using external CO_2 standard (1000 ppm). GC conditions were as follows: Supelco 80/100 alumina F1 column, 65 °C oven temperature, 35 °C detector temperature, 45 mL min^{-1} hydrogen flow, 400 mL min^{-1} dry air flow during the 4 min analysis time. The GC was calibrated with external ethylene standard (25 ppm) for ethylene production. Then, similarly, 1 mL gas sample injected to flame ionization detector following GC conditions: GS-GasPro 113-4362 capillary column, 130 °C oven temperature, 275 °C detector temperature, 35 mL min^{-1} hydrogen flow, 350 mL min^{-1} dry air flow during the 2 min analysis time.

Respiration rate and ethylene production were calculated using Equation (1):

$$Respiration\ rate\ or\ ethylene\ production = A \times (V_j - V_p)/(t \times g) \quad (1)$$

where A is the sample area (ppm)/standard area (ppm), V_j is the jar volume (L), V_p is the product volume (L), t is the time (hour) and g is the fruit weight (kg).

The contents of sugar (glucose and fructose) and organic acid (citric and malic) were determined using the same extracts. For this purpose, fruit were crushed with hand blender and 5 g of sample was immersed in 45 mL of distilled water and homogenized with an ultraturrax (T-25, Ika-Labortechnik, Staufen, Germany). This mixture was extracted for 30 min in an ultrasonic water bath at 40 °C. The samples were centrifuged at 4000× g for 5 min at room temperature after incubation. Finally, the supernatant was put into vials after passing through a 0.45 µm membrane filter (Macherey-Nagel, Düren, Germany). The sugar contents were determined according to a method described by Sturm et al. [17] using high performance liquid chromatography (HPLC) (Shimadzu, LC-20 AD, Kyoto, Japan) equipped with Refractive Index Detector (RID). The analysis was carried out using a Rezex RCM-monosaccharide column (300 × 7.8 mm; Phenomenex, Torrance, CA, USA) with a flow of 0.6 mL min^{-1}, and column temperature was maintained at 65 °C. Twice distilled water was used as mobile phase. Organic acids were determined using ICE ORH-801-organic acid column (300 × 6.5 mm, Transgenomics, Ohama, USA) with a flow 0.6 mL min^{-1}, and column temperature was maintained at 65 °C. For the mobile phase, 0.125 mM H_2SO_4 was used.

2.3. Experimental Design and Statistical Analysis

The experiment was conducted using completely randomized design with three replicates, and each replication included 22 fruits. The data were analyzed by using the SAS 9.0 statistical software (SAS Inst., Cary, NC, USA) and means were compared using Duncan's multiple range test ($p \leq 0.05$). A comparison of taste scores was performed using a nonparametric ANOVA and multiple comparison of Kruskal–Wallis test ($p \leq 0.05$).

3. Results and Discussion

3.1. Weight Loss

Weight loss gradually increased with storage time in all atmospheric compositions. Minimum weight loss was determined in PA-1 (2.68%) followed by PA-2 (2.73%), PA-3 (3.00%), and RA (3.64%) after 28 days of cold storage (Figure 2).

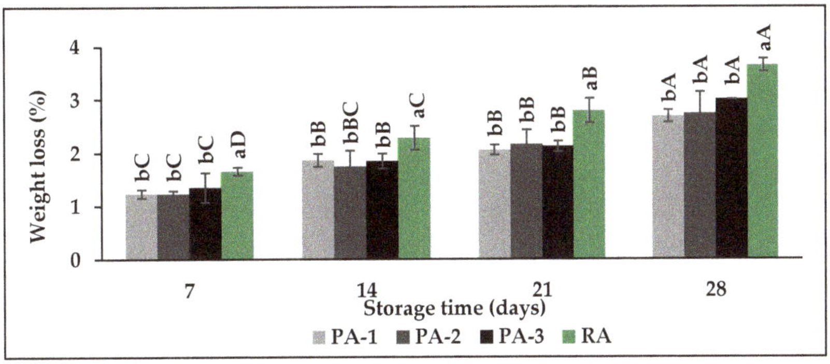

Figure 2. Effects of different atmospheric compositions on fruit weight loss in fresh fig cv. Bursa Siyahi. Different uppercase letters indicate significant differences for each atmosphere composition during storage and different lowercase letters indicate significant differences between atmosphere compositions for each sampling date at $p \leq 0.05$ by Duncan's multiple range test. Data are the mean ± SE of three replicates. PA-1: 3% O_2 + 10% CO_2; PA-2: 3% O_2 + 15% CO_2; PA-3: 3% O_2 + 20% CO_2; RA: 21% O_2 + 0.03% CO_2.

These findings clearly illustrated that storage atmospheres except for RA have positive effects to inhibit weight loss; however, there were no statistical differences among atmospheric compositions. The potential of CA storage including low O_2 and high CO_2 level is effective in preventing weight loss due to its slowdown effects on fruit metabolism [8]. Similarly, CA storage slowed down weight loss compared to control in the Ottomonit fig variety stored for 30 days. Although the weight loss reached 14.4% in control fruit, 5% O_2 + 5% CO_2 limited this weight loss in 4.4% [15]. Considering the previous studies, it can be said that the palliflex storage system for fig fruit inhibited weight loss compared to RA. This inhibiting effect can be caused by cultivar, maturity stage, and the pallet cover, which is not permeable to water and gas.

3.2. Fruit Firmness

Fruit firmness is an extremely important characteristic for consumers' acceptance of fig fruit, and the soft ones are generally preferred by consumers; however, too-soft fruit may not be appropriate for handling and transportation. The initial fruit firmness was 6.18 N at harvest. Fruit firmness gradually decreased with storage time in all atmospheric compositions. During the first 7 days of storage, PA-1 and PA-2 maintained firmness while PA-3 and RA stored fruit softened faster (Figure 3). At the end of the storage, the highest decrease in firmness reached up to 52.3% in PA-3 followed by 46.8% in RA in comparison to initial firmness. The highest fruit firmness was determined in PA-1 (3.82 N) and PA-2 (3.37 N) at the end of the storage. Fruit softening is linked to cell-wall carbohydrate metabolism and pectin substrate, and the polygalacturonase (PG) enzyme is important in this process [18]. A similar mechanism has also been molecularly revealed in fig fruit [19]. Teixeira et al. [20] reported that CA storage with 5% O_2 and 20% CO_2 increased fruit softening in guava fruit during 28 days of storage.

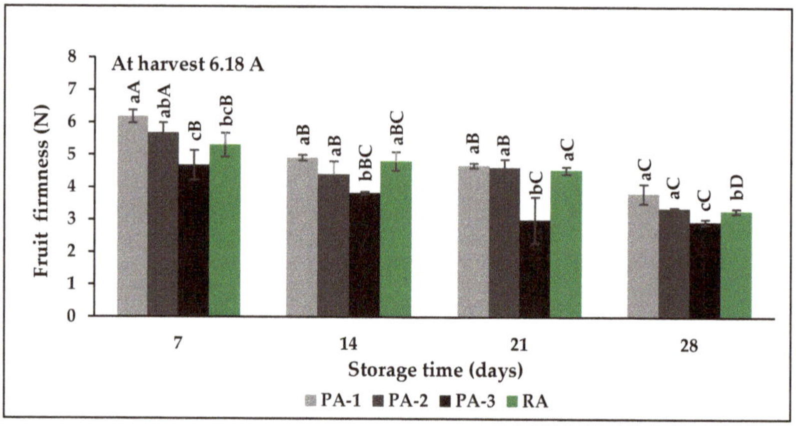

Figure 3. Effects of different atmospheric compositions on fruit firmness in fresh fig cv. Bursa Siyahi. Different uppercase letters indicate significant differences for each atmosphere composition during storage and different lowercase letters indicate significant differences between atmosphere compositions for each sampling date at $p \leq 0.05$ by Duncan's multiple range test. Data are the mean ± SE of three replicates. PA-1: 3% O_2 + 10% CO_2; PA-2: 3% O_2 + 15% CO_2; PA-3: 3% O_2 + 20% CO_2; RA: 21% O_2 + 0.03% CO_2.

This situation was explained by a negative correlation between high CO_2 and solubilization of pectic compounds. Villalobos et al. [21] stated that fruit softening varies according to the atmospheric compositions, and modified atmosphere packaging (MAP) slowed softening in the Cuello Dama Blanco, Cuello Dama Negro, and San Antonio fig cultivars. Similarly, an atmosphere of 5% CO_2 is more effective than 10% and 15% CO_2 to delay softening in Ottomonit fig [15]. On the other hand, there were no significant differences between the atmospheric compositions (RA and 6% O_2 + 17% CO_2) on the firmness of Brown Turkey and Kadota, while a significant effect was found in Mission fig [14].

3.3. Decay Rate and Total Microbial Count

Fresh fig is a highly perishable fruit due to its pomological characteristics, and the postharvest decay rate can reach up to 50% in a short period of time under unfavorable storage conditions. Therefore, most fig fruits are generally consumed as fresh in regions close to the producing areas. Fresh figs are highly affected by pathogens, especially in unsuitable storage conditions [22]. Successful postharvest management in fresh fig is dependent on avoiding mechanical damage and as rapidly as possible getting into the cooling chain. In this study, the palliflex storage was highly efficient to control deterioration during the first 2 weeks of storage and there was no decay observed. After this period, the decay rate was lower on the fruit stored at different atmospheric compositions compared to fruit stored RA. In both 21 and 28 days of storage, the highest decay rate was on the fruit stored under RA. Microbial activity increased during the storage in all atmospheres and the highest total microbial count was recorded on the fruit stored under RA with 14.87 cfu mL^{-1}. However, PA-2 and PA-3 were found highly effective to control microbial count (Table 1). Fungal decay caused by some pathogens such as *Alternaria* spp., *Botrytis* spp., *Fusarium* spp., *Aspergillus* spp., *Penicillium* spp., and *Rhizopus* spp. [2,4,5] is the main cause of rapid and extensive postharvest losses in fresh fig. Lowering O_2 and increasing CO_2 level are known as an effective treatment to control anaerobic and aerobic microbial growth due to their toxicity effects. The efficiency of inhibition by high CO_2 on microorganisms has also been reported for strawberries [23].

Table 1. Effects of different atmospheric compositions on decay rate and microbial count of fresh fig fruit (cv. Bursa Siyahi).

Atmospheric Compositions	Tested Parameters	Storage Time (days)				
		0	7	14	21	28
PA-1	Decay rate (%)	**	**	**	6.06 bB	18.18 bA *
PA-2		**	**	**	7.58 bB	13.64 bcA
PA-3		**	**	**	1.52 cB	9.09 cA
RA		**	**	**	18.18 aB	30.30 aA
PA-1	Microbial count (cfu)	0.95 aE	2.60 bD	7.16 bC	9.47 bB	12.12 aA
PA-2		0.95 aB	2.78 bB	5.60 cA	6.16 cA	7.13 bA
PA-3		0.95 aD	2.68 bC	6.22 cB	5.63 cB	8.90 bA
RA		0.95 aE	6.66 aD	9.77 aC	11.85 aB	14.87 aA

* Different capital letters show significant differences for each composition during storage and different lowercase letters show significant differences among compositions for each sampling date at $p < 0.05$ by Duncan's multiple range test. ** No decay was observed. PA-1: 3% O_2 + 10% CO_2; PA-2: 3% O_2 + 15% CO_2; PA-3: 3% O_2 + 20% CO_2; RA: 21% O_2 + 0.03% CO_2.

Many studies have revealed that CA storage can inhibit decay development and microbial count during the storage. Serradilla et al. [24] reported that 5% O_2+ 10% CO_2 and 8% O_2 + 10% CO_2 inhibited the growth of bacteria and molds in cherries stored for 15 days. Colelli et al. [10] reported that 15 and 20% CO_2 enriched atmospheres resulted in reduction of decay incidence in Black Mission fig. The inhibition effects of modified atmosphere with low O_2 and high CO_2 on decay incidence and microbial count were also observed on fresh fig [25]. Similarly, microperforated films delayed fungal deteriorations and physiological disorders in San Antonio and Banane figs [26]. In other studies, it was reported that the low O_2 level can maintain firmness and decrease respiration and ethylene production, and thus control the deterioration of fresh figs during storage [11,12].

3.4. Total Soluble Solids (TSS), Titratable Acidity (TA) and Taste

The effects of different atmospheric compositions and storage times on the total soluble solid content (TSS) were found statistically significant ($p \leq 0.05$). The TSS contents of fresh fig on the 14th day in storage were higher than their initial value at harvest (Table 2). However, TSS content decreased during the storage period due to the fruit senescence and respiration. At the end of the storage, PA-1 and PA-2 had higher TSS content compared to other storage atmospheres (Table 2). A similar trend was also reported for long-term storage of Bursa Siyahi fig fruit during CA storage at 0 °C [13]. Villalobos et al. [26] reported that fruits covered with microperforated film, which can provide a low O_2 high CO_2 atmosphere, resulted in higher TSS content than control treatment due to decreasing respiration rate and delaying ripening. Titratable acidity (TA) also showed fluctuations during the storage period. Palliflex storage atmospheres (PA-1, PA-2 and PA-3) provided a slight increase in TA content during the first 14 days of storage while RA did not show any changes. After this storage period, TA content decreased in all storage atmospheres. At the end of storage, TA contents in the PA-2 (0.25%) and PA-3 (0.25%) atmospheres were higher than PA-1 (0.23%) and RA (0.19%) stored fruit. After comparison to the initial TA content, it can be clearly said that there were no changes on the fruit stored in PA-2 and PA-3. In a study, chitosan-coated fresh fig had higher TA content than the uncoated fruit due to the effects, which create a gas barrier around the fruit and so slow down metabolism [27]. On the other hand, low O_2 and high CO_2 caused no off flavor and they provided more stable taste than the fruit stored in RA. Fruit stored under controlled atmospheres had a higher taste score than the fruit under RA after 7 days of cold storage; however, there were no significant differences among PA-1, PA-2, and PA-3. The taste score of the fruit stored in RA decreased after 14 days of storage while the taste score of fruit stored in PA-3 decreased at the end of the storage. PA-2 and PA-1 maintained their initial taste score during the storage period. Waghmare and Annapure [28] reported that irradiation and MAP (5% O_2 + 10% CO_2) and their combinations positively affected the taste of Poona fig fruit stored at 5 °C for 15 days.

Table 2. Effects of different atmospheric compositions on total soluble solid (TSS), titratable acidity (TA), and taste [1] of fresh fig fruit (cv. Bursa Siyahi).

Atmospheric Compositions	Tested Parameters	Storage Time (days)				
		0	7	14	21	28
PA-1	TSS (%)	15.03 aB	15.07 abB	16.50 aA	13.83aC	12.83 abD *
PA-2		15.03 aB	14.63 bBC	16.17 abA	14.17 aCD	13.73 aD
PA-3		15.03 aB	14.83 bB	15.90 abA	12.77 bC	12.53 bcC
RA		15.03 aB	15.67 aA	15.50 bAB	11.67 cC	11.73 cC
PA-1	TA (%)	0.24 aBC	0.26 bB	0.28 aA	0.24 bBC	0.23 bD
PA-2		0.24 aB	0.28 aA	0.29 aA	0.25 aB	0.25 aB
PA-3		0.24 aB	0.29 aA	0.28 aA	0.26 abBC	0.25 aBC
RA		0.24 aA	0.25 bA	0.24 bA	0.21 cB	0.19 cB
PA-1	Taste [1]	5.00 aA	5.00 aA	4.94 aA	4.83 aA	4.67 aA
PA-2		5.00 aA	5.00 aA	4.94 aA	4.83 aA	4.67 aA
PA-3		5.00 aA	5.00 aA	4.89 aAB	4.78 aAB	4.61 aB
RA		5.00 aA	4.67 bA	3.89 bB	3.61 bB	3.50 bB

* Different capital letters show significant differences for each composition during storage and different lowercase letters show significant differences among compositions for each sampling date at $p \leq 0.05$ by Duncan's multiple range test. [1] The taste analysis was conducted on the base of a 5-point hedonic scale, where. 1 = very poor (not marketable); 2 = poor (limit of acceptability); 3 = good; 4 = very good; 5 = excellent. The means of taste score were compared by Kruskal–Wallis test ($p \leq 0.05$). PA-1: 3% O_2 + 10% CO_2; PA-2: 3% O_2 + 15% CO_2; PA-3: 3% O_2 + 20% CO_2; RA: 21% O_2 + 0.03% CO_2.

Similarly, Ayhan and Karacay [29] reported that active MAP, including 20% CO_2 with low and high O_2, had a higher taste score and overall appearance in Bursa Siyahi fig fruit compared to the control. MAP was effective to maintain initial taste and texture in Black Mission fig at both 1 °C and 25° C [30].

3.5. Sugar and Organic Acid Contents

The fructose is the major sugar in fresh fig followed by glucose and sucrose. At harvest, the fructose and glucose contents of fig fruit were determined as 9.65 and 7.43 g 100 g^{-1}, respectively (Table 3). Similarly, Çalişkan and Polat [31] reported that fructose and glucose are the dominant sugars in fresh fig fruit grown in the Mediterranean basin. Fructose and glucose content gradually increased during the first 14 days of storage in all atmospheres and then decreased at the end of storage. The fructose contents of the fig fruit stored under PA-1 and PA-3 were higher than the fruit stored under PA-2 and RA. While glucose content of the fruit stored under RA did not change at 7 and 14 days of cold storage, the glucose contents of fruits stored under different atmospheres increased after 7 days of storage. After this period, glucose content decreased on the 14th and 21st days, depending on atmospheric compositions. At the end of 28 days of storage, the glucose contents of the fruit stored under atmospheric compositions were higher than RA; however, no statistical differences were found among the atmospheric compositions. In this study, sugar content first increased and then decreased. This decrease can be the result of senescence and sugar used in respiratory metabolism. The glucose and fructose contents can show different patterns based on cultivars, maturity stage, and storage conditions. Fructose and glucose content increased during the 30 days of cold storage at 0.5 °C plus 1 day at 20 °C in Masui Dauphine and Banane fig cultivars, whereas Bongresi showed similar pattern with our results [32].

Table 3. Effects of different atmospheric compositions on fructose, glucose, citric and malic acid content of fresh fig fruit (cv. Bursa Siyahi).

Atmospheric Compositions	Tested Parameters	Storage Time (Days)				
		0	7	14	21	28
PA-1	Fructose (g 100 g^{-1})	9.65 aA	8.09 bB	9.85 aA	7.69 aB	8.08 aB *
PA-2		9.65 aA	10.29 aA	9.43 aA	6.52 bB	6.49 bB
PA-3		9.65 aA	9.91 aA	8.06 bB	7.14 abB	7.86 aB
RA		9.65 aA	8.41 bB	9.03 abAB	6.49 bC	6.43 bC
PA-1	Glucose (g 100 g^{-1})	7.43 aB	8.38 bcA	8.02 aA	5.92 aC	5.72 aC
PA-2		7.43 aB	10.22 aA	7.66 aB	5.32 aC	5.58 aC
PA-3		7.43 aB	9.44 abA	6.47 bC	5.98 aC	5.42 abD
RA		7.43 aA	7.34 cA	7.92 aA	5.52 aB	5.19 bB
PA-1	Citric acid (mg 100 g^{-1})	340.8 aA	343.0 aA	354.6 aA	345.3 abA	324.8 bA
PA-2		340.8 aA	350.7 aA	363.3 aA	372.7 aA	359.4 aA
PA-3		340.8 aB	344.3 aB	348.5 aAB	373.3 aA	351.3 abAB
RA		340.8 aA	356.8 aA	327.3 aA	324.3 bA	266.3 cB
PA-1	Malic acid (mg 100 g^{-1})	161.0 aC	180.5 aB	195.5 aA	158.0 aC	139.7 bD
PA-2		161.0 aC	171.8 aBC	193.0 aA	178.2 aB	168.9 aBC
PA-3		161.0 aBC	175.3 aB	194.4 aA	170.2 aB	149.7 bC
RA		161.0 aB	176.5 aA	180.8 bA	163.6 aB	122.4 cC

* Different capital letters show significant differences for each composition during storage and different lowercase letters show significant differences among compositions for each sampling date at $p \leq 0.05$ by Duncan's multiple range test. PA-1: 3% O_2 + 10% CO_2; PA-2: 3% O_2 + 15% CO_2; PA-3: 3% O_2 + 20% CO_2; RA: 21% O_2 + 0.03% CO_2.

Citric and malic acids are the most abundant organic acids in fig fruit. There were no significant differences among the atmospheric compositions during the 7, 14, and 21 days of storage. At the end of the 28 days of storage, the highest citric acid content was found in PA-2 and PA-3, whereas the lowest citric acid was in RA. Malic acid content showed fluctuation during the storage period. It decreased in PA-1 and RA compared to their initial concentration at harvest; however, malic acid content in PA-2 and PA-3 did not change during the storage period. At the end of the storage, the highest malic acid content was in PA-2 and the lowest was in RA. Trad et al. [33] reported that citric acid is the most abundant organic acid following malic acid in fresh fig fruit as determined in this study

The decrease in malic and citric acid contents of medlar fruit was significantly prevented by CA storage [16].

3.6. Respiration Rate and Ethylene Production

The respiration rate of fresh fig at harvest was 37.38 mL CO_2 kg^{-1} h^{-1} and sharply decreased in all storage atmospheres after 7 days of storage (Figure 4a). This situation can be the result of low temperature effects on the respiration rate. Allegro et al. [34] reported that the initial respiration rate of Dottato figs was 28.0 ± 2.5 mL CO_2 kg^{-1} h^{-1} and decreased to 6.8 ± 1.8 mL CO_2 kg^{-1} h^{-1} in the fruit stored at 4 °C after 72 h. A similar decreasing trend after storage was also reported in fig fruit belonging to the cultivars Cuello Dama Blanco, Cuello Dama Negro, and San Antonio [21]. In this study, at the 14th and 21st days of storage, the highest respiration rate was found in the fruit stored in RA, while the lowest rates were found those stored in PA-1 and PA-2. At the end of the storage period, fig fruit stored under different atmospheric compositions had lower respiration rates compared to the fruit stored in RA; however, there were no significant differences among the PA-1, PA-2, and PA-3.

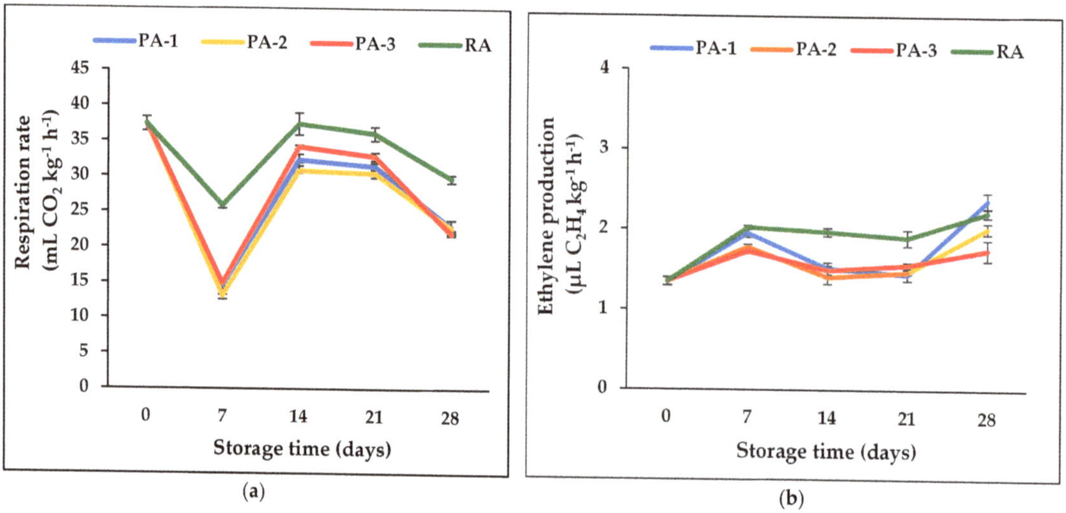

Figure 4. Effects of different atmospheric compositions on respiration rate (**a**) and ethylene production (**b**) in fresh fig cv. Bursa Siyahi. Data are the mean ± SE of three replicates. PA-1: 3% O_2 + 10% CO_2; PA-2: 3% O_2 + 15% CO_2; PA-3: 3% O_2 + 20% CO_2; RA: 21% O_2 + 0.03% CO_2.

The ethylene production of the Bursa Siyahi fig fruit was 1.35 µL C_2H_4 kg^{-1} h^{-1} and increased after 7 days of storage in all tested atmospheres (Figure 4b). While the highest ethylene production was in fruit stored at RA (2.2 µL C_2H_4 kg^{-1} h^{-1}), the lowest was in PA-2 and PA-3 (1.78 and 1.73 µL C_2H_4 kg^{-1} h^{-1}), respectively. After 14 and 21 days of storage, ethylene production slightly decreased. At the end of the storage, ethylene productions increased with senescence in all atmospheres, but PA-3 inhibited this increase.

In previous studies, it was reported that CA storage slowed down respiration rate and ethylene production in strawberry [35] and blueberry [36]. Modified atmosphere decreased respiration rate and ethylene production, and thus storage life extended [37]. In other studies, enriched CO_2 levels suppressed ethylene production in Masui Dauphine [38] and Mission [10] fig cultivars.

4. Conclusions

Fresh fig has a limited postharvest life and commonly suffers from diseases, softening, and weight loss. A high level of CO_2 has been effective in preventing weight loss, decay development, and micro-organism growth in figs as well as slowing down metabolic activities such as respiration rate and ethylene production. However, an atmosphere containing 3% O_2 + 20% CO_2 accelerated the softening. It was concluded that a 3% O_2 + 15% CO_2 containing atmosphere can be used to maintain postharvest fruit quality and to ensure reduced product loss in the Bursa Siyahi fig cultivar under palliflex storage. Depending on pre- and postharvest factors, fig varieties give different responses to atmospheric compositions. For this reason, the biochemical properties and molecular mechanisms of the fig fruit stored under variable atmospheric composition should be investigated in the future.

Author Contributions: Conceptualization, A.D. and M.E.; methodology, A.D. and M.E.; software, A.D.; formal analysis, A.D.; investigation, A.D.; writing—original draft preparation, A.D.; writing—review and editing, M.E.; visualization, A.D.; supervision, M.E.; project administration, M.E.; funding acquisition, M.E. All authors have read and agreed to the published version of the manuscript.

Funding: The chemicals and necessary equipment were funded by TUBITAK, grant number 217O016 and The Scientific Research Projects Coordination Unit of Akdeniz University, grant number FBA-2017-2608.

Data Availability Statement: The data presented in this study are available on request from the corresponding author.

Acknowledgments: The authors gratefully thank the support of Alanar Fruit Company (Turkey) for providing the fruit materials.

Conflicts of Interest: The authors declare no conflict of interest.

References

1. Food and Agriculture Organization Statistics Database. Available online: https://www.fao.org/faostat/en/#data/QCL (accessed on 25 December 2022).
2. Doğan, A.; Erkan, M. *Ficus carica: Production, Cultivation and Uses*; Zeynel, D., Ed.; Nova Science Publishers: New York, NY, USA, 2022; pp. 185–208.
3. Crisosto, H.; Ferguson, L.; Bremer, V.; Stover, E.; Colelli, G. Fig (*Ficus carica* L.). In *Postharvest Biology and Technology of Tropical and Subtropical Fruits*; Yahia, E., Ed.; Woodhead Publishing: Cambridge, UK, 2011; pp. 134–158.
4. Doster, M.A.; Michailides, T.J. Fungal decay of first-crop and main crop figs. *Plant Dis.* **2007**, *91*, 1657–1662. [CrossRef]
5. Cantín, C.M.; Palou, L.; Bremer, V.; Michailides, T.J.; Crisosto, C.H. Evaluation of the use of sulfur dioxide to reduce postharvest losses on dark and green figs. *Postharvest Biol. Technol.* **2011**, *59*, 150–158. [CrossRef]
6. Dogan, A.; Cat, A.; Catal, M.; Erkan, M. First report of *Alternaria alternata* causing postharvest decay in fig (*Ficus carica* L. cv. Bursa Siyahi) fruit in Turkey. *J. Biotechnol.* **2018**, *280*, S84. [CrossRef]
7. Karabulut, O.A.; Ilhan, K.; Arslan, U.; Vardar, C. Evaluation of the use of chlorine dioxide by fogging for decreasing postharvest decay of Fig. *Postharvest Biol. Technol.* **2009**, *52*, 313–315. [CrossRef]
8. Thompson, A.K.; Prange, R.K.; Bancroft, R.D.; Puttongsiri, T.; Beaudry, R.M. *Controlled Atmosphere Storage of Fruit and Vegetables*, 3rd ed.; CAB International: Oxfordshire, UK, 2008; ISBN 9781786393739.
9. Cantín, C.M.; Minas, I.S.; Goulas, V.; Jiménez, M.; Manganaris, G.A.; Michailides, T.J.; Crisosto, C.H. Sulfur dioxide fumigation alone or in combination with CO_2-enriched atmosphere extends the market life of highbush blueberry fruit. *Postharvest Biol. Technol.* **2012**, *67*, 84–91. [CrossRef]
10. Colelli, G.; Mitchell, F.G.; Kader, A.A. Extension of postharvest life of Mission figs by CO_2 enriched atmospheres. *Hortscience* **1991**, *26*, 1193–1195. [CrossRef]
11. Dogan, A. Effects of different oxygen levels with high-carbon dioxide atmosphere on postharvest quality of fresh fig under palliflex storage systems. *Horticulturae* **2022**, *8*, 353. [CrossRef]
12. Tsantil, E.; Karaiskos, G.; Pontikis, C. Storage of fresh figs in low oxygen atmosphere. *J. Hortic. Sci. Biotechnol.* **2003**, *78*, 56–60. [CrossRef]
13. Türk, R.; Eriş, A.; Özer, M.H.; Tuncelli, E.; Henze, J. Research on the CA storage of Fig cv. Bursa Siyahı. *Acta Hortic.* **1994**, *368*, 830–839. [CrossRef]
14. Crisosto, G.; Bremer, V.; Dollahite, S.; Crisosto, C.H.; Stover, E.; Ferguson, L. Effect of Controlled Atmosphere Storage on the Quality of Three Fresh Fig Cultivars. 2009. Available online: http://figs4fun.com/Links/FigLink291.pdf (accessed on 26 December 2022).

15. Bahar, A.; Lichter, A. Effect of controlled atmosphere on the storage potential of Ottomanit fig fruit. *Sci. Hortic.* **2018**, *227*, 196–201. [CrossRef]
16. Selcuk, N.; Erkan, M. The effects of modified and palliflex controlled atmosphere storage on postharvest quality and composition of 'Istanbul' medlar fruit. *Postharvest Biol. Technol.* **2015**, *99*, 9–19. [CrossRef]
17. Sturm, K.; Koron, D.; Stampar, F. The composition of fruit of different strawberry varieties depending on maturity stage. *Food Chem.* **2003**, *83*, 417–422. [CrossRef]
18. Giongo, L.; Poncetta, P.; Loretti, P.; Costa, F. Texture profiling of blueberries (*Vaccinium* spp.) during fruit development, ripening and storage. *Postharvest Biol. Technol.* **2013**, *76*, 34–39. [CrossRef]
19. Cui, Y.; Wang, Z.; Chen, S.; Vainstein, A.; Ma, H. Proteome and transcriptome analyses reveal key molecular differences between quality parameters of commercial-ripe and tree-ripe fig (*Ficus carica* L.). *BMC Plant Biol.* **2019**, *19*, 146. [CrossRef] [PubMed]
20. Teixeira, G.H.A.; Júnior, L.C.C.; Ferraudo, A.S.; Durigan, J.F. Quality of guava (*Psidium guajava* L. cv. Pedro Sato) fruit stored in low-O_2 controlled atmospheres is negatively affected by increasing levels of CO_2. *Postharvest Biol. Technol.* **2016**, *111*, 62–68. [CrossRef]
21. Villalobos, M.C.; Serradilla, M.J.; Martín, A.; López, C.M.; Pereira, C.; Córdoba, M.D.G. Preservation of different fig cultivars (*Ficus carica* L.) under modified atmosphere packaging during cold storage. *J. Sci. Food Agric.* **2016**, *96*, 2103–2115. [CrossRef]
22. Üstün, H.; Doğan, A. The effect of different UV-C illumination doses on postharvest quality of fresh Fig. *Gıda* **2022**, *47*, 744–753.
23. Allende, A.; Marín, A.; Buendía, B.; Tomás-Barberán, F.; Gil, M.I. Impact of combined postharvest treatments (UV-C light, gaseous O_3, super atmospheric O_2 and high CO_2) on health promoting compounds and shelf-life of strawberries. *Postharvest Biol. Technol.* **2007**, *46*, 201–211. [CrossRef]
24. Serradilla, M.J.; Villalobos, M.C.; Hernández, A.; Martín, A.; Lozano, M.; de Guía Córdoba, M. Study of microbiological quality of controlled atmosphere packaged 'Ambrunés' sweet cherries and subsequent shelf-life. *Int. J. Food Microbiol.* **2013**, *166*, 85–92. [CrossRef]
25. Villalobos, M.C.; Ansah, F.; Amodio, M.L.; Serradillas, M.J.; Colelli, G. Application of modified atmosphere packaging with moisture absorber to extend the shelf life of 'Domenico Tauro' breba fruit. *Acta Hortic.* **2017**, *1173*, 365–370. [CrossRef]
26. Villalobos, M.C.; Serradilla, M.J.; Martín, A.; Ruiz-Moyano, S.; Pereira, C.; Córdoba, M.G. Use of equilibrium modified atmosphere packaging for preservation of 'San Antonio' and 'Banane' breba crops (*Ficus carica* L.). *Postharvest Biol. Technol.* **2014**, *98*, 14–22. [CrossRef]
27. Adiletta, G.; Zampella, L.; Coletta, C.; Petriccione, M. Chitosan coating to preserve the qualitative traits and improve antioxidant system in fresh figs (*Ficus carica* L.). *Agriculture* **2019**, *9*, 84. [CrossRef]
28. Waghmare, R.B.; Annapure, U.S. Integrated effect of radiation processing and modified atmosphere packaging (MAP) on shelf-life of fresh Fig. *J. Food Sci. Technol.* **2018**, *55*, 1993–2002. [CrossRef] [PubMed]
29. Ayhan, Z.; Karaçay, E. Preservation of the Bursa Siyahı fresh fig under modified atmosphere packaging (MAP) and cold storage. *Int. J. AgriSci.* **2011**, *1*, 1–9.
30. Martínez-Damián, M.T.; Cruz-Arvizu, O.; Cruz-Alvarez, O. Effect of modified atmosphere packaging on nutraceutical quality and overall appearance of figs stored at 1 °C. *Not. Bot. Horti. Agrobot. Cluj-Napoca* **2020**, *48*, 2292–2305. [CrossRef]
31. Çalişkan, O.; Polat, A.A. Phytochemical and antioxidant properties of selected fig (*Ficus carica* L.) accessions from the eastern Mediterranean region of Turkey. *Sci Hortic.* **2011**, *128*, 473–478. [CrossRef]
32. Byeon, S.E.; Lee, J. Differential responses of fruit quality and major targeted metabolites in three different cultivars of cold-stored figs (*Ficus carica* L.). *Sci. Hortic.* **2020**, *260*, 108877. [CrossRef]
33. Trad, M.; Bourvellec, L.B.; Gaaliche, B.; Renard, C.M.G.C.; Mars, M. Nutritional compounds in figs from the Southern Mediterranean region. *Int. J. Food Prop.* **2014**, *17*, 491–499. [CrossRef]
34. Allegra, A.; Sortino, G.; Inglese, P.; Settanni, L.; Todaro, A.; Gallotta, A. The effectiveness of *Opuntia ficus-indica* mucilage edible coating on post-harvest maintenance of 'Dottato' fig (*Ficus carica* L.) fruit. *Food Packag. Shelf Life* **2017**, *12*, 135–141. [CrossRef]
35. Almenar, E.; Hernández-Muñoz, P.; Lagarón, J.M.; Catalá, R.; Gavara, R. Controlled atmosphere storage of wild strawberry fruit (*Fragaria vesca* L.). *J. Agric. Food Chem.* **2006**, *54*, 86–91. [CrossRef]
36. Falagan, N.; Miclo, T.; Terry, L.A. Graduated controlled atmosphere: A novel approach to increase "Duke" blueberry storage life. *Front. Plant Sci.* **2020**, *11*, 221. [CrossRef] [PubMed]
37. Bouzo, C.A.; Travadelo, M.; Gariglio, N.F. Effect of different packaging materials on postharvest quality of fresh fig fruit. *Int. J. Agric. Biol.* **2012**, *14*, 821–825.
38. Mathooko, F.M.; Sotokawa, T.; Kubo, Y.; Inaba, A.; Nakamura, R. Retention of freshness in fig fruit by CO_2 enriched atmosphere treatment or modified atmosphere packaging under ambient temperature. *J. Jpn. Soc. Hortic. Sci.* **1993**, *62*, 661–667. [CrossRef]

horticulturae

Article

Foliar Application of Chitosan Accelerates Wound Periderm Formation with an Intensified Deposition of Suberin Polyphenolic and Lignin in the Wounds of Potato Tubers

Yan Zhu [1,*], Wei Zhang [1], Yuanshou Zhao [2], Fu Chen [3], Fupeng Zhu [1], Jipeng Yang [1], Xiangzhen Ge [1], Wenjuan Gong [1], Ruitong Cui [1] and Wenjun Sheng [1]

1 College of Food Science and Engineering, Gansu Agricultural University, Lanzhou 730070, China
2 Lanzhou Industrial Research Institute, Lanzhou 730050, China
3 Institute of Potato, Gansu Academy of Agricultural Sciences, Lanzhou 730070, China
* Correspondence: zhuyan0102@163.com; Tel./Fax: +86-0931-7631201

Abstract: Potato tubers are susceptible to wounding during post-harvest processes, leading to quality decline, perishability and large economic losses. In this study, the potato cultivar, 'Longshu No.7', was foliar-sprayed with 3% chitosan (*w/v*) three times during the pre-harvest period after flowering to evaluate the effect of foliar spraying with chitosan on suberization processing in the wounds of harvested potato tubers. Our results demonstrated that foliar sprayed with chitosan significantly reduced wound-induced fresh weight loss and dry rot disease index by 37.34% and 41.60% on day 28 after wounding, respectively. Foliar sprayed with chitosan accelerated the deposition of suberin polyphenolic and lignin at the wound sites of potato tubers with the formation of thicker cell layers. This occurred with increased localized activities of key enzymes in the suberin polyphenolic and lignin pathways, including phenylalanine ammonia lyase, 4-coumaryl-coenzyme A ligase, cinnamyl alcohol dehydrogenase and peroxidase (33.90–64.32%), as well as the contents of cinnamic acid, sinapic acid, flavonoids, lignins and total phenolics (19.70–23.46%) in the wounded sites of potato tubers on day 7 after wounding. Our results indicated that foliar application of chitosan accelerated wound-induced suberization of potato tubers and could mitigate post-harvest product damages.

Keywords: chitosan; foliar application; lignin; potato; suberin polyphenolic; wound-induced suberization

Citation: Zhu, Y.; Zhang, W.; Zhao, Y.; Chen, F.; Zhu, F.; Yang, J.; Ge, X.; Gong, W.; Cui, R.; Sheng, W. Foliar Application of Chitosan Accelerates Wound Periderm Formation with an Intensified Deposition of Suberin Polyphenolic and Lignin in the Wounds of Potato Tubers. *Horticulturae* **2023**, *9*, 663. https://doi.org/10.3390/horticulturae9060663

Academic Editor: Elazar Fallik

Received: 9 May 2023
Revised: 31 May 2023
Accepted: 1 June 2023
Published: 3 June 2023

1. Introduction

Potato (*Solanum tuberosum* L.) is the world's fourth largest food crop and an important raw material in food industrial processing, with annual production reaching 18 million tons in China (potato grain production is converted to 20% of the fresh weight of potatoes as the standard yield; data from the 2021 statistics of the National Bureau of China). About 70–80% of the total production of potato tubers is stored after harvest for consumption over the following year [1]. However, the primary periderm of potato tubers can be fragile and susceptible to peeling, scraping and friction injuries during harvesting, post-harvest transportation and storage operations [2]. Epidermal wounds are the main channels for pathogen infection and water loss, leading to a decline in product quality and increased product decay during storage [3]. In China, the incidence of potato tuber rot during storage was reported to be on average between 10 and 30% and up to 60%, which represents huge economic losses [4]. Injury to the tuber periderm leads to secondary periderm formation as protection against pathogen infestations and water loss [5]. This healing process involves wound-induced suberization at the wound site, which occurs with the localized synthesis and deposition of suberin polyphenolic (SPP) and lignin. However, this process can take 2–3 weeks to complete, which is sufficient for substantial water loss, opportunistic pathogen entry, disease incidence and product loss during storage [5]. Therefore, the study of methods to accelerate wound-induced suberization is of economic importance [3,6].

Chitosan, a deacetylation product of chitin (poly-β-(1,4)-N-acetyl-D-glucosamine), is widely found in the exoskeletons of crustaceans, insects and the cell walls of fungi. Chitosan functions as an elicitor in the plant system for the induction of antimicrobial activities and innate disease resistance [7,8]. Chitosan is non-toxic, edible and biodegradable, and has been commercially registered for production in some countries for use in food preservation [9]. Reports have shown that soaking or coating treatments of plant products by chitosan can effectively control the occurrence of post-harvest decay in some fruits and vegetables [10–12]. Significantly, some studies have shown that pre-harvest foliar application of chitosan can also effectively enhance plant stress resistance and reduce the occurrence of field diseases in wheat and maize [13,14]. For fruit and vegetable preservation, Nia et al. (2021) found that foliar spraying of table grapes with chitosan improved fruit disease resistance and reduced the incidence of fruit during the post-harvest period [15]. Li et al. (2021) reported that pre-harvest spraying of chitosan promoted the accumulation of SPP in muskmelon fruits and enhanced the rate of wound healing at the fruit surface [16]. Chitosan treatment also induced lignin production in suspension cells of *Pinus elliottii* and callus of *Oryza sativa* [17,18]. However, the effect of foliar application of chitosan on the wound-induced suberization of potato tubers after harvest has not been reported. In this study, *S. tuberosum* L. cv. Longshu No.7 was used to explore the effects of pre-harvest foliar spraying of chitosan on the suberization of wounded tubers after harvest and the underlying biochemical mechanism involved and to provide a theoretical basis for the use of this treatment to maintain product quality during commercial potato tuber storage.

2. Materials and Methods

2.1. Materials

Chitosan (poly-β-(1,4)-2-amino-D-glucose) with a degree of deacetylation of ≥90% was purchased from WN Group of Publishers Ltd. (Mansouriah block1, Paris, France). 1,3,5-Trihydroxybenzene (No. Y93552), standards for cinnamic acid (No. B21082) and sinapic acid (No. B25310) were purchased from Yuanye Biotechnology Co., Ltd. (Shanghai, China).

Fusarium sambucinum was isolated from potato tissues exhibiting typical symptoms of dry rot. The identity of the pathogen was verified from its rDNA-ITS sequence and the pathogenicity was verified in potato tubers according to Koch's Postulates [19]. The pathogen was preserved in Potato Dextrose Agar (PDA) medium at 4 °C and cultured on a PDA medium for 7 days at 28 °C before use.

2.2. Experimental Design

Virus-free seeds of *S. tuberosum* L. cv. Longshu No.7 potatoes, provided by the Potato Institute of the Gansu Academy of Agricultural Sciences (Lanzhou, China), were planted in an open field on 20 April 2019 in Nangou Village, Huichuan Town, Weiyuan County, Dingxi City, Gansu Province (35°06′30″ N, 103°58′15″ E, 2260 m above sea level). Potatoes were planted in six rows per plot, with an inter-row spacing of 60 cm and plant spacing of 30 cm, and 210 plants were sown in each plot. Potato plants were sprayed evenly (1 L/30 plants) using a hand sprayer with either water (control) or 3% (*w/v*) chitosan at the flowering stage, the tuber bulking stage and 2 weeks before harvesting, respectively. The experiment was conducted in a completely randomized group arrangement with three replications [20]. Tubers were developed to maturity and harvested on 10 October 2019, dried in the sun for 4 h, and then packed into standard corrugated cartons. The tubers used for physiological and biochemical analyses were stored refrigerated (5 ± 2 °C, RH 80–90%) until further use.

2.3. Methods

2.3.1. Artificial Wounding of Potato Tubers

Three batches of 300 potato tubers from each of the chitosan and control treatments of similar size, free of mechanical damage, pests and diseases were selected, rinsed under running water and soaked in 1% sodium hypochlorite for 10 min. After air drying, the periderm was cut along the equatorial plane with a sterile scalpel to a depth of 0.3–0.5 mm according to the method of Zheng et al. (2020) [6]. The wounded tubers were then stored in a polyethylene preservation box, which had moist sterilized filter paper inside. The box was wrapped with a perforated black bag in the dark at 20 ± 2 °C with an RH of 75–85%. The images of the artificial wounding of potato tuber and storing procedures after injury are shown in Figure 1.

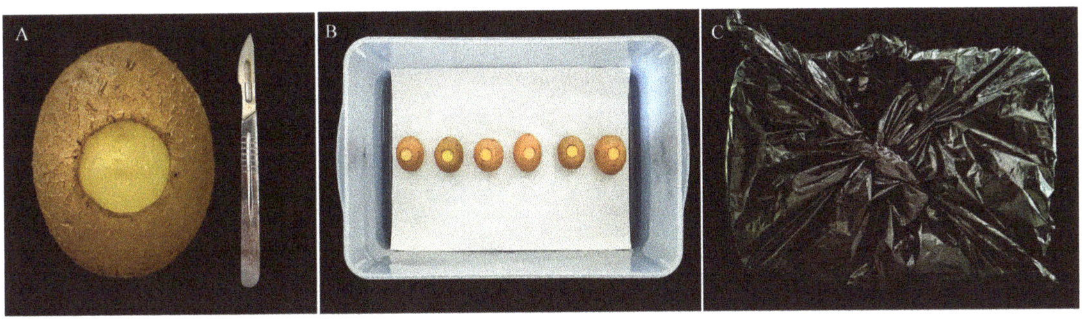

Figure 1. Artificial wounding and storing procedures of potato tubers. (**A**) Wounded tubers and sterile scalpel; (**B**) polyethylene preservation box; (**C**) wrapped black bags.

2.3.2. Determination of Dry Rot Disease Index and Loss of Fresh Weight in Wounded Potato Tubers

The loss of tuber fresh weight and the disease index of dry rot were determined according to the method of Zheng et al. (2020) [6]. Using sterile water containing 0.01% (*w*/*v*) Tween 80, spore suspensions (1×10^6 spores/mL) were prepared from 7-day-old *F. sambucinum* solid PDA cultures. On days 7, 14 and 28 after injury, 20 µL of the *F. sambucinum* spore suspension was evenly applied at the surface of wounded tubers, which were air-dried, placed in polyethylene preservation boxes wrapped with perforated black bags and stored in a ventilated storage room in the dark (20 ± 2 °C, RH 75–85%) for 7 days before determining the disease incidence [20]. Ninety tubers were used for each treatment and randomly divided into three groups of thirty categorized as three biological replicates. The severity of the dry rot was graded based on the percentage of tubers displaying visible fungi on the tuber surface [20]. The disease index was then calculated using the following equation with three biological replicates.

$$\text{Disease index} = \frac{\sum(\text{Number of diseased tubers} \times \text{Relative level value})}{(\text{Total number of tubers} \times \text{High estrepresentative value})} \times 100 \quad (1)$$

To determine the fresh weight loss, ninety artificially wounded tubers were selected from each treatment group and randomly divided into three biological replicates of thirty tubers for storage under the conditions described above. The fresh weight loss was calculated according to the following equation on days 7, 14 and 28 after wounding. The experiment was repeated three times.

$$\text{Fresh weight loss}(\%) = \frac{\text{Fresh weight before wounded} - \text{Fresh weight after wounded}}{\text{Fresh weight before wounded}} \times 100 \quad (2)$$

2.3.3. Microscopic Observation of SPP and Lignin Deposition at the Wound Site of Potato Tubers

The vertical wound surface of the tuber was cut into 0.2–0.3 mm slices of ca. 1 cm^2 using a sterile scalpel, which were washed three times with distilled water to remove starch particles. Lignin was stained by the phloroglucinol–hydrochloric acid method [6] and observed using a microscope (CX21FS1C Olympus, Tokyo, Japan). The deposition of SPP was monitored from its autofluorescence using a fluorescence microscope (Shimadzu RF-5301 PC, Tokyo, Japan), with the excitation wavelength and the emission wavelength at 340–390 nm and 420 nm, respectively [21]. The IS Capture software (Tucsen, Fujian, China) was used to determine the thickness of the cell layer of lignin and SPP [6].

2.3.4. Sample Collection

The suberized tissue sampling was collected from the wounded site after 0, 1, 3, 5 and 7 days of artificial injury according to the method of Ge et al. (2021) with some modifications [20]. Briefly, the sample was taken 3 mm around and below the wound site. Samples were snap-frozen in liquid nitrogen and ground into a powder with a grinding mill (IKA M20, IKA-Werke GmbH & Co., KG, Staufen im Breisgau, Germany), and then stored at −80 °C until further analysis.

2.3.5. Determination of Enzyme Activities in SPP and Lignin Anabolism

For the determination of the relative activities of phenylalanine aminolase (PAL), 4-coumaroyl-coenzyme A ligase (4CL), cinnamyl alcohol dehydrogenase (CAD), peroxidase (POD), and 0.5 g (FW) of sampled tissue were homogenized on ice with 2 mL of extraction reagents and centrifuged at 4 °C and 10,000× *g* for 10 min; the supernatant was used as the enzyme crude extract. The activities of the enzyme were analyzed using microplate assay kits as the manufacturer's instructions (Comin Bio. Co., Ltd., Suzhou, China). Crude extracts were placed on ice and tested within 20 min. A total of 10 µL crude extract and 190 µL reaction reagent were added and mixed as a reaction system for measuring the enzyme activity. The extraction and reaction reagents were provided in the kits. Colorimetric determination was made by ultraviolet and visible spectrophotometers (Shimadzu UV-2450, Tokyo, Japan). Using BSA as a standard, the protein content of the tissue extracts ($mg \cdot mL^{-1}$) was determined using the Bradford assay [22]. The activity of the enzyme was expressed as U activity mg^{-1} protein. One unit of PAL and POD enzymes was defined as that required to achieve an increase in the absorbance of 290 nm of 0.1 $min^{-1} \cdot mL^{-1}$. One unit of 4CL was defined as that required for the production of one nmoL of 4-coumaryl coenzyme A per minute and one unit of CAD enzyme activity was defined as that required for the generation of one nmol of Nicotinamide adenine dinucleotide phosphate per minute. All enzyme activities presented were determined from three biological replicates.

2.3.6. Determination of Metabolic Contents of Suberization

The content of phenolic acid monomers (cinnamic acid and sinapic acid) was determined according to the method of Gruz et al. (2008) with minor modifications [23]. Briefly, 1 g of the frozen tissue homogenate was extracted by ultrasonication in 3 mL of 70% (v/v) methanol for 30 min and centrifuged twice at 8000× *g* for 20 min. The supernatant was concentrated in a vacuum concentrator (EYELA UT2000, Tokyo, Japan) and dissolved in 1 mL of a 70:30:1 mixture of methanol, ultrapure water and glacial acetic acid, respectively, and filtered using a 0.22 µm nylon filter membrane (Biosharp, Hefei, China). Quaternary gradient ultrafast liquid chromatography (ACQUITY Arc, Waters, Milford, MA, USA) and Symmetry® C18 column (4.6 mm × 250 mm, 5 µm) were used to analyze the filtrate. The analysis conditions used were as described by Zhu et al. (2022) [21], where cinnamic and sinapic acids were detected at 276 nm and 325 nm, respectively, and identified from their retention times relative to those of their pure standards. The content of these phenolic acid monomers was calculated from their standard curves and expressed as μg·g^{-1} (FW). The relative contents of lignin, total phenols and flavonoids were measured according to the method of Ge et al. (2021) [20], where the contents of lignin and total phenols were expressed as OD_{280}·g^{-1} (FW), and the content of flavonoid was expressed as OD_{325}·g^{-1} (FW).

2.4. Statistical Analysis

All data were presented as the average of at least three biological replicates and ± standard error. Significance testing was performed using Duncan's multiple difference with SPSS 19.0 (Chicago, IL, USA), and with $p < 0.05$ as the threshold. All graphs were generated using Origin 2023 (OriginLab, Northampton, MA, USA).

3. Results

3.1. Foliar Spraying of Chitosan Reduces the Effects of Wounding on Tuber Fresh Weight Loss and Dry Rot Development

Both fresh weight loss and dry rot disease development were promoted by tuber wounding. The loss of fresh weight in the wounded tubers gradually increased with increasing healing time. However, the fresh weight loss in the wounded tubers of the control group was significantly higher (5.49%) than that in the chitosan-sprayed group (3.44%) after 28 days of storage, representing a ca. 37.34% reduction in fresh weight loss after wounding ($p < 0.05$; Figure 1A). Typical dark brown spots symptomatic of dry rot were observed on the surfaces of the injured tubers inoculated with *F. sambucinum*. In the control group, the disease index of dry rot increased rapidly from the 7th (ca. 10%) to the 28th day of incubation (ca. 73%). In contrast, the disease index in the wounded tubers of the chitosan-sprayed group was consistently and significantly lower ($p < 0.05$) than that in the control group by 28.80–41.60% (Figure 2B). These results demonstrate that the foliar spraying of chitosan significantly reduced fresh weight loss and the development of dry rot in wounded potato tubers after harvesting.

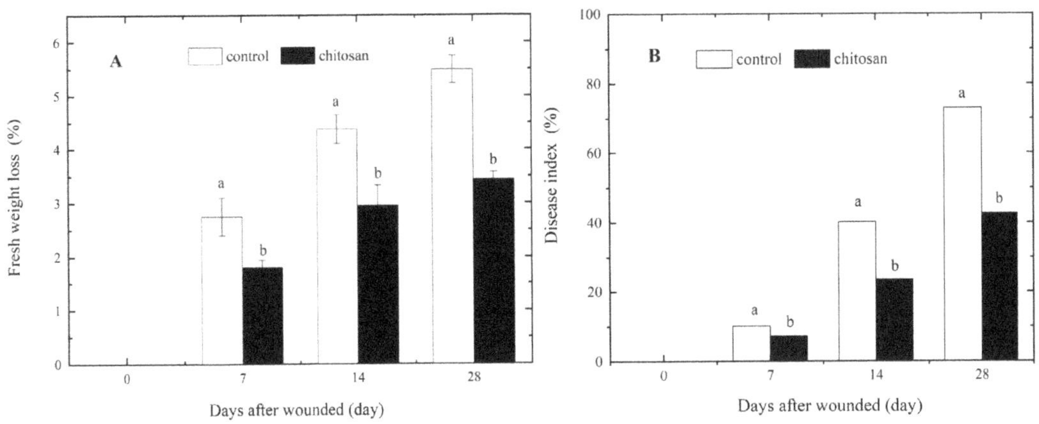

Figure 2. Effects of the foliar spraying of chitosan on (**A**) the fresh weight loss and (**B**) the development of dry rot disease in wounded potato tubers. Lowercase letters indicate a significant difference between the different treatment groups at the same time point after injury ($p < 0.05$).

3.2. Effect of Foliar Spraying of Chitosan on SPP and Lignin Accumulation at the Wound Site of Potato Tuber

The SPP and lignin layers are important components of the wound periderm in wounded potato tubers. As shown in Figure 3A, tubers from the foliar spraying of the chitosan group showed an increased deposition of SPP at the wound site relative to the control group as early as 3 days after injury. After 7 days of injury, tubers from the chitosan treatment group showed larger increases in SPP fluorescence intensity and fluorescent cell layer thickness (27.27%; Figure 4A) at the wound site relative to the control tubers. Similarly, tuber wounding induced the localized deposition of lignin, which increased over 7 days (Figure 3B). After 7 days of injury, the lignin content of the treated group increased by 23.46% (Figure 4C) and the thickness of the cell layer by 26.54% compared to the control group (Figure 4B). These results indicated that the foliar spraying of chitosan significantly accelerated the deposition of SPP and lignin at the wounds of potato tubers.

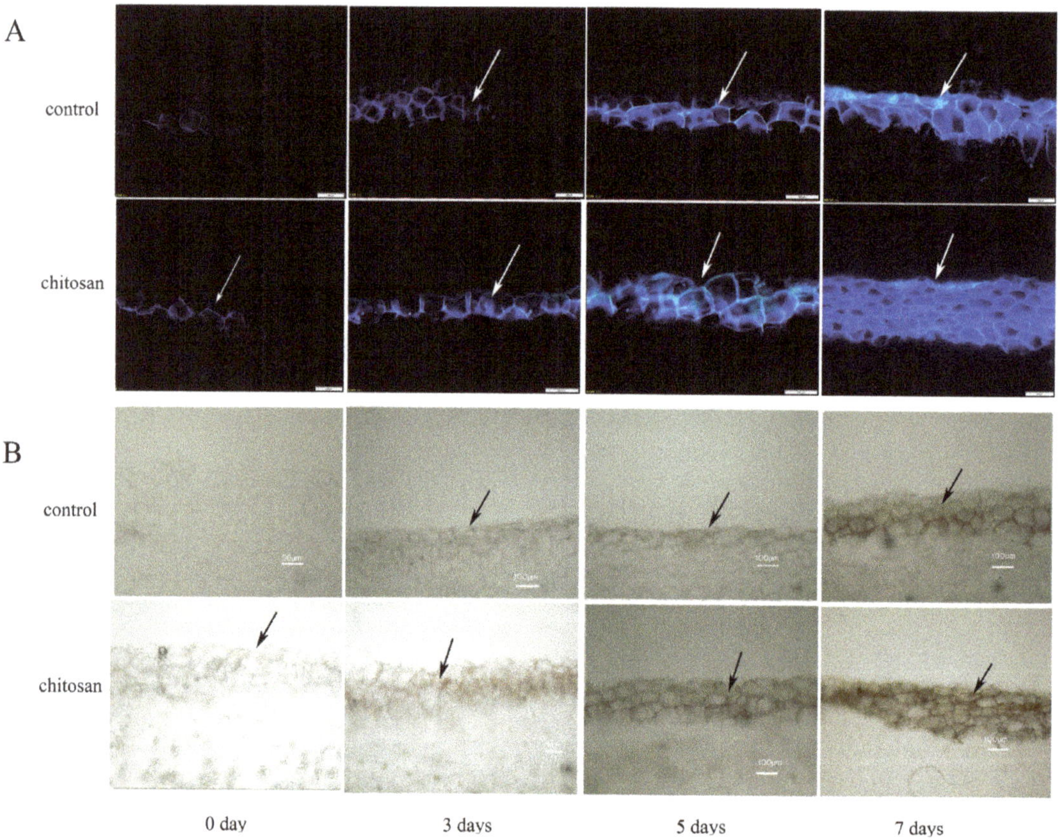

Figure 3. Effect of foliar spraying of chitosan on the localized accumulation of (**A**) SPP and (**B**) lignin in wounded potato tubers. White arrows indicate the deposition of SPP at the wound sites and black arrows indicate the deposition of lignin at the wound sites, respectively.

3.3. Effect of Foliar Spraying of Chitosan on Key Enzyme Activities of SPP and Lignin Anabolism during Wound-Induced Suberization

The activities of key enzymes in both the SPP and lignin synthesis pathways were determined. Prior to wounding, there were no significant differences ($p \geq 0.05$) in the activities of PAL, 4CL or POD between the control and chitosan treatment groups. However, the activity of CAD was comparatively 17.61% higher in tubers of the chitosan treatment group at the time of harvest (Figure 5).

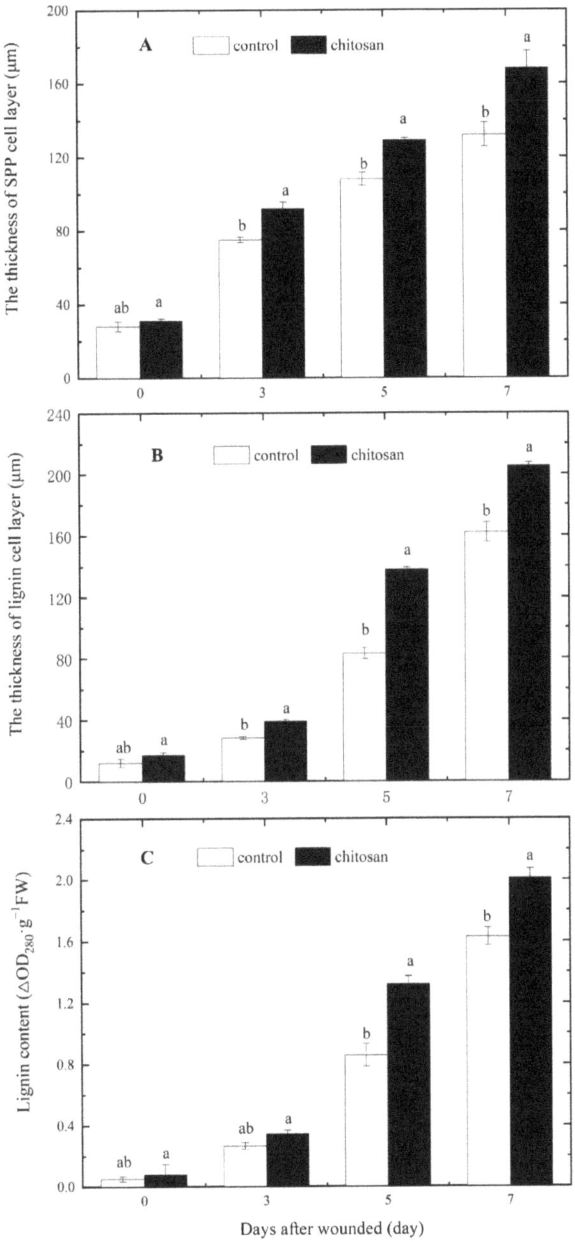

Figure 4. Effect of foliar spraying of chitosan on (**A**) the thickness of the SPP cell layer at the wound site, (**B**) the thickness of the lignin cell layer at the wound site, and (**C**) the content of lignin. Lowercase letters indicate a significant difference between the different treatment groups ($p < 0.05$).

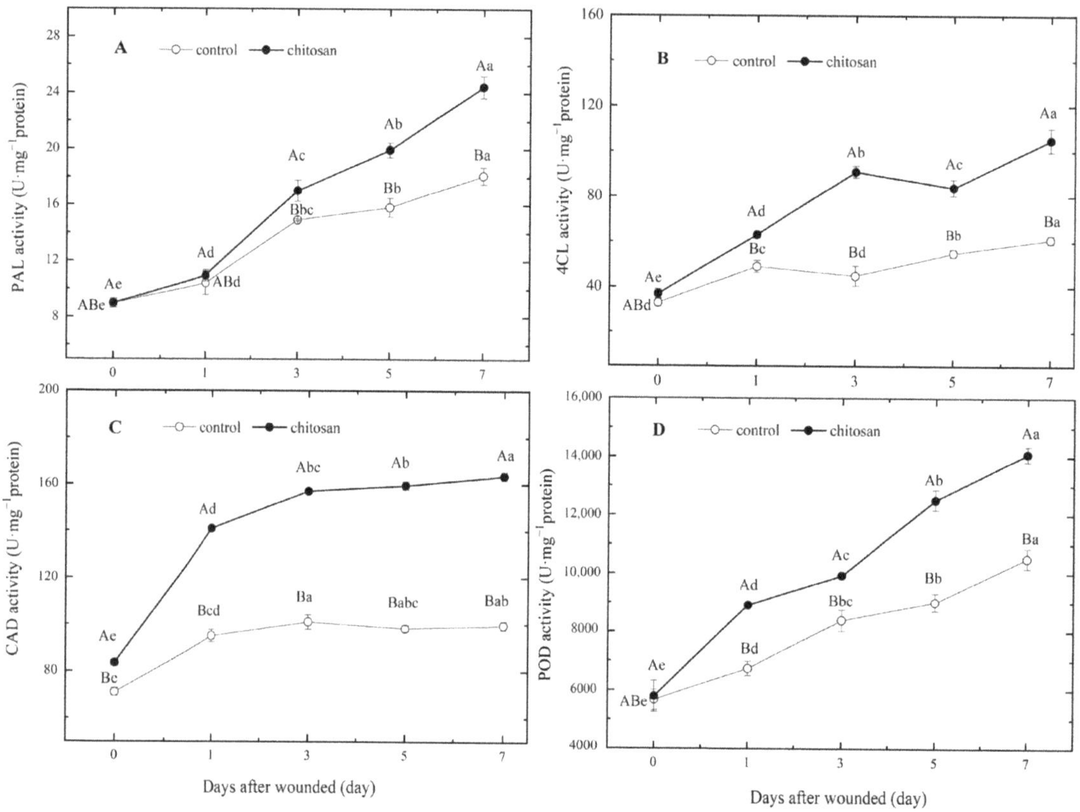

Figure 5. Effects of the foliar spraying of chitosan on enzyme activities related to SPP and lignin synthesis at the wound site of potato tubers. The activities of (**A**) PAL, (**B**) 4CL, (**C**) CAD and (**D**) POD are shown. Capital letters indicate significant differences between the control and chitosan treatments ($p < 0.05$). Lowercase letters indicate significant differences between time points in the same treatment group ($p < 0.05$).

Following wounding, the activities of the four key enzymes showed a gradual increase in the control group. However, these enzyme activities in the tubers of the chitosan treatment group were all higher than those of the control group (Figure 5). Notable relative differences included a larger increase in PAL activity 0–3 days post-injury (Figure 5A) and two peaks of 4CL activity on days 3 and 7 after wounding (Figure 5B). The activity of CAD at the wound site was 48.42% higher than that of the control after 1 day of wounding and maintained higher activities over the following 6 days (Figure 5C). POD activities showed a continual increase after wounding, with the highest activity on day 7 (Figure 5D). Relative to the control group, the activities of PAL, 4CL, CAD and POD of the chitosan treatment group were 35.02%, 55.74%, 64.32% and 33.9% higher, respectively, after 7 days of wounding (Figure 5). These results indicated that the foliar spraying of chitosan significantly enhanced the localized mobilization of PAL, 4CL, CAD and POD activities in response to the wounding of potato tubers after harvest.

3.4. Effects of the Foliar Spraying of Chitosan on the Contents of Phenolic acid Monomers, Total Phenols and Flavonoids in Tubers during Wound-Induced Suberization

SPPs are polymerized from different phenolic acid monomers, of which cinnamic acid and sinapic acid are the main constituents. Total phenols and flavonoids are the sum of phenolic acid in tubers. The data on day 0 showed that the foliar spraying of chitosan had no significant effect on the contents of cinnamic acid, sinapic acid, or total phenols (Figure 6A–C), but resulted in a significantly higher flavonoid content relative to the control ($p < 0.05$; Figure 6D). Following tuber wounding, the contents of phenolic acids and flavonoids showed an increasing trend (Figure 6), and the contents of sinapic acid and cinnamic acid increased rapidly, reaching a maximum value on days 5 and day 7, respectively. In the chitosan-treated group, the contents of both phenolic acid monomers were significantly higher relative to the control group (Figure 6A,B). In tubers of the chitosan treatment group, the total phenolic content was significantly higher than that of the control group from days 3 to 7 after wounding (Figure 6C), whereas the flavonoid content was not only relatively higher directly after harvesting but was also induced to higher levels after wounding with further increases over time (Figure 6D). Compared to the control group, the content of cinnamic acid, sinapic acid, total phenols and flavonoids in the tubers from the chitosan-treated group increased by 20.34%, 20.50%, 22.86% and 19.70%, respectively, on day 7 after wounding (Figure 6). The above results indicated that the foliar spraying of chitosan promoted a higher rate of synthesis of cinnamic acid, sinapic acid, total phenols and flavonoids in potato tubers after harvest during wound-induced suberization.

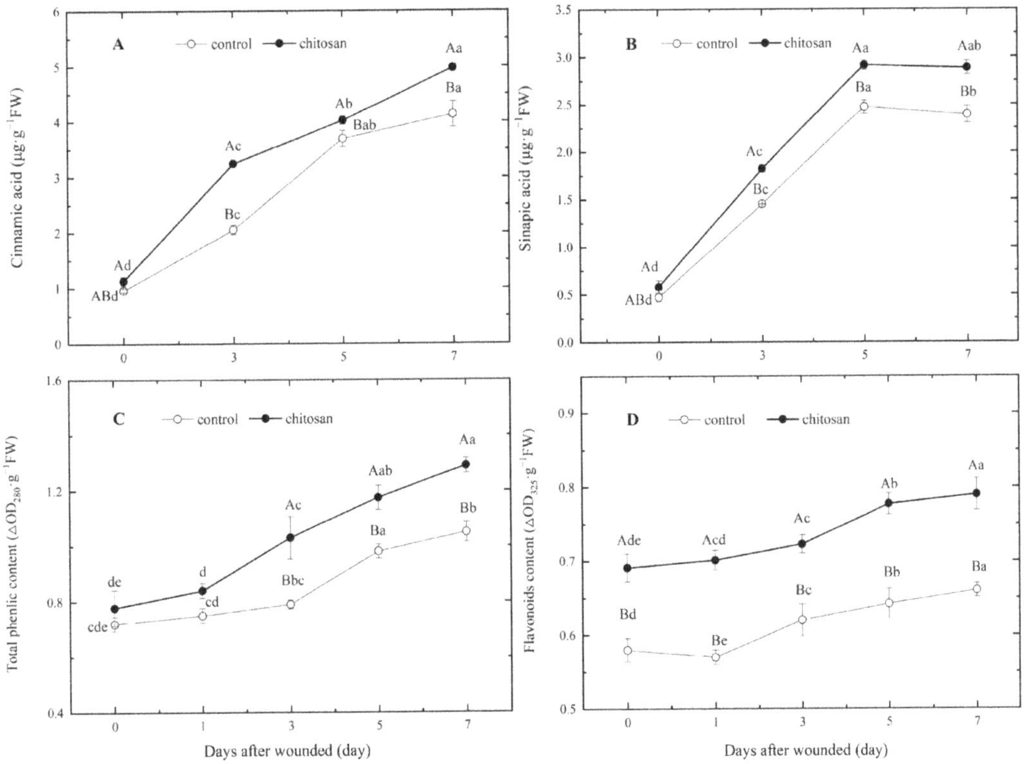

Figure 6. Effect of foliar spraying of chitosan on the contents of (**A**) cinnamic acid, (**B**) sinapic acid, (**C**) total phenolics and (**D**) flavonoids at the wound site of potato tubers. Capitals letters indicate significant ($p < 0.05$) differences between chitosan and control treatments. Lowercase letters indicate significant ($p < 0.05$) differences between time points in the same treatment group.

4. Discussion

Pre-harvest treatments, including seed dipping or foliar spraying with selected agents, have been shown to enhance stress resistance of fruit and vegetable crops during plant cultivation, reduce damage to product quality from aspects of post-harvest processes, and improve post-harvest preservation of the freshness of fruits and vegetables, all of which have great potential for application in agricultural production [24–28]. Many agents have been reported to improve the post-harvest quality of fruits and vegetables, such as brassinosteroid, salicylic acid, sorbitol, sodium nitroprusside, sodium silicate, chitosan, etc., during the pre-harvest or post-harvest period [21,24,29–32]. Chitosan is generally recognized as safe (GRAS) and can be applied to many crops during pre-harvest [10]. It is known that potato tubers decay easily in water, so harvested tubers are not suitable for preservative treatment with water solution after harvest, as potato tubers are not suitable for post-harvest treatments involving their immersion, and there is a lack of fumigation treatments in current potato production practices. Therefore, the pre-harvest foliar application is a good strategy for improving potato tuber quality after harvest.

Our results showed that pre-harvest foliar spraying of chitosan three times during potato tuber development was effective in reducing fresh weight loss and dry rot disease development in wounded tubers during the post-harvest period (Figure 2). The reduced fresh weight loss is likely a direct result of the accelerated formation of the wound periderm [20]. Potato tubers are susceptible to *Fusarium* spp. and *F. sambucinum*, which are the dominant causal pathogens of dry rot during potato storage [4]. Pre-harvest foliar spraying of chitosan significantly impaired the development of dry rot in injured tubers, which can be related to the accelerated suberization of the wound periderm and its provision of an enhanced protective barrier against pathogen infection. Similar post-harvest protective effects have been reported for potatoes using a stroby (kresoxim-methyl) [20].

Wound-induced tissue suberization is a complex biological process involving an increase in precursor phenylpropanoid synthesis, SPP and lignin formation in potentially dedicated pathways [5,16,33]. The phenylpropanoid pathway is an important secondary metabolic pathway closely related to the plant immune system, and large amounts of phenolic acids are biosynthesized via this pathway [34]. PAL is the key rate-limiting enzyme in the phenylpropanoid pathway and catalyzes the deamination of phenylalanine to produce trans-cinnamic acid [35], which undergoes a series of catalytic transformations to generate many phenolic compounds, including *p*-coumarate, caffeic, ferulic acid and sinapic acid [33]. Phenolic acid monomers are catalyzed to hydroxylated phenolic acids, which are polymerized in the presence of POD to produce SPP [31]. Hydroxylated phenolic acids are also catalyzed by 4CL to acetylate and produce *p*-coumaroyl-CoA, feruloyl-CoA, sinapic-CoA, etc., which are further catalyzed in the presence of CAD to produce monolignols, including *p*-coumaryl, coniferyl and sinapyl alcohols, which are subsequently polymerized by POD into lignin [31,33,35]. The increased availability of phenolic substrates has been associated with an accelerated deposition of SPP and lignin during the wound-induced suberization process [4,21]. Phenolic acids provide precursor substrates for the synthesis of SPP and lignin, which provide important waterproofing properties and a protective barrier against pathogen infection in the periderm of potato tubers [36]. It has been shown that chitosan treatment can enhance fruit resistance by activating the activity of defense enzymes related to the lignin synthesis pathway [15,16,24], promote lignin synthesis and prolong fruit shelf life in citrus [37,38], pears [39], grapes [40], muskmelons [16] and bamboo shoots [41]. These effects occurred during the foliar spraying of chitosan in our study. Our results showed that pre-harvest foliar spraying of chitosan enhanced the wound-induction of PAL, 4CL, CAD and POD activities (Figure 5), and promoted increased contents of two phenolic acid monomers in the wounded sites of potato tubers (Figure 6A,B). Key enzyme activities of the phenylpropanoid pathway, SPP and lignin anabolism were activated, which were associated with an accelerated SPP and lignin deposition in the suberizing wound periderm (Figures 3 and 4).

Many phenolic acids and the derivatives of the phenylpropanoid pathway contribute to the total phenolic content of plant tissues [31,35]. The total synthesized phenols and flavonoids can have antioxidant and antimicrobial activities, which inhibit pathogen expansion in the host [7,24,33,42]. The accumulation of total phenols and flavonoids can contribute to improved fruit and vegetable storage performance through potential antimicrobial and antioxidant activities. Li et al. (2021) found that the pre-harvest chitosan treatment increased the content of flavonoids during wound healing of post-harvest melon fruits and reduced the disease index of fruit [16]. Cui et al. (2020) reported that pre-harvest chitosan spray promoted the post-harvest synthesis of phenolic acids in apricot with the enhancement of the fruit's antioxidant capacity [24]. Potato varieties with high contents of total phenols tend to have a higher disease resistance [6]. Pre-harvest chitosan spray induced the accumulation of total phenols and flavonoids in muskmelons and apricots, together with an enhanced fruit resistance to infection and extended shelf-life [16,24]. Flavonoids have high antioxidant capacity and antimicrobial activity in plant products, which can effectively improve plant resistance to infection by pathogens [43,44]. In this study, the foliar spraying of chitosan led to an increased content of flavonoid in tubers during harvest and a relatively larger increase in flavonoid levels after tuber wounding (Figure 6D), as well as an increased content of total phenols after wounding in potato tubers (Figure 6C), which are consistent with an improvement in antioxidant capacity, inhibition of pathogen infection and a reduction in disease index.

The pre-harvest chitosan treatment presented here involves foliar spraying of potato plants with 3% (*w/v*) chitosan during the flowering period, tuber enlargement period and two weeks before tuber harvest. The efficacy of this treatment in improving post-harvest potato tuber performance was successfully field-tested in 2017 and 2018. A similar method was developed for the pre-harvest spraying of muskmelon with chitosan in [16]. As for the regulatory mechanism underlying the beneficial effects of the foliar spraying of chitosan on the wound-induced suberization of potato tubers, further research at the gene level of priming is needed.

5. Conclusions

In summary, foliar spraying of chitosan on potato plants activated PAL and increased cinnamic acid, sinapic acid, total phenolics and flavonoids in tuber wounds. Foliar spraying of chitosan activated 4CL, CAD and POD and increased SPP and lignin on wounded sites of potato tubers. The accelerated deposition of SPP and lignin domains of the periderm was associated with an enhanced wound periderm formation in the wounds of potato tubers, resulting in alleviated wound-facilitated fresh weight loss and dry rot disease index.

Based on our findings, we hypothesize that the reduced fresh weight loss and disease index of dry rot result from the accelerated ability of wound periderm formation. Pre-harvest application of chitosan elicited immune activity in potato tubers after post-harvest and activated PAL, 4CL, CAD and POD, increased phenolic compounds, resulting in intensified deposition of SPP and lignin, and accelerated wound periderm formation in the wounds of potato tubers. A possible model of foliar spraying of chitosan to accelerate wound periderm formation of potato tubers is illustrated in Figure 7. The foliar spraying of chitosan is a good strategy to improve the potato tuber quality. Our study provides an alternative and eco-friendly treatment for the preservation of commercial potato tuber qualities after harvest.

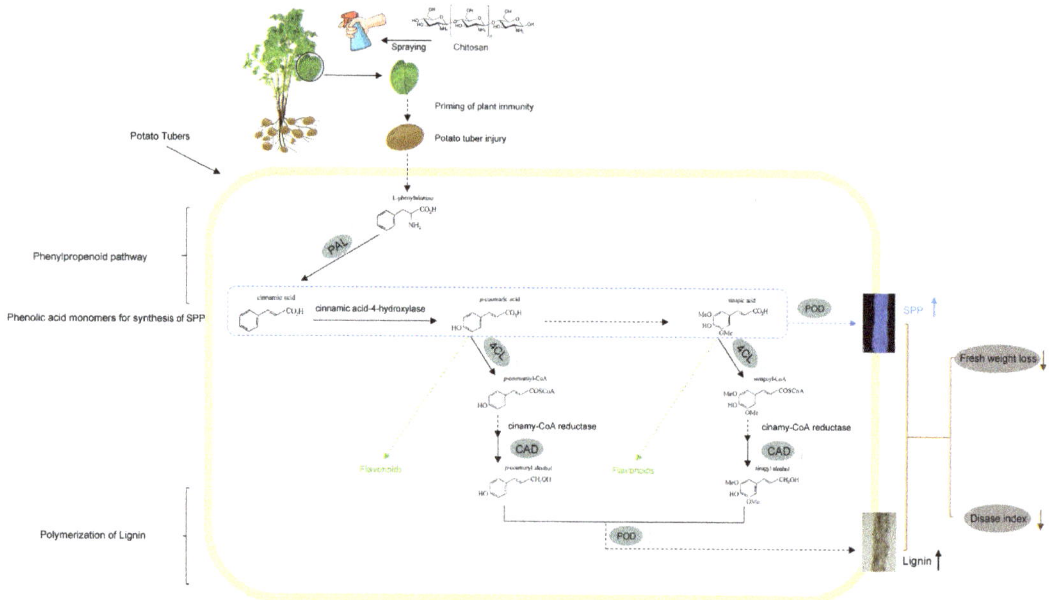

Figure 7. Foliar spraying of chitosan accelerated the deposition of suberin polyphenolic and lignin in the wounds of potato tubers by eliciting phenylpropanoid metabolism via the polymerization of SPP and lignin. Reactions denoted by solid lines are known, whereas those denoted by dashed lines are hypothetical metabolic steps catalyzed by multiple enzymes. Black arrows represent the flow of material that synthesizes lignin, blue dashed lines in the box and blue arrows represent the flow of material that synthesizes SPP, and green arrows represent the synthesis of flavonoids, respectively. The black brackets and words on the left side of the model diagram show different metabolic pathways, and the brown brackets and ellipses on the right side of the model diagram show the variation in fresh weight loss and disease index of potato tubers after wounding. PAL, phenylalanine ammonialyase; 4CL, 4-coumaric acid coenzyme A ligase; CAD, cinnamoyl alcohol dehydrogenase; POD, peroxidase; SPP, suberin polyphenolic.

Author Contributions: Conceptualization, Y.Z. (Yan Zhu) and Y.Z. (Yuanshou Zhao); methodology, W.Z., F.Z. and J.Y.; software, W.S. and W.G.; validation, X.G. and R.C.; formal analysis, F.C. and Y.Z. (Yuanshou Zhao); investigation, X.G. and W.G.; resources, F.C.; data curation, W.S. and J.Y.; writing—original draft preparation, Y.Z. (Yan Zhu) and W.Z.; writing—review and editing, Y.Z. (Yan Zhu) and W.S.; visualization, R.C. and F.Z.; project administration, Y.Z. (Yan Zhu); funding acquisition, Y.Z. (Yan Zhu) and F.C. All authors have read and agreed to the published version of the manuscript.

Funding: This research was funded by the Young Doctor Fund Project of the Gansu Provincial Education Department grant number 2022QB-077, the National Natural Science Foundation of China, grant number 31760473, and the Excellent Postgraduate Innovation Star Project of Gansu Province grant number 2023CXZX-687.

Data Availability Statement: No new data were created or analyzed in this study. Data sharing does not apply to this article.

Conflicts of Interest: The authors declare that they have no known competing financial interests or personal relationships that could have influenced the work reported in this paper.

References

1. Sun, J.; Wang, X.; Huang, Z.; Sun, H.; Cheng, Q.; Zhu, M. Classification and integration of storage and transportation engineering technologies in potato producing areas of China. *J. Agric. Sci. Technol.* **2017**, *18*, 710–718.
2. Campilho, A.; Nieminen, K.; Ragni, L. The development of the periderm: The final frontier between a plant and its environment. *Curr. Opin. Plant Biol.* **2020**, *53*, 10–14. [CrossRef] [PubMed]
3. Singh, B.; Bhardwaj, V.; Goutam, U. Potato periderm is the first layer of defence against biotic and abiotic stresses: A Review. *Potato Res.* **2021**, *64*, 131–146. [CrossRef]
4. Du, M.R.; Ren, X.Y.; Sun, Q.H.; Wang, Y.; Zhang, R.F. Characterization of *Fusarium* spp. causing potato dry rot in China and susceptibility evaluation of Chinese potato germplasm to the pathogen. *Potato Res.* **2012**, *55*, 175–184. [CrossRef]
5. Lulai, E.C.; Neubauer, J.D. Wound-induced suberization genes are differentially expressed, spatially and temporally, during closing layer and wound periderm formation. *Postharvest Biol. Technol.* **2014**, *90*, 24–33. [CrossRef]
6. Zheng, X.Y.; Jiang, H.; Bi, Y.; Wang, B.; Wang, T.L.; Li, Y.C.; Gong, D.; Wei, Y.N.; Li, Z.C.; Prusky, D. Comparison of wound healing abilities of four major cultivars of potato tubers in China. *Postharvest Biol. Technol.* **2020**, *164*, 111167. [CrossRef]
7. Al-Mokadem, A.Z.; Alnaggar, A.E.-A.M.; Mancy, A.G.; Sofy, A.R.; Sofy, M.R.; Mohamed, A.K.S.H.; Abou Ghazala, M.M.A.; El-Zabalawy, K.M.; Salem, N.F.G.; Elnosary, M.E.; et al. Foliar application of chitosan and phosphorus alleviate the *potato virus* Y-induced resistance by modulation of the reactive oxygen species, antioxidant defense system activity and gene expression in potato. *Agronomy* **2022**, *12*, 3064. [CrossRef]
8. Duan, C.; Meng, X.; Meng, J. Chitosan as a preservative for fruits and vegetables: A review on chemistry and antimicrobial properties. *J. Bioresour. Bioprod.* **2019**, *4*, 11–21. [CrossRef]
9. Miranda-Castro, S.P. Application of chitosan in fresh and minimally processed fruits and vegetables. In *Chitosan in the Preservation of Agricultural Commonities*; Bautista-Baños, S., Romanazzi, G., Jiménez-Aparicio, A., Eds.; Academic Press: Cambridge, MA, USA, 2016; pp. 67–113.
10. Romanazzi, G.; Sanzani, S.M.; Bi, Y.; Tian, S.; Martinez, P.G.; Alkan, N. Induced resistance to control postharvest decay of fruit and vegetables. *Postharvest Biol. Technol.* **2016**, *122*, 82–94. [CrossRef]
11. Niu, X.D.; Zhu, L.; Xi, L.J.; Guo, L.; Wang, H.S. An antimicrobial agent prepared by N-succinyl chitosan immobilized lysozyme and its application in strawberry preservation. *Food Control.* **2020**, *108*, 106829. [CrossRef]
12. Romanazzi, G.; Moumni, M. Chitosan and other edible coatings to extend shelf life, manage postharvest decay, and reduce loss and waste of fresh fruits and vegetables. *Curr. Opin. Biotechnol.* **2022**, *78*, 102834. [CrossRef] [PubMed]
13. Deshaies, M.; Lamari, N.; Ng, C.K.Y.; Ward, P.; Doohan, F.M. The impact of chitosan on the early metabolomic response of wheat to infection by *Fusarium graminearum*. *BMC Plant Biol.* **2022**, *22*, 73. [CrossRef]
14. Hassan, E.O.; Shoala, T.; Attia, A.M.F.; Badr, O.A.M.; Mahmoud, S.Y.M.; Farrag, E.S.H.; EL-Fiki, I.A.I. Chitosan and nano-chitosan for management of *Harpophora maydis*: Approaches for investigating antifungal activity, pathogenicity, maize-resistant lines, and molecular diagnosis of plant infection. *J. Fungi* **2022**, *8*, 509. [CrossRef]
15. Nia, A.E.; Taghipour, S.; Siahmansour, S. Pre-harvest application of chitosan and postharvest *Aloe vera gel* coating enhances quality of table grape (*Vitis vinifera* L. cv. 'Yaghouti') during postharvest period. *Food Chem.* **2021**, *347*, e129012.
16. Li, Z.C.; Xue, S.L.; Xu, X.Q.; Wang, B.; Zheng, X.Y.; Li, B.J.; Xie, P.D.; Bi, Y.; Prusky, D. Preharvest multiple sprays with chitosan accelerate the deposition of suberin poly phenolic at wound sites of harvested muskmelons. *Postharvest Biol. Technol.* **2021**, *179*, 111565. [CrossRef]
17. Lesney, M.S. Growth responses and lignin production in cell suspensions of *Pinuse elliottii* 'elicited' by chitin, chitosan or mycelium of *Cronartium quercum* f.sp. *fusiforme*. *Plant Cell Tissue Org.* **1989**, *19*, 23–31. [CrossRef]
18. Notsu, S.; Saito, N.; Kosaki, H.; Inui, H.; Hirano, S. Stimulation of phenylalanine ammonia-lyase activity and lignification in rice callus treated with chitin, chitosan, and their derivatives. *Biosci. Biotechnol. Biochem.* **1994**, *58*, 552–553. [CrossRef]
19. Todd, O.A.; Peter, B.M. *Candida albicans* and *Staphylococcus aureus* pathogenicity and polymicrobial interactions: Lessons beyond Koch's Postulates. *J. Fungi* **2019**, *5*, 81. [CrossRef]
20. Ge, X.Z.; Zhu, Y.; Li, Z.C.; Bi, Y.; Yang, J.; Zhang, J.L.; Prusky, D. Preharvest multiple fungicide stroby sprays promote wound healing of harvested potato tubers by activating phenylpropanoid metabolism. *Postharvest Biol. Technol.* **2021**, *171*, 111328. [CrossRef]
21. Zhu, Y.T.; Zong, Y.Y.; Liang, W.; Kong, R.; Gong, D.; Han, Y.; Li, Y.C.; Bi, Y.; Prusky, D. Sorbitol immersion accelerates the deposition of suberin polyphenolic and lignin at wounds of potato tubers by activating phenylpropanoid metabolism. *Sci. Hortic.* **2022**, *297*, 110971. [CrossRef]
22. Bradford, M.M. A rapid and sensitive method for the quantitation of microgram quantities of protein utilizing the principle of protein-dye binding. *Anal. Biochem.* **1976**, *72*, 248–254. [CrossRef] [PubMed]
23. Gruz, J.; Novak, O.; Strnad, M. Rapid analysis of phenolic acids in beverages by UPLC-MS/MS. *Food Chem.* **2008**, *111*, 789–794. [CrossRef]
24. Cui, K.B.; Shu, C.; Zhao, H.D.; Fan, X.G.; Cao, J.K.; Jiang, W.B. Preharvest chitosan oligochitosan and salicylic acid treatments enhance phenol metabolism and maintain the postharvest quality of apricots (*Prunus armeniaca* L.). *Sci. Hortic.* **2020**, *267*, 109334. [CrossRef]

25. Benkeblia, N.; Tennant, D.P.F.; Jawandha, S.K.; Gill, P.S. Preharvest and harvest factors influencing the postharvest quality of tropical and subtropical fruits. In *Postharvest Biology and Technology of Tropical and Subtropical Fruits*; Yahia, E.M., Ed.; Woodhead Publishing Limited: Cambridge, UK, 2011; pp. 112–141.
26. Wójcik, P. Effects of preharvest sprays of iodine, selenium and calcium on apple biofortification and their quality and storability. *PLoS ONE* **2023**, *18*, e0282873. [CrossRef] [PubMed]
27. Zheng, J.R.; Chen, H.; Wang, T.L.; Mustafa, G.; Liu, L.H.; Wang, Q.M.; Shao, Z.Y. Quality improvement of tomato fruits by preharvest application of chitosan oligosaccharide. *Horticulturae* **2023**, *9*, 300. [CrossRef]
28. Fu, E.; Zhang, Y.; Li, H.; Wang, X.; Zhang, H.; Xiao, W.; Chen, X.; Li, L. Chitosan reduces damages of strawberry seedlings under high-temperature and high-light stress. *Agronomy* **2023**, *13*, 517. [CrossRef]
29. Han, Y.; Yang, R.R.; Zhang, X.J.; Wang, Q.H.; Wang, B.; Zheng, X.Y.; Li, Y.C.; Prusky, D. Brassinosteroid accelerates wound healing of potato tubers by activation of reactive oxygen metabolism and phenylpropanoid metabolism. *Foods* **2022**, *11*, 906. [CrossRef]
30. Ge, Y.H.; Chen, Y.R.; Li, C.Y.; Zhao, J.R.; Wei, M.L.; Li, X.H.; Yang, S.Q.; Mi, Y.T. Effect of sodium nitroprusside treatment on shikimate and phenylpropanoid pathways of apple fruit. *Food Chem.* **2019**, *290*, 263–269. [CrossRef]
31. Han, Y.; Yang, R.R.; Wang, Q.H.; Wang, B.; Prusky, D. Sodium silicate promotes wound healing by inducing the deposition of suberin polyphenolic and lignin in potato tubers. *Front. Plant Sci.* **2022**, *13*, 942022. [CrossRef]
32. Wang, B.; Bi, Y. The role of signal production and transduction in induced resistance of harvested fruits and vegetables. *Food Qual. Saf.* **2021**, *5*, 1–8. [CrossRef]
33. Lulai, E.C.; Campbell, L.G.; Fugate, K.K.; Mccue, K.F. Biological differences that distinguish the 2 major stages of wound healing in potato tubers. *Plant Signal. Behav.* **2016**, *11*, 1256531. [CrossRef]
34. Dong, N.Q.; Lin, H.X. Contribution of phenylpropanoid metabolism to plant development and plant-environment interactions. *J. Integr. Plant Biol.* **2021**, *63*, 180–209. [CrossRef] [PubMed]
35. Vogt, T. Phenylpropanoid biosynthesis. *Mol. Plant* **2010**, *3*, 2–20. [CrossRef] [PubMed]
36. Bernards, M.A.; Razem, F.A. The poly (phenolic) domain of potato suberin: A non-lignin cell wall bio-polymer. *Phytochemistry* **2000**, *57*, 1115–1122. [CrossRef] [PubMed]
37. Zhao, Y.J.; Deng, L.L.; Zhou, Y.H.; Ming, J.; Yao, S.X.; Zeng, K.F. Wound healing in citrus fruit is promoted by chitosan and *Pichia membranaefaciens* as a resistance mechanism against *Colletotrichum gloeosporioides*. *Postharvest Biol. Technol.* **2018**, *145*, 134–143. [CrossRef]
38. Gao, Y.; Kan, C.N.; Wan, C.P.; Chen, C.Y.; Chen, M.; Chen, J.Y. Quality and biochemical changes of navel orange fruits during storage as affected by cinnamaldehyde-chitosan coating. *Sci. Hortic.* **2018**, *239*, 80–86. [CrossRef]
39. Kou, X.H.; Guo, W.L.; Guo, R.Z.; Li, Z.H. Effects of chitosan, calcium chloride, and pullulan coating treatments on antioxidant activity in pear cv."Huang guan" during storage. *Food Bioprocess Technol.* **2014**, *7*, 671–681. [CrossRef]
40. Hu, B.; Lin, L.; Fang, Y.J.; Zhou, M.; Zhou, X.Y. Application of chitosan-lignosulfonate composite coating film in grape preservation and study on the difference in metabolites in fruit wine. *Coatings* **2022**, *12*, 494. [CrossRef]
41. Yang, H.Q.; Zheng, J.Y.; Huang, C.H.; Zhao, X.F.; Chen, H.Y.; Sun, Z.D. Effects of combined aqueous chlorine dioxide and chitosan coatings on microbial growth and quality maintenance of fresh-cut bamboo shoots *(Phyllostachys praecox* f. *prevernalis.*) during storage. *Food Bioprocess Technol.* **2015**, *8*, 1011–1019. [CrossRef]
42. Hahlbrock, K.; Grisebach, H. Enzymic controls in the biosynthesis of lignin and flavonoids. *Ann. Rev. Plant Physiol.* **1976**, *30*, 105–130. [CrossRef]
43. Cascaes, M.M.; Guilhon, G.M.S.P.; Zoghbi, M.d.G.; Andrade, E.H.A.; Santos, L.S.; Silva, J.K.R.; Uetanabaro, A.P.T.; Araújo, I.S. Flavonoids, antioxidant potential and antimicrobial activity of *Myrcia rufipila* mcvaugh leaves (myrtaceae). *Nat. Prod. Res.* **2021**, *35*, 1717–1721. [CrossRef] [PubMed]
44. Górniak, I.; Bartoszewski, R.; Króliczewski, J. Comprehensive review of antimicrobial activities of plant flavonoids. *Phytochem. Rev.* **2019**, *18*, 241–272. [CrossRef]

horticulturae

Review

Microbiological Activity Affects Post-Harvest Quality of Cocoa (*Theobroma cacao* L.) Beans

Edy Subroto *, Mohamad Djali, Rossi Indiarto, Elazmanawati Lembong and Nur Baiti

Department of Food Industrial Technology, Faculty of Agro-Industrial Technology, Universitas Padjadjaran, Bandung 45363, Indonesia; djali@unpad.ac.id (M.D.); rossi.indiarto@unpad.ac.id (R.I.); elazmanawati.lembong@unpad.ac.id (E.L.); nur17003@mail.unpad.ac.id (N.B.)
* Correspondence: edy.subroto@unpad.ac.id

Abstract: Cocoa beans are the basic ingredient to produce chocolate and its derivatives, including cosmetics, foods, and pharmaceutical products. The quality of cocoa beans is greatly affected by post-harvest handling, especially by microbial activity involved in pre-conditioning after they are harvested, including fermentation, drying, and storage. This review aims to provide various factors that affect each stage of post-harvest cocoa beans, process mechanisms, and various latest technologies that can be used to improve the quality of cocoa beans. Microorganisms could be involved in each post-harvest stage and affect the cocoa beans' quality. However, fermentation was one of the keys to determining the quality of cocoa beans because fermentation involved various microorganisms, such as yeast, lactic acid bacteria, and acetic acid bacteria, which were interrelated primarily to produce precursor flavor compounds. The drying and storage processes were decisive in maintaining quality, especially in preventing mold growth and other microbial contaminants. Various technologies could improve the quality of cocoa beans during post harvest, especially by adding microbial starters during fermentation. Using several technologies of vacuum drying and a controlled atmosphere during storage could maintain the quality of the cocoa beans. However, many challenges must be faced, especially those related to controlling microbial activity during post-harvest. Therefore, post-harvest technology needs to be continuously developed, especially in controlling microbiological activities to improve the quality of cocoa beans effectively.

Keywords: cocoa beans; microorganism; post harvest; quality; fermentation

Citation: Subroto, E.; Djali, M.; Indiarto, R.; Lembong, E.; Baiti, N. Microbiological Activity Affects Post-Harvest Quality of Cocoa (*Theobroma cacao* L.) Beans. *Horticulturae* **2023**, *9*, 805. https://doi.org/10.3390/horticulturae9070805

Academic Editors: Yang Bi, Yongcai Li and Di Gong

Received: 5 May 2023
Revised: 10 July 2023
Accepted: 12 July 2023
Published: 13 July 2023

1. Introduction

The quality of cocoa beans depends on several factors, including the variety or type of cocoa, the growing environment, and the post-harvest handling of the cocoa beans [1–3]. Proper post-harvest handling can improve quality and prevent product loss. Post-harvest of cocoa beans includes sorting, ripening the fruit, breaking/splitting the fruit, fermenting, drying, and storage [4–6]. Each post-harvest stage can involve both beneficial and detrimental microbial activity. Therefore, controlling the factors and microbial activity during post-harvest handling is necessary. The important factor that must be considered, especially in fermenting cocoa beans. The method or conditions of the fermentation process that is carried out determines the final quality, especially in the formation of flavor precursors in cocoa beans. Fermentation can also reduce bitterness, change the color of the beans to black-brown, and change the texture of the beans hardens like a shell [7–10].

The low quality can be caused by the cocoa beans not being fermented properly or intentionally not fermented. Some farmers only soak fresh cocoa beans using water, which aims to remove the pulp, and then is immediately carried out the drying process without any fermentation process, even though fermentation is an important aspect of producing good quality cocoa beans [7,11,12]. The fermentation process involves the activity of many microorganisms. However, fermentation involves beneficial and harmful microorganisms [13,14]. The role of beneficial microorganisms is to assist in the fermentation process by breaking down carbohydrates into simple compounds in the form of flavor precursors. In contrast, harmful microorganisms contaminate and can make cocoa beans rot or spoil [15,16].

The fermentation of cocoa beans occurs naturally with the help of microbes and lasts 6 days [17,18]. Microorganisms in fermentation will produce ethanol, acetic acid, and lactic acid compounds. Acetic acid and alcohol diffuse into the interior of the cocoa bean, followed by an increase in temperature, causing the seeds to die [18–20]. The presence of *Lactobacillus lactis* and *Acetobacter aceti* can shorten fermentation time [21,22]. Microorganisms that have a role in the early stages of fermentation are dominated by yeast, then followed by the growth of lactic acid bacteria, and then ended by the growth of acetic acid bacteria [7,23–25]. The mold growth during the fermentation needs to be avoided because the mold can produce mycotoxins and cause cocoa to have a bitter taste [21,26].

There are two methods of post-harvest processing of cocoa beans: dry cocoa beans with fermentation and dry cocoa beans without fermentation [27]. Non-fermented dry cocoa beans produce low-quality products. The quality of the non-fermented dry cocoa beans can be improved by the dry bean fermentation method. Still, the fermentation that is carried out must be optimized and conditioned as well as possible so that it can run well [12,28]. Various technologies have been developed to improve the quality of cocoa beans, especially by utilizing and controlling the role of microorganisms during post harvest.

Therefore, this review discusses in more depth the activities of microorganisms involved in the post-harvest handling process of cocoa beans. This review also discusses the effects of the activities of these microorganisms on the quality, as well as efforts to improve the quality of cocoa beans, especially by adding starter culture in the fermentation process and maintaining the quality of cocoa beans by applying drying and storage technology.

2. Microbiological Activity in Post-Harvest Handling of Cocoa Beans

Cocoa beans were essentially sterile while still inside the fruit, but the pulp often became contaminated by various microorganisms after being exposed. Acetic acid bacteria, yeast, lactic acid bacteria, as well as molds were commonly found in commercial cocoa beans [6]. Furthermore, unwanted contamination could occur during the processing and transportation of these beans [26,29]. Therefore, each stage in the post-harvest of cocoa beans requires process control, both microbiological and environmental controls. The stages of post harvest of cocoa beans and the factors that need to be considered schematically can be seen in Figure 1.

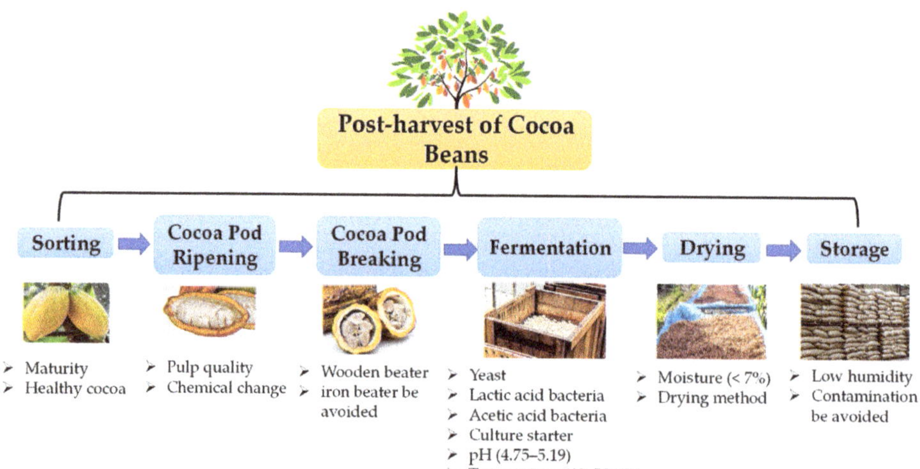

Figure 1. The stages of post harvest of cocoa beans and the factors that need to be considered.

Post-harvest handling is a crucial problem that must be addressed, as it greatly affected the final yield of cocoa beans produced. Therefore, a chain strategy for post-harvest handling was needed, where the potential for the emergence of *Salmonella* and mycotoxins in cocoa derivative products was the major focus, as followed in the HACCP, as well as in microbiological quality guidelines issued [29,30]. According to Lima et al. [6], bacteria from the genus *Bacillus* were the main microorganisms found in this plant. Some of these *Bacillus* species could form spores that were highly resistant to heat, thereby surviving the roasting process [31,32].

Based on the conditions required for the cocoa beans fermentation process, microbial ecological studies focused on the constituent mycotoxins and the derivative products [33]. Other studies had also focused on the viability of *Salmonella* during the handling of cocoa beans and derivative products [34,35]. These studies were not only relevant in terms of quality and safety but also provided an opportunity to identify microorganisms with new physiological characteristics [16,36].

Potentially, pathogenic microorganisms could also grow on this plant, such as *Aspergillus*, *Penicillium*, and *Fusarium*. These fungi often posed a risk to public health due to their ability to produce toxins [37]. Furthermore, their existence had the potential to cause mycotoxin contamination, which could endanger human health. The types of mycotoxins often found in cocoa beans included aflatoxins and ochratoxins, which were produced from *Aspergillus* and *Penicillium* species [34,38].

Cocoa could be the origin of chocolate contamination by bacteria, such as *Salmonella* due to poor hygiene practices during the processing [39]. However, mycotoxin-producing molds were a prominent concern in post-harvest handling because they determine the quality of the product cocoa beans. The optimal temperature of *Aspergillus* sp. for the formation of aflatoxin was 5 °C–45 °C with 80% minimum humidity and 5.5–7.0 pH [40]. Several types of mold that commonly contaminate cocoa included *Aspergillus flavus*, *Aspergillus niger*, and *Penicillium* [33,34,38].

Contamination from mold could cause weathering, reduced nutrition, and the presence of mycotoxins, thereby causing health problems [33]. The growth of these fungi was due to hydration in foodstuffs, problems during transportation, and poor storage [41]. The existence of toxins contained in cocoa beans was very detrimental because it led to a decrease in quality and could interfere with the health of consumers [42,43].

Therefore, several factors that influenced microbes in the post harvest of cocoa beans must be considered, including (1) cocoa genotype, which will affect the type and amount of polyphenols, carbohydrates, and proteins that were used as substrates by microbes during post harvest; (2) pod storage, which reduced pulp volume and increased micro-aeration which then reduced the formation of acetic acid and alcohol during fermentation; (3) de-pulping which reduced pulp sugar content which then reduced acid production; (4) fermentation conditions, which included cleanliness of equipment, environment, the addition of starter, reversal of cocoa beans, and length of fermentation; (5) drying, which included weather, technique, and drying equipment; and (6) microbial contamination, especially pathogenic bacteria and fungi, which produced harmful toxins.

2.1. Sorting

Harvested cocoa pods must be sorted to select good quality pods because the initial condition or quality determined the process conditions at later stages. Sorting determined the uniformity of the process in the next stage, namely, ripening. Furthermore, excessively ripe, damaged, or diseased cocoa pods could not be ripened because it often caused rotting and contamination [44,45]. The sorting process was often performed to separate healthy cocoa pods from the damaged variant [46].

This process must also be carried out to determine the quality of the product by separating cocoa beans from the impurities, followed by grouping based on quality and physical appearance [47]. The aspects that need to be considered during sorting include clipped, germinated, raw, flat, unfermented, purple, broken, dirt, insect, and moldy beans [48].

The moldy type was caused by wet cocoa beans that germinated and became dry, but the product overgrew, leading to the appearance of mold [49].

The criteria often considered during the sorting stage included size, color, shape, and health. The size of cocoa beans could be divided into 3 classes: large, medium, and small. Meanwhile, the color was typically distinguished by brown, purple, and black. The shape of cocoa beans could either be oval, round, or flat. Healthy beans were also separated from unhealthy, defective, or imperfect types [50,51].

2.2. *Cocoa Pod Ripening*

The ripening of cocoa pods typically involved storage for a certain period before breaking. Furthermore, this process had a significant effect on the product in terms of the chemical content and was useful for developing the flavor [52,53]. Ripening aimed to facilitate the formation of empty spaces in cocoa heaps, thereby aiding the penetration of oxygen during fermentation. The process also helped in reducing the pulp layer, speeding up fermentation, as well as decreasing water content, acidity, and polyphenols levels. Cocoa pods with poor quality, such as overripe, damaged, and diseased types could not be ripened due to rotting and contamination [54].

Changes in the properties of the pulp could occur during the ripening process of cocoa pods, including a reduction of the volume of pulp per bean due to water evaporation and conversion of sucrose, a decrease in the total sugar content, an increase in micro-aeration in the pulp, and a reduction of the production of alcohol and acetic acid [2]. The presence of excess seed pulp could affect the fermentation process, leading to a sour taste. Based on previous studies, products with low pulp volume and water content often experienced faster fermentation compared to others [55].

The process of ripening cocoa pods could be carried out In an open room or directly in the garden to prevent the proliferation of mold [48]. Hamdouche et al. [56] reported that several microorganisms were found in products aged 8 days, including *Acinetobacter* sp., *Klebsiella pneumoniae*, and *Bacillus* spp. These bacteria commonly proliferate on the surface during the ripening stage. *Acinetobacter* sp. caused nosocomial infections, which were widely spread in health facilities and could be transmitted through contaminated objects, touch, and saliva. These infections often occurred in the intestine and urinary tract, with several symptoms, such as diarrhea, fever, abdominal cramps, and lower abdominal pain. Several studies had shown that *Acinetobacter* sp. could grow on almost all types of surfaces, including cocoa pods [57,58]. *Klebsiella pneumoniae* was bacteria known for its ability to produce enzymes, leading to resistance to antibiotics. *Klebsiella pneumoniae* caused nosocomial infections, such as *Acinetobacter* sp. [58]. According to a previous study, it also caused pneumonia, which began with symptoms of fever, dry cough, and malaise [59]. Meanwhile, *Bacillus* spp. was a plant-grown promotion *rhizobacteria (PGPR)* bacteria, which could increase growth and production in plants. This bacteria had been reported to have the possibility of causing endocarditis and endophthalmitis [60]. The aforementioned bacteria had been used as biocontrol agents, phosphate solvents, nitrogen fixatives, and phytopathogens [61,62].

The time required for the ripening process of cocoa pods could significantly affect the organic matter and moisture content. The longer the time used for ripening, the greater the percentage of organic matter and the lower the water content [54].

2.3. *Cocoa Pods Breaking*

Fruit breaking referred to the process of removing and separating cocoa beans from shells. The breaking was often carried out using tools, such as sickles and paddles, by hitting the pods with each other or with a de-podding machine [63,64]. Furthermore, the cracking process was rarely performed mechanically because the tools used were usually not developed into commercial tools, and it was difficult to separate the fresh cocoa beans from the broken shells [65].

The best breaking method involved the use of a wooden beater. Cracking cocoa pods using an iron beater must be avoided because it could lead to black coloration, iron smell, as well as reduced aroma and taste. Meanwhile, the use of iron cutting tools could cause an oxidation reaction of phenolic compounds in cocoa beans, and the cut scars were easily overgrown by fungi [65]. The existence of mold on cocoa beans was not allowed because it reduced the quality of the product [14,49].

2.4. Fermentation

Fermentation referred to activity carried out by microorganisms either aerobically or anaerobically, which could cause changes in complex compounds into simpler types. Furthermore, the success of fermentation depended on the presence of microorganisms, and each of these microbes had different living conditions, such as pH, temperature, humidity, substrate, oxygen content, and others [16,66]. This process involved the decomposition of sugar components and citric acid compounds present in the pulp into organic acids by microbes [67]. The microbial activity present during this process facilitated the formation of flavor precursor compounds by changing the carbohydrate substrates into acetic acid, lactic acid, and ethanol [68]. Fermentation of cocoa beans was generally carried out for 5–6 days, and the first inversion treatment was conducted on the 2nd day or after 48 h, with further inversion being performed once every 24 h [69]. Fermentation occurred naturally without the addition of a starter culture due to the presence of glucose, sucrose, citric acid, and fructose in the pulp, leading to the growth of microorganisms [70,71].

The pulp of fresh cocoa beans contained water (80–90%), sugar (10–15%), and protein (0.5–0.7%). This indicated that it was an important source of macronutrients for microbial growth [7]. Furthermore, the water activity (Aw) of the pulp ranged from 0.98 to 0.99 and these conditions supported the spontaneous growth of microorganisms to carry out fermentation, such as lactic acid bacteria (LAB) and acetic acid bacteria (AAB) [72]. The pulp was very suitable for the growth of microorganisms, and during the fermentation process, the activity of microbes led to the production of organic acids and alcohol, as well as the release of heat (exothermic reaction) [73]. The exothermal reaction caused the diffusion of metabolites into the beans, leading to their death. This was followed by enzymatic reactions to form flavors, colors, and aromas that determined the quality of cocoa beans [74].

Cocoa bean fermentation was assisted by several types of microorganisms, namely, yeast, lactic acid bacteria, and acetic acid bacteria [12,14]. Yeast was pioneering microorganisms in the process of fermenting cocoa beans. This microbe could facilitate the production of proto-pectinase enzymes, leading to the breakdown of pectin compounds into pectin acids and alcohols, and pectinase enzymes. Subsequently, the pectinase enzyme catalyzed the breakdown of pectin acids into arabinose, galactose, and acetic acid [7,14]. The pulp was then crushed and released due to the decomposition of pectin [25]. Yeast and lactic acid bacteria were microorganisms involved in the decomposition of citric acid during the fermentation of cocoa beans [14,75].

The main aim of fermentation was to remove the mucus and kill cocoa beans, leading to the occurrence of changes, such as the formation of flavor and color precursors [69,76]. Factors affecting this process included the time or length of the procedure, aeration, fruit ripeness, quantity, uniformity of the speed of turning/stirring, climate, and the container [77]. An extended duration could increase the number of germinating and moldy cocoa beans, while a short time led to the emergence of slaty samples (unfermented beans) [27].

The stages of fermenting cocoa beans involved the opening of the pods, followed by the removal of beans still covered with pulp, and collection in a container [69,77,78]. Fermentation of cocoa beans was carried out in aerobic and anaerobic [77,79]. The presence of citric acid compounds made the pulp environment acidic and initiated the growth of yeast, followed by the occurrence of the anaerobic process. Furthermore, aerobic fermentation was carried out by acetic acid bacteria and lactic acid bacteria. Fermentation also involved enzyme activity, including endoproteases, carboxypeptidases, aminopeptidases, glycosidases, polyphenol oxidases, and invertases [80]. The process produced good quality cocoa beans, which was indicated by the color of beans being changed from brown to slightly purple brown, slightly crumbly, bitter, and slightly astringent taste, and a strong chocolate odor [47].

The fermentation was often initiated by the growth of yeast colonies, followed by lactic acid bacteria, acetic acid bacteria, and mold [20,56,81]. However, bacterial succession in cocoa bean fermentation can be affected by the type of yeast and filamentous fungi present. The succession led to the growth of *Enterobacteriaceae*, LAB, and AAB, which can generate a wide range of genetic metabolic potentials related to carbohydrate and protein metabolism. Based on in silico evidence, the interspecific quorum sensing (QS) arsenal found the genera *Bacillus*, *Lactobacillus*, *Enterobacter*, and *Pantoea*, while the potential for intraspecific QS was found in the genera *Lactobacillus*, *Komagataeibacter*, *Enterobacter*, *Bacillus*, and *Pantoea*. On the other hand, quorum quenching potential (QQ) was detected in the *Lactobacillus* and AAB groups. These QS and QQ can modulate the dominance of bacteria during cocoa bean fermentation, which is also affected by cross-feeding, over a long period of time [82].

Microorganisms in the fermentation process caused several biochemical changes, including the hydrolysis of sucrose in the pulp into fructose and glucose during the first 24 h [7,19,83]. The growth of microbes on this sugar caused an increase in temperature. The pulp had a high sugar content of 10–15% sugar, thereby stimulating the growth of yeast, which could convert these molecules into alcoholic compounds under anaerobic conditions and hydrolyze pectin compounds covering cocoa beans. The process was then accompanied by the death of yeast due to the presence of these alcohol compounds, leading to a higher temperature change. Furthermore, *Streptococcus* and *Lactobacillus lactic* acid bacteria were able to grow, and then the pulp was stirred for aeration purposes. The presence of oxygen and low pH caused acetic acid bacteria (*Acetobacter* and *Gluconobacter*) to grow [13,84]. The main stages of cocoa bean fermentation involving yeast, lactic acid bacteria, and acetic acid bacteria are presented schematically in Figure 2.

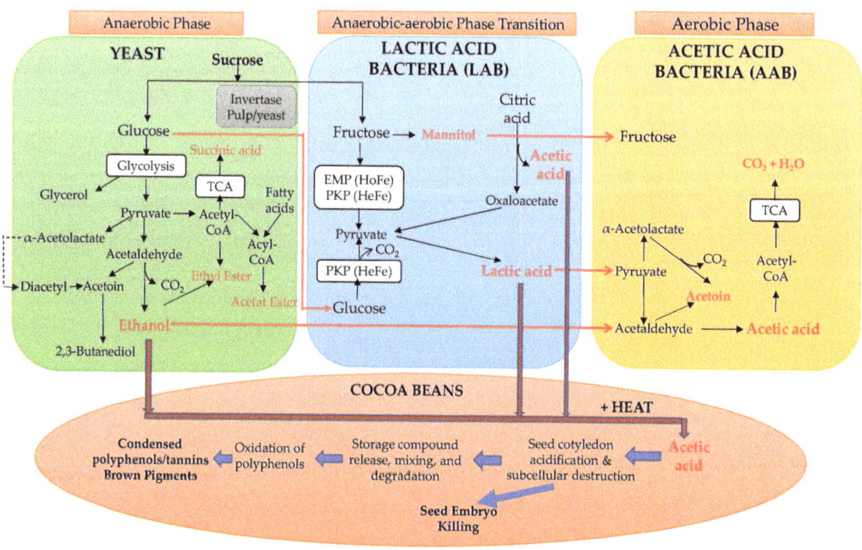

Figure 2. The main stages of cocoa beans' fermentation mechanism involve yeast in the anaerobic phase, lactic acid bacteria in the anaerobic–aerobic phase transition, and acetic acid bacteria in the aerobic phase.

During cocoa fermentation, there were approximately 100 million microbes per gram after 5–6 days. The process was then stopped and cocoa beans were dried because the continuation of fermentation could cause an unwanted odor due to the growth of molds (*Aspergillus*, *Mucor*, and *Penicillium*), which hydrolyzed lipids [27]. Furthermore, some yeasts that grew earlier played a role in degrading the pulp and converting sugars into alcohol compounds. The process was carried out exothermically, with heat being generated, leading to an increase in the temperature of the cocoa mass that was being fermented [19,85]. The types of yeast involved could differ in each region, as shown in Table 1.

Table 1. Several types of yeast are involved in fermentation process of cocoa beans in several countries.

Country	Yeast Species	Characteristics	References
Ecuador	*S. cerevisiae*, *R. minuta*, *P. manshurica*, *P. kudriavzevii*, *P. kluyveri*, *K. marxianus*, *H. opuntiae*, *C. tropicalis*, *C. sorbosivorans*-like, and *T. delbrueckii*	*P. manshurica*, *P. kudriavzevii*, and *S. cerevisiae* were the dominant ethanol producers.	[86]
Brazil	*P. kundriavzevii*, *C. orthopsilosis*, *K. ohmeri*, *D. etchellsii*, *I. orientalis*, *H. uvarum*, *P. kluyveri*, and *S. cerevisiae*	These species adapted well to both fermentation box and the stainless steel used in large-scale fermentation.	[87]
	P. Kluyver, *C. magnoliae*, and *S. cerevisiae*		[76]
Cuba	*T. delbruekii*, *P. terricola*, *C. ortopsilosis*, *P. occidentalis*, *C. tropicalis*, *P. kluyveri*, *P. kundriavzevii*, *H. opuntiae*, and *P. manshurica*	*Pichia kudriavzevii* was the most common. Some yeasts undergo mutations caused by natural processes, such as transposons genetic recombination, changes in ploidy, and sexual reproduction.	[67,88]
Dominican Republic	*C. zeylanoides*, *Y. lipolytica*, *H. guillermondii*, and *C. inconspicua*	*Candida inconspicua* was the most common and dominant because it could survive up to the 36th hour of fermentation.	[89]
Indonesia	*S. Cerevisiae*, *Kloeckera* sp., *S. fibuligera*, *C. tropicalis*, and *C. krusei*	Yeast found in Indonesia could generally live in tropical environments.	[90]
	C. tropicalis, *S. cerevisiae*, and *Kl. apis*	*Saccharomyces cerevisiae* and *Candida tropicalis* were resistant to high temperatures (>40 °C).	[25]
Ivory Coast	*G. geotrichum*, *W. anomalus*, *P. galeiforms*, *P. kudriavzevii*, *C. tropicalis*, *S. cerevisiae*, *P. kluyveri*, and *P. kundriavzevii*	*Pichia kudriavzevii*, *Pichia kluyveri*, and *Saccharomyces cerevisiae* were the most common and had intraspecific diversity.	[91]
	P. fermentans, *P. klyvera*, *Candida* sp., *C. insectorum*, *P. kudriavezii*, *I. hanoiensis*, *P. sporocuriosa*, *P. manshurica*, and *H. opuntiae*		[92]
	P. kudriavezii, *I. hanoiensis*, *P. sporocuriosa*, *P. manshurica*, and *H. opuntiae*		[56]
Ghana	*H. guilliermondii*, *P. membranifaciens*, *Sc. cerevisiae*, *S. crataegensis*, *P. Pijperi*, *I. Hanoiensis*, *C. zemplinina*, *C. michaelii*, *C. diversa*, *C. ethanolica*, *Schiz. pombe*, and *I. orientalis*	*H. guilliermondii* was the most common species at the beginning of fermentation (0–24 h), while *P. membranifaciens* was the dominant species at the end of fermentation (36–144 h).	[70]

Table 1. *Cont.*

Country	Yeast Species	Characteristics	References
	P. manshurica M.(P.) carribica, K. ohmeri, C. orthopsilosis, C. carpophila, H. opuntiae, S. cerevisiae, P. kundriavzevii	*S. cerevisiae* and *P. kundriavzevii* were the most common species. *Hanseniaspora opuntiae* was able to live at a fairly low pH.	[93]
	Saccharomyces cerevisiae, Kluyveromyces lactis, Candida glabrata,	*S. cerevisiae* and *K. lactis* were the most common species.	[13]
Mexico	*H. guilliermondii, S. crataegensis, S. cerevisiae*, and *P. kundriavzevii*	*S. cerevisiae* was the most important species because it had the best survival.	[94]

The fermentation process was followed by the growth of several types of bacteria. Furthermore, bacteria belonging to the Beta genus, which were heterofermentative and involved in cocoa bean fermentation, were capable of producing acetic acid and lactic acid [95,96]. Some acetic acid bacteria also played a role in the oxidation of alcohol compounds to acetic acid. The microbes involved in the fermentation of cocoa beans included lactic acid bacteria and acetic acid bacteria [97]. The types of acetic acid bacteria and lactic acid bacteria implicated in fermentation were different in each region [81,97,98]. The diversity of these microorganisms was caused by different factors in each region, such as oxygen levels, temperature, and fermentation techniques used [20,96]. The various species present in each country are presented in Table 2.

Table 2. Several types of bacteria species are involved in cocoa bean fermentation in several countries.

Country	Bacterial Species	Characteristics	References
Ecuador	*Lb. fermentum, A. pasteurianus, Leu. pseudomesenteroides, Lb. plantarum, A. fabarum, F. tropaeoli*-like, *Lb fabifermentans, Lac. lactis, Lb. nagelii, Lb. cacaonum, E. casseliflavus, A. peroxydans, A. cibinongensis,* and *A. malorum/indonesiensis*	*Lb. fermentum, A. pasteurianus,* and *Leu. Pseudomesenteroides* were the most commonly found in fermented cocoa beans.	[86]
Indonesia	*B. licheniformis, B. pumilus, A. pasteurianus L. plantarum,* and *L. cellobiosus*	*L. plantarum* was the most consistent bacteria, while lactic acid bacteria had a dominant role in the microbial ecology of cocoa beans fermentation.	[25]
Ghana	*Ent. faecium, Ent. casseliflavus, W. ghanensis, Leuc. Mesenteroides, Leuc. pseudomesenteroides, L. mali, L. brevis, L. plantarum,* and *L. Fermentum*	*L. plantarum* was the most commonly found in fermented cocoa beans. *Weisella ghanensis* was known as the first-line divergent in the genus *Weisella.*	[99]
	Lactiplantibacillus plantarum, Lactobacillus nagelii, Liquorilactobacillus cacaonum, Limosilactobacillus fermentum, and *Leuconostoc pseudomesenteroides*		[100]
	Lb. plantarum, A. pasteurianus, Leu. mesenteroides, G. oxydans, G. diazotrophicus, G. hansenii, Leu. citreum, Lb. fermentum, Lb. brevis, and *E. coli*		[13]

Table 2. *Cont.*

Country	Bacterial Species	Characteristics	References
	Lb. Plantarum, *Pd. acidilactici*, *Lb. hilgardii*, *Lc. pseudoficulneum*, *Lb. fermentum*, *G. oxydans*, *A. malorum*, *A. tropicalis*, *A. syzygii*, and *A. pasteurianus*		[70]
	A. tropicalis-like, *A. tropicalis*, *A. syzygii*-like, *A. senegalensis*, *A. senegalensis*, and *A. pasteurianus*		[21]
Dominican Republic	*L. paracasei* subsp. *paracasei*, *L. brevis*, *L. pentosus*, and *L. plantarum*	*L. plantarum* was most commonly found in fermented cocoa beans.	[89]
Ivory Coast	*W. cibaria*, *W. paramesenteroide*, *L. casei*, *F. pseudoficulneus*, *Ent. faecium*, *L. curieae*, *Leuc. mesenteroides*, and *L. plantarum*,	*Leuconostoc mesenteroides* was most commonly found in fermented cocoa beans. *Leuconostoc mesenteroides* could catabolize citrate more efficiently.	[101]
	A. malorum, *A. ghanensis*, *A. okinawensis*, *A. tropicalis*, *A. pasteurianus*, and *G. oxydans*	*A. pasteurianus*, *A. okinawensis*, and *A. tropicalis* were the most commonly found acetic acid bacteria in fermented cocoa beans.	[96]
	Lactobacilli sp., *Lactococci* sp.	*Lactobacilli* and *lactococci* could metabolize sucrose, fructose, and glucose during fermentation. *Lactobacilli* strains were unable to metabolize citrate, while *lactococci* strains could use citrate as a carbon source.	[75]
Brazil	*Oenococcus oeni*, *P. acidilactici*, *S. salivarius*, *F. pseudoficulneus*, *Lc. mesenteroides*, *Lc. lactis*, *L. reuteri*, *L. amylovo-rus*, *P. dextrinicus*, *L. brevis*, *L. acidophilus*, *L. delbrueckii*, *L. lactis*, *L. rhamnosus*, *L. casei*, *L. fermentum*, and *L. plantarum*.	*L. plantarum* was the most commonly found in fermented cocoa beans. *L. plantarum* adapted well to cocoa ecosystem by responding to changes in ethanol concentration, temperature, and acid stress.	[92]
	Acetobacter senegalensis, *Bacillus subtilis*, *Limosilactobacillus fermentum*, *Brevundimonas*, *Pseudomonas*, and *Kozakia baliensis*.		[100]
	G. saccharivorans, *G. xylinus*, *Ga. oxydans*, *A. peroxydans*, *A. cerevisiae*, *A. malorum*, *A. indonesiensis*, *A. fabarum*, *A. lovaniensis*, *A. senegalensis*, *A. ghanensis*, *A. pasteurianus*, and *A. aceti*		[102]
Ecuador	*W. fabaria*, *W. cibaria*, *L. satsumensis*, *F. ficulneus*, *E. saccharolyticus*, *L. amylovorus*, *L. cacaonum*, *L. nagelii*, *Lc. lactis* subsp. *lactis*, *L. fabifermentans*, *F. tropaeoli*-like, *Leuc. pseudomesenteroides*, and *L. fermentum*,	*L. fermentum* was the dominant and widely studied lactic acid bacteria. *L. fermentum* lived at the beginning of fermentation of cocoa beans and could change citrate. Assimilation of citric acid increased the pH levels, thereby allowing the growth of less acid-fast lactic acid bacteria species, facilitating acetic acid bacteria growth, and optimizing the expression of some microbial activity, such as pectinolytic activity by yeast.	[86]

Lactic acid bacteria were generally gram-positive bacteria and could be categorized into two groups, namely, homofermentative and heterofermentative. The homofermentative group only produced the final product in the form of lactic acid, while the heterofermentative group produced the final product of lactic acid and other compounds, such as acetaldehyde, carbon dioxide, and ethanol [103]. Furthermore, lactic acid bacteria had a crucial role in the fermentation of cocoa beans, especially in breaking down the sugar in the pulp through the homofermentative and heterofermentative pathways [81,104]. The activity of these bacteria produced acidic compounds, which diffused into beans, causing a decrease in the pH to a range of 4.0 to 5.0 [105].

Some of the lactic acid bacteria involved in cocoa bean fermentation are classified as probiotics, which are defined as "live microorganisms that, when administered in adequate amounts, confer a health benefit on the host" [106]. Therefore, the aspect of pro and postbiotics in fermented cocoa beans is one of the interesting potentials to be developed in cocoa and its processed products. One of the probiotics found in fermented cocoa beans is *Lactobacillus plantarum* [107]. Several other probiotics, such as *Bifidobacterium lactis, Lactobacillus rhamnosus, Lactobacillus acidophilus, Lactobacillus paracasei*, and *Lactobacillus casei* can also be added to the fermentation process or to some chocolate processing processes which can increase the probiotic content in the product [108,109]. *L. plantarum* is known to produce postbiotic metabolites in the form of bacteriocins, including RG14, RG11, RI11, RS5, TL1, and UL4, which have cytotoxicity capabilities against various types of cancer cells [110]. Foong et al. [107] reported that they succeeded in isolating *L. plantarum* from fermented cocoa beans and applying it to dark chocolate to produce a product that has the potential to be very good for health.

The growth activity of yeast, acetic acid bacteria, and lactic acid bacteria affected the concentration of lactic acid, acetic acid, and ethanol produced. *Saccharomyces cerevisiae* grew dominantly at the 24th hour of fermentation, followed by the optimum growth of *Lactobacillus lactis* at the 48th hour. Furthermore, *Acetobacter aceti* was already active at 24 h, but its maximum number was reached at 72 h [25]. Lactic acid bacteria converted sugar into lactic acid under anaerobic conditions, and acetic acid bacteria converted alcohol into acetic acid under aerobic conditions [111,112]. The activity of microorganisms implicated in the cocoa beans fermentation process and their roles can be seen in Table 3.

Table 3. Microorganisms' activity is implicated in cocoa bean fermentation.

Microorganism	Lifetime	pH	Temperature	Role	References
Yeast	Lived at the beginning of fermentation, and then the population increased in the 24th hour.	3.1–3.3	30–35 °C	Yeast converted glucose from the pulp into ethanol. Its decomposed pectin compounds into pectin acids and alcohols in the presence of proto-pectinase enzymes, then decomposed pectin acids into arabinose, galactose, and acetic acid using pectinase enzymes. Converted citric acid contained in the pulp.	[20,75]
Lactic acid bacteria	Grew from the beginning of fermentation, and then became dominant at 36 to 72 h.	3.3–4.0	30–40 °C	Broke down sugar into lactic acid, pyruvate, and mannitol, and then lowered the pH.	[20,81,105]
Acetic acid bacteria	Grew from the beginning of fermentation, and then became dominant at 72 h.	4.0–5.0	28–30 °C	Played a role in the process of oxidation of alcohol compounds (ethanol) to acetic acid.	[20,113]
Mold	Grew at moisture content > 8%	2.0–8.5	25–30 °C	It caused the rotting of cocoa beans and produced toxins and other secondary metabolites.	[114,115]

Contamination by other microorganisms, such as mold, was also often found during the fermentation of cocoa beans. Copetti et al. [33] reported that several molds were identified during the process for 6 days, including *Absidia corymbiera, Geotrichum candidum,*

Penicillium paneum, and *Monascus ruber*. The physiological aspects of molds caused them to be found in the fermentation of the products [14,114,116].

The presence of molds must be avoided due to their ability to cause a bitter taste and an off flavor in the product due to the production of several undesirable flavor compounds [14,116]. These microbes often proliferate due to contaminants from the surrounding environment. The presence of mold on the surface of cocoa beans' shells did not cause significant losses, but their penetration into the seeds caused damage to the color and flavor [117].

2.5. Drying

The drying process was a continuation of the oxidation stage of fermentation, which could reduce the bitterness and chelation of cocoa beans. Furthermore, drying produced dry cocoa beans with good quality in terms of physical characteristics, including strong flavor and aroma precursor. In conditions where the drying process was carried out slowly, this situation could be detrimental because it excited the growth of molds and their penetration into cocoa beans [118]. A relatively fast drying process could interfere with the oxidation reaction and cause excessive acidity. An increase in the temperature also caused an increment in acidity and chelation; hence, the maximum threshold was 70 °C [3]. Hayati et al. [119] reported that cocoa beans treated with fermentation and dried with 12 treatment combinations had the lowest water content obtained with a drying temperature of 60 °C for all treatments. This was due to the material's ability to release moisture from the surface and this was directly proportional to the increase in temperature.

Drying was often carried out after fermentation and was commonly useful for reducing the moisture content of cocoa beans to 6–8% [27]. Apart from using sunlight, this could also be conducted using a drying machine, such as an oven blower, especially when the weather was not sunny [120,121]. This alternative was utilized to prevent damage to cocoa beans. A moisture content of less than 6% could cause excessive brittleness, thereby complicating handling and further processing. Meanwhile, a content of >9% could cause weathering of beans due to the growth of fungi. Drying could be conducted by utilizing sunlight, which required approximately 10–14 days [3].

The drying process must only be carried out on cocoa beans that had already undergone fermentation. Freshly harvested cocoa beans often had a very high moisture content of approximately 51–60%, making them more prone to damage or rot due to the growth of microorganisms [27]. The drying was expected to reduce the water content in the product to around 7.5%, thereby preventing microorganisms from growing and prolonging the shelf life of cocoa beans. This process also made it easier to remove nibs from cocoa shells [3,27].

Salazar et al. [122] conducted observations of microbiological properties in post-harvest handling of cocoa beans, where 4 types of treatment were carried out, namely, treatment A and A-farmer, treatment B and B-farmer. Treatments A and A-farmer (performed by farmers directly in the field) were carried out by drying cocoa beans through an artificial process, while treatments B and B-farmers (carried out by farmers directly in the field) were performed using sunlight. The observation results showed that the highest total aerobic mesophilic (TAM) occurred in fresh samples, and *Salmonella* was not found in all treatments. The highest TAM and yeast values were also found in the fresh samples because cocoa beans in the pods were still sterile after being cut. The moisture and high sugar content in fresh cocoa beans were used for the growth and development of microorganisms [7,122].

The maximum allowable value of mold on cocoa beans in the Venezuelan standard was 3.00–4.00 log CFU/g [122,123]. Meanwhile, the international standard ISO 2451:2017 for the cocoa beans—specification and quality requirements have set a maximum limit for moldy dry cocoa beans, which is 3% for grade 1, and 4% for grade 2 [124]. Samples below this threshold still met the good manufacturing practice (GMP) standards. GMP standards had recommended a drying process for cocoa beans to avoid contamination, especially by fungi, including: (a) fermented cocoa beans were immediately dried using direct sunlight or artificial methods until the moisture content was less than 8%; (b) the drying area must be protected from sources of contaminants and thickness of the stack of cocoa beans must be less than 6 cm; (c) the samples were turned over several times (5–10 times per day) to achieve total dryness; (d) cocoa beans must be protected from animals to avoid biological contamination; and (e) equipment must be cleaned regularly [125,126]. Microbes growing on dried cocoa beans could also vary, which was mainly influenced by the drying process and the conditions of the area where the samples were obtained. Furthermore, Delgado-Ospina et al. [127] stated that drying 18 samples of cocoa beans from different plantations led to the presence of some groups of microorganisms, such as yeast, mold, *Lactobacillus* spp., *Lactococcus* spp., total aerobic thermophiles, total aerobic mesophiles, and *Enterobacter* spp. The differences in microbes were caused by several factors, such as variations in the drying techniques used, the maturity level of the samples, storage, and environmental factors, including field conditions, weather, and the initial condition of cocoa beans [16].

Thermophilic bacteria could grow optimally at a high-temperature range of 40–80 °C. These microbes were also found in extreme environmental conditions, such as a pH of more than 10 or less than 2, high salinity (saturated NaCl), and substrate pressure [128]. Meanwhile, mesophilic bacteria could live at an optimum temperature range of 25–37 °C [129]. *Enterobacter* sp. had been reported to have various enzyme activities, including proteolytic properties, and it acted as an opportunistic pathogen [130]. One of the microbiological characteristics of the drying process of cocoa beans after fermentation is presented in Figure 3.

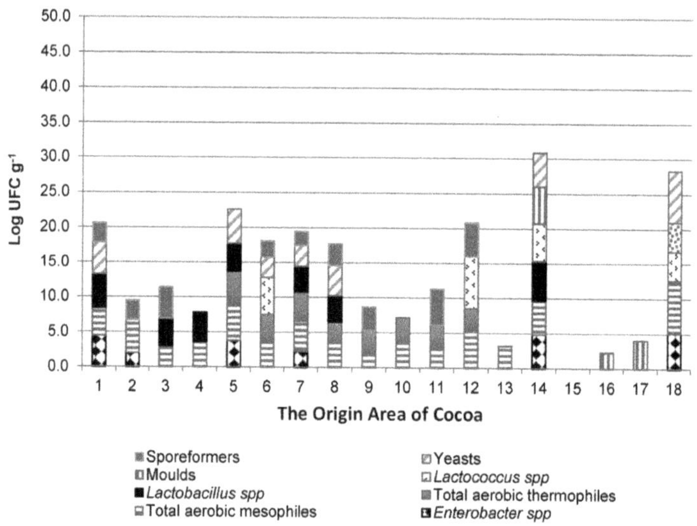

Figure 3. Microbiological characteristics of the drying process of Colombian Criollo cocoa beans (the origin area of cocoa, Valle del Cauca = 1−4, 6−8, 11−13, 15, 17, 18; Cauca = 5, 9, 10, 14; Narino = 16) [127].

Several species of mold were sometimes found during the drying of cocoa beans, such as *Absida corymbifera*, *Penicillium paneum*, *Aspergillus parasiticus*, *Aspergillus candidus*, *Aspergillus flavus*, *Aspergillus niger*, and *Eurotium chebalieri*. Furthermore, *Aspergillus flavus* had great potential as a producer of aflatoxins and ochratoxins [33,34,131]. *P. paneum* was a species that could grow in low oxygen conditions, 4–5 pH, and high CO_2 levels [132].

2.6. Storage

Dried cocoa beans must be stored under certain conditions to maintain their quality. During the storage period, several strains of mold could be found [116,133]. Copetti [33] succeeded in identifying several strains, including *Aspergillus penicillioides*, *Eurotium rubrum*, *Eurotium chevalieri*, *Eurotium amstelodami*, and *Absida corymbifera*. These molds were a class of xerophilic fungi that could live in dry environmental conditions, aw, and low moisture content conditions. Furthermore, they were often found in dry products, such as spices, powdered herbs, and nuts. Xerophilic fungi had the potential to produce mycotoxins, which were harmful to health [134].

Based on previous studies, it was not advisable to store cocoa beans for a long time to avoid damage. This was because dry products were hygroscopic and could easily absorb moisture until they reach equilibrium conditions [135]. Under aw conditions and low moisture content, mold spores could survive for a long time. Poor storage, such as high humidity levels, caused spore germination, leading to rot and the presence of toxins. The relative humidity for storing cocoa beans was recommended to be around 68–70% or lower with a temperature of 25–30 °C [27,42,115]. Several types of fungi, such as *Aspergillus restrictus* and *Aspergillus glaucus*, could grow in environments with relative humidity of >70%. These two groups of fungi species were considered dominant in initiating damage to agricultural products, including cocoa beans. Samples with moisture of 8% or more were more likely to become moldy [27]. Sánchez-Hervás et al. [136] explored 9 samples of cocoa beans originating from Sierra Leone, Ecuador, and Guinea. The results showed that the dominant molds during the storage period included *Aspergillus flavus* and *Aspergillus niger*. Storage under controlled atmospheric conditions was also recommended because it could prevent deterioration due to environmental factors. The use of inert gases, such as carbon dioxide or nitrogen was also effective in maintaining the quality of stored agricultural products, including coffee and cocoa beans [137–140].

3. Effect of Microorganism Activity on Quality of Cocoa Beans

3.1. Physical Properties

The effect of microbiological activity on the quality of cocoa beans could be observed visually through the physical changes on the surface of the samples. Fermented samples had a rather dark brown color and a hollow texture. Physical assessment of the effect of microorganisms' activity on cocoa beans was assessed in more detail through a cut test. This split test was used as a standard for determining the degree of fermentation of the samples [55]. Furthermore, the principle of the test was that cocoa beans were split lengthwise using a knife to display the appearance of the cotyledon surface. The color of the two halves of beans was then observed visually, and the color was classified as gray (slaty), purple (violet), and fully brown for unfermented, partially fermented, and fully fermented samples [21]. Based on physical observation, slaty beans had a firm texture, did not produce a distinctive taste, had excessive astringent and bitter taste, low aroma quality (still smells of alcohol), and purple color, especially in Linda cocoa [141].

3.2. Chemical Properties

Chemical influences were also found on cocoa beans due to the activity of microorganisms in post-harvest handling, which included pH, total acid, ethanol content, and reduced sugar content [21]. Kouamé et al. [142] reported that during fermentation, there was an alteration in pH or acidity. The initial pH of cocoa beans before the process was 4.2, but decreased to 3.6 after 24 h. However, then there was an increase in indigo pH at the 60th hour to 5.8 and 7.3 at the end of the process.

Organic acids were formed and this occurred simultaneously with the production of ethanol by yeast at the beginning of fermentation during the anaerobic phase [7,143]. The ethanol produced was then used by lactic acid bacteria to produce lactic acid. Ethanol was produced at the beginning of the fermentation process by yeast due to the breakdown of sugar [144]. The most widely available reducing sugars included fructose and glucose, but there were also non-reducing sugars, such as sucrose. The sugar content, especially reducing sugars, decreased as fermentation progressed. This was because sugar was used as energy for physiological activity and seed metabolism [17].

The compounds produced during the process could be used as a determining parameter known as the fermentation index. This index was used as a benchmark for assessing the progress of fermentation in cocoa beans [145] and was also considered to have more objective results compared to the cut test. Furthermore, it compared the absorbance at a wavelength between 460 to 540 nm. Fermentation was considered good if the index produced was close to 1 [12].

The index value was determined based on the absorbance level of the fermented compounds and their formation [146,147]. Compounds that changed during the process included polyphenolic compounds, such as flavonoids. Based on previous studies, flavonoids could significantly decrease during fermentation up to 115–43 mg/g compared to non-fermented cocoa beans, which contained higher levels of these compounds [148–150]. Compounds produced from fermentation included tannin complexes that were brown in color and had maximum absorbance values at a wavelength of 460 nm. Some other compounds were reduced due to the process, including anthocyanins, with purple color and maximum absorbance values at a wavelength of 530. Potential flavors from the seeds of cocoa could be seen from the quality of its fermentation through fermentation index value [2].

Certain bacteria starters could also be added to fermented cocoa beans to facilitate the rate of the process. Kresnowati et al. [151] added *L. plantarum* as a starter and this led to an increase in the rate of sugar consumption by 64%, while its absence produced a 46% increment. The concentration of metabolites after the addition of the *L. Plantarum* starter culture also increased by 5–6 times for lactic acid, 4 times for acetic acid, and 50 times for ethanol compared to standard fermentation.

Fermented ethanol was useful for stopping seed germination and accelerating the death of cocoa beans. This was because the heat produced during the formation of ethanol did not damage the structure of the seed cells, thereby facilitating the diffusion of various substances out or into the seeds. *L. plantarum* was one of the microorganisms that produced ethanol in cocoa bean fermentation [152]. Furthermore, its addition was useful as a trigger to increase the growth rate of heterofermentative lactic acid bacteria and prevent its production by yeast. An increase in ethanol levels also affected the increment in organic acid levels in dry cocoa beans [9,71,152]. The pH value of cocoa beans' pulp tended to increase at the end of fermentation along with the extended duration. The activity of bacteria implicated in the process was generally a reaction that produced heat (exothermic) [12]. In spontaneously fermented cocoa beans, there was the growth of acidic bacteria, which showed a normal growth curve. This could be seen by the adaptation, exponential, stationary, optimum growth, and death phases at 0 to 4 h, 8 h, 8 to 12 h, 12 h, and 16 to 20 h, respectively [113].

4. Quality Improvement of Cocoa Beans

4.1. Post-Harvest Treatments for the Quality Improvement of Cocoa Beans

The relatively low quality of conventional cocoa beans could be improved using various methods. Quality improvement could be carried out starting from crop modeling to post-harvest handling of the product [5,153]. This was because post-harvest handling was still not optimal, thereby causing quality defects and non-compliance with the requirements [12,154]. Farmers with smallholder plantations did not often carry out fermentation properly, naturally, or with the addition of inoculum. These farmers only rinsed the samples using water and then dried them in the sun, leading to the brief occurrence of spontaneous fermentation or total absence [12,155].

The common method used to improve the quality of cocoa beans was through controlled fermentation [85,143]. The biochemical processes that occurred during the process using microbes produced several flavor precursors compounds. Furthermore, these compounds could improve or increase the quality of cocoa beans. Fermentation was carried out by several microbes, including lactic acid bacteria (*Lactobacillus* sp. *KSL2*), acetic acid bacteria (*Acetobacter* sp. *KLK1*), and yeast (*Candida* sp. *KLK4*) [16,19,20]. Various methods or post-harvest treatments to improve the quality of cocoa beans are presented in Table 4.

Table 4. Various methods or post-harvest treatments to improve the quality of cocoa beans.

Post-Harvest Stage	Treatments	Conditions	Characteristics	References
Sorting and fermentation	Determining cocoa pod maturity and fermentation time to increase healthful bioactive compounds.	Pod harvest: mature and ripe. Post fermentation: 1, 3, 5 days.	Mature cocoa pods and fermentation for 3 days could produce cocoa beans with a high content of bioactive compounds, high antioxidant activity, and the desired flavor.	[52]
Fermentation	Fermentation by administering anti-fungal strains.	The anti-fungal strain of *L. fermentum* and *S. cerevisiae* were added to 180 kg box.	Culture of *L. fermentum 223* and *S. cerevisiae H290* had good anti-fungi activity and produced less off-flavor, a good percentage of fermented beans, less astringency, and the best cocoa taste.	[156]
Fermentation	Fermentation with the addition of a mixture of LAB and AAB starter cultures	Starter culture: *A. pasteurianus 386B*, *L. fermentum 222*, *S. cerevisiae H5S5K23* (3 heaps and 1 box).	The addition of lactic acid bacteria and acetic acid bacteria starter culture mixture accelerated carbohydrate fermentation by increasing the conversion of lactic acid and citric acid and was proven to improve the chocolate flavor produced.	[105]
Fermentation	Controlled temperature and pH during fermentation.	Temperature: 12–28 °C, 20–26 °C, and 16–28 °C. RH: 80–85%, 80–85%, and 70–75%.	Controlled fermentation of cocoa beans at a pH between 4.75 to 5.19 and a temperature below 40 °C could optimize activity of microbes and enzymes forming flavor compounds, as well as other sensory attributes.	[85]
Fermentation	Cocoa bean turning start times on fermentation	Cocoa beans: Criollo, turning start times: 24 and 48 h).	Turning start of 48 h could stimulate flavor-forming microbes, such as *M. carpophila*, *P. manshurica*, and *H. opuntiae* in cocoa beans fermentation.	[143]
Fermentation	Addition of acetic acid bacterial culture starter in fermentation	Culture starter: 130 AAB from 3 countries (French Guiana, Ivory Coast, and Mexico).	The addition of *A. pasteurianus* starter culture increased the conversion of ethanol and lactic acid into acetoin or acetic acid, thereby improving quality of cocoa beans.	[84]
Fermentation	Addition of LAB culture starter in fermentation to inhibit the growth of fungi.	Starter LAB: *L. fermentum* and *L. plantarum*. Fermentation periods: 5 days.	The addition of *L. fermentum*, *L. plantarum*, and a combination of *L. plantarum* with *A. aceti* and *S. cerevisiae* provided suitable pH and temperature, and inhibited the growth of fungi.	[116]
Fermentation	Addition of indigenous LAB, yeast, AAB in fermentation to inhibit mycotoxins produced by fungi.	Starter: indigenous *Acetobacter* spp. *HA-37*, *L. plantarum HL-15*, *C. famata HY-37* at concentration of 10^9 CFU/mL. Fermentation periods: 5 days.	The use of indigenous *L. plantarum HL-15* or in combination with *Acetobacter* spp. and *C. famata* inhibited ochratoxin-A produced by fungi.	[14]
Fermentation	Addition of BAL starter *L. plantarum* in the fermentation of cocoa beans.	Starter *L. plantarum*: 10^3 CFU per gram cocoa beans.	The addition of BAL starter accelerated the growth of lactic acid bacteria and acetic acid bacteria. There was also an increase in the amount of acetic acid, lactic acid, and ethanol produced. Fermentation index increased and the time was shorter.	[151]
Fermentation and drying	Determining fermentation and drying times in the rainy season.	Fermentation periods: 5–8 days. Drying: 4–6 days.	Fermentation for 8 days and drying for 6 days using sunlight in the rainy season produced the best quality cocoa beans.	[27]
Drying	Drying cocoa beans with adsorption, vacuum drying, and freeze drying to get high antioxidant cocoa beans.	Pressure of freeze dryer: 0.015 mbar, condenser dimensions: 31.3 cm × 34.5 cm × 46.0 cm.	Freeze-drying produced cocoa beans with the highest antioxidant activity (71.8 mg Trolox/g) and polyphenol content (126.3 mg GAE/mg).	[157]
Drying	Drying cocoa beans with solar power equipped with a heat pump.	Temperature: 32–48 °C and RH: 35–80%.	A solar dryer with a heat pump can speed up the drying of cocoa beans from 6 days to 5 days.	[118]

Table 4. *Cont.*

Post-Harvest Stage	Treatments	Conditions	Characteristics	References
Storage	Re-fermentation of dry non-fermented cocoa beans by administering pure cultures to fermented the non-fermented of dry cocoa beans.	Moisture: 15%. Fermentation periods: 120 h.	The addition of pure culture of *A. aceti*, *L. lactis*, and *S. cerevisiae* improved quality of dry cocoa beans by facilitating the fermentation process and increasing the fermentation index up to 1.03.	[12]

Fermentation modification involving various microbial starter cultures had been proven to improve the quality of cocoa beans, but the process was prone to contamination. Romanens et al. [156] reported that the spontaneous fermentation of cocoa beans led to the growth of mold, causing a decrease in quality. Controlled fermentation by administering two anti-fungi strains improved the anti-fungi activity in fermenting and drying to improve quality. The anti-mold cultures used included *S. cerevisiae H290* and *L. fermentum* M017 cultures and *L. fermentum 223* and *S. cerevisiae H290 B* cultures. The dominance of *L. fermentum M017* and *223* caused an increase in mannitol originating from the ability of Lb. Culture B *L. fermentum 223* and *S. cerevisiae H290* showed the best results with less off-flavor, a good percentage of fermented beans, less astringency, and the best taste.

Several studies had also extensively explored fermentation modification, including Rahayu et al. [14], who added indigenous lactic acid bacteria, yeast, and acetic acid bacteria to inhibit mycotoxins produced by fungi. The use of indigenous *L. plantarum HL-15* or in combination with *Acetobacter* spp. and *C. famata* inhibited ochratoxin-A produced by fungi. Marwati et al. [116] also reported that the addition of *L. fermentum*, *L. plantarum*, and a combination of *L. plantarum* with *A. aceti* and *S. cerevisiae* provided suitable pH and temperature, as well as inhibited the growth of fungi.

Improving the quality of cocoa beans could also be done on non-fermented dry samples, which were known to be low quality. Improvements were made by administering pure cultures of *A. aceti*, *L. lactis*, and *S. cerevisiae* to fermented and non-fermented dry cocoa beans, as reported by Apriyanto et al. [12]. The results showed that fermentation treatment with the addition of pure culture gradually improved the quality of the samples. This was because the treatment facilitated fermentation and the role of microorganisms during the process could be controlled based on the succession of each microorganism.

The activity of microorganisms during post-harvest handling of cocoa beans had various beneficial effects, especially on fermentation. However, some risks or adverse effects could occur due to the activity of microorganisms. Some of the advantages and disadvantages of microorganisms' activity in the post-harvest handling of cocoa beans are presented in Table 5.

Table 5. The advantages and disadvantages of microorganisms' activity in post-harvest handling of cocoa beans.

Advantages			
No.	**Role**	**Microorganism**	**References**
1.	It decomposed pectin compounds into pectin acids and alcohol, and the pulp was crushed and released due to the decomposition of pectin. Sugars were then converted into alcohol compounds and citric acid was broken down.	Yeast	[25,75]

Table 5. *Cont.*

2.	Broke down citric acid in cocoa fermentation sugar in the pulp through homofermentative and heterofermentative pathways.	Lactic acid bacteria	[75,105]
3.	Played a role in the process of oxidation of alcohol compounds to acetic acid.	Acetic acid bacteria	[19,20]
4.	Killed the seeds to ensure changes, such as the formation of color and flavor precursors in fermentation.	Acetic acid bacteria, yeast, and lactic acid bacteria	[158]
	Disadvantages		
No.	**Role**	**Microorganism**	**References**
1.	The growth of mold was a risk to public health due to the toxins it produced.	*Aspergillus*, *Penicillium*, and *Fusarium*	[159,160]
2.	It caused typhus with symptoms of diarrhea, nausea, and dizziness if cocoa beans were contaminated.	*Salmonella*	[39]
3.	The cause of weathering, reduced nutrition, and the presence of mycotoxins, which could cause health problems in cocoa beans.	Mold	[34]
4.	Contamination by bacteria could cause nosocomial infections when cocoa pods were ripened.	*Acinetobacter* sp., *Klebsiella pneumoniae*	[57]
5.	There was damage to the color and flavor of cocoa beans.	Kapang	[161]

Some of the microorganisms' activities during post harvest were beneficial because they contributed to improving the quality of cocoa beans, including (1) yeast, which converted sugar into alcohol and citric acid; (2) lactic acid bacteria (LAB), which converted citric acid into lactic acid; and (3) acetic acid bacteria (AAB), which oxidized alcohol to acetic acid, which then played a role in killing the seeds, forming a brown color, and forming flavor precursor compounds. However, some microorganism activities could contaminate, which harmed or reduced the quality of cocoa beans, and were dangerous because they produced toxins. These microorganisms included (1) *Salmonella*, which caused diarrhea and typhus; (2) *Acinetobacter* sp. and *Klebsiella pneumoniae*, which caused nosocomial infections; and (3) *Aspergillus*, *Penicillium*, and *Fusarium*, which produced toxins that were harmful to health. Therefore, microbial contamination must be prevented and controlled during post harvest to maintain the quality of the cocoa beans.

4.2. Technological Difficulties in Maintaining the Postharvest Quality of Cocoa Beans

Post-harvest handling greatly affects the quality of the cocoa beans. However, many of the cocoa beans circulating in the market are of low quality due to improper post-harvest handling [162,163]. Various technological difficulties are still encountered in maintaining the post-harvest quality of cocoa beans, especially during the fermentation, drying, and storage processes. The process of fermenting cocoa beans takes quite a long time, around 5–6 days, causing some farmers not to ferment their cocoa beans or to ferment them impatiently and incorrectly so that the quality of the cocoa beans is low. Until now, fermentation technology has been developed with various methods and various additions of microbial starters, but the fermentation time cannot be significantly shortened [84,116,156]. Shortened fermentation of cocoa beans can improve the activity of antioxidants and other bioactive compounds but cannot produce good flavor precursor compounds [52]. Dry cocoa beans that are not fermented are also a challenge because they are low quality. Several re-fermentation technologies have been applied to dry unfermented cocoa beans and can slightly improve the quality of cocoa beans, but not as good as cocoa beans fermented from scratch. This is because the substrate available for re-fermentation cannot support the type and number of microbial populations that should be involved in the fermentation process [12].

Drying technology is also still experiencing some difficulties. This is because the drying process of cocoa beans requires sufficient time to obtain a moisture content of less than 8% with good quality. Drying can be accelerated using various drying machines, but drying too fast can increase the acidity in the cocoa beans, while drying for too long can cause mold growth which can produce mycotoxins and reduce the quality of the cocoa beans [134]. Cocoa bean storage technology also has challenges in maintaining the quality of cocoa beans over a long period of time. Controlled atmosphere technology, such as nitrogen or carbon dioxide, is quite effective in maintaining the quality of cocoa beans, but it requires quite complicated installation and is expensive [139,140]. Using preservatives to prevent microbial growth also carries risks related to their toxicity and food safety. Therefore, it is still necessary to develop other storage innovations that are cheaper, more effective, and more efficient.

5. Conclusions and Future Research

Post-harvest handling of cocoa beans needs attention, starting from picking the pods, sorting, ripening the pods, fermenting, drying, and storing. The activity of microorganisms in each post-harvest stage affects the quality of cocoa beans. Fermentation is one of the important stages involving microorganisms such as acetic acid bacteria, lactic acid bacteria, and yeast which play a major role in producing precursor flavor compounds in cocoa beans. However, harmful microbes such as pathogenic bacteria and fungi can contaminate and reduce the quality of cocoa beans. Mold contamination causes weathering, reduced nutrition, and the presence of mycotoxins. Improvement of cocoa bean quality can be carried out by adding a starter culture in fermentation. The addition of these cultures can increase the fermentation rate and improve the biochemical process in cocoa beans. Proper drying processes and controlled storage can maintain the fermented cocoa beans' quality. However, many difficulties and challenges remain, especially those related to controlling microbial activity during post harvest. Post-harvest technology needs to be continuously developed, especially in controlling microbiological activities to improve the quality of cocoa beans.

Various studies to develop post-harvest handling technology had been carried out, especially to improve the quality of the product. Various fermentation techniques continued to be developed to create ideal conditions for microbiological activity. Various types of starters and various substrates were also added to the fermentation process to increase the speed and improve the quality of the products. Various efforts had also been made to improve the quality of non-fermented cocoa beans by further fermentation or by approaching the chemical compounds produced during the process using synthetic materials. Various technologies had also begun to be developed to increase the bioactive compounds in cocoa beans to provide high antioxidant activity, which was beneficial for health. Furthermore, several combinations of modification technologies were carried out to obtain quality cocoa beans, especially from the desired biochemical compound content and flavor precursors.

Author Contributions: Conceptualization, E.S. and E.L.; methodology, M.D. and N.B.; software, R.I.; validation, M.D., R.I. and E.L.; formal analysis, N.B.; investigation, E.S. and N.B.; resources, E.S. and N.B.; data curation, E.S. and N.B.; writing—original draft preparation, E.S. and N.B.; writing—review and editing, M.D., R.I. and E.L.; visualization, R.I.; supervision, M.D.; project administration, E.L.; funding acquisition, E.S. All authors have read and agreed to the published version of the manuscript.

Funding: This article was supported by Universitas Padjadjaran, Indonesia, grant number: 1549/UN6.3.1/PT.00/2023.

Data Availability Statement: Not applicable.

Acknowledgments: Thank you to the Rector of Universitas Padjadjaran and The Ministry of Education, Culture, Research, and Technology of the Republic of Indonesia.

Conflicts of Interest: The authors declare no conflict of interest.

References

1. Mihai, R.A.; Landazuri Abarca, P.A.; Tinizaray Romero, B.A.; Florescu, L.I.; Catană, R.; Kosakyan, A. Abiotic Factors from Different Ecuadorian Regions and Their Contribution to Antioxidant, Metabolomic and Organoleptic Quality of *Theobroma cacao* L. Beans, Variety "Arriba Nacional". *Plants* **2022**, *11*, 976. [CrossRef]
2. Kongor, J.E.; Hinneh, M.; de Walle, D.V.; Afoakwa, E.O.; Boeckx, P.; Dewettinck, K. Factors influencing quality variation in cocoa (*Theobroma cacao*) bean flavour profile—A review. *Food Res. Int.* **2016**, *82*, 44–52. [CrossRef]
3. Dzelagha, B.F.; Ngwa, N.M.; Nde Bup, D. A Review of Cocoa Drying Technologies and the Effect on Bean Quality Parameters. *Int. J. Food Sci.* **2020**, *2020*, 8830127. [CrossRef] [PubMed]
4. Forte, M.; Currò, S.; Van de Walle, D.; Dewettinck, K.; Mirisola, M.; Fasolato, L.; Carletti, P. Quality Evaluation of Fair-Trade Cocoa Beans from Different Origins Using Portable Near-Infrared Spectroscopy (NIRS). *Foods* **2023**, *12*, 4. [CrossRef] [PubMed]
5. Streule, S.; Freimüller Leischtfeld, S.; Galler, M.; Miescher Schwenninger, S. Monitoring of cocoa post-harvest process practices on a small-farm level at five locations in Ecuador. *Heliyon* **2022**, *8*, e09628. [CrossRef] [PubMed]
6. Lima, L.J.R.; Almeida, M.H.; Nout, M.J.R.; Zwietering, M.H. *Theobroma cacao* L. "The Food of the Gods": Quality Determinants of Commercial Cocoa Beans, with Particular Reference to the Impact of Fermentation. *Crit. Rev. Food Sci. Nutr.* **2011**, *51*, 731–761. [CrossRef] [PubMed]
7. Gutiérrez-Ríos, H.G.; Suárez-Quiroz, M.L.; Hernández-Estrada, Z.J.; Castellanos-Onorio, O.P.; Alonso-Villegas, R.; Rayas-Duarte, P.; Cano-Sarmiento, C.; Figueroa-Hernández, C.Y.; González-Rios, O. Yeasts as Producers of Flavor Precursors during Cocoa Bean Fermentation and Their Relevance as Starter Cultures: A Review. *Fermentation* **2022**, *8*, 331. [CrossRef]
8. Lembong, E.; Djali, M.; Zaida; Utama, G.L. The potential of dry fermented cocoa (*Theobroma cacao* L.) variety Lindak bean shell treated at different degrees of roasting as a functional food. *IOP Conf. Ser. Earth Environ. Sci.* **2021**, *733*, 12068. [CrossRef]
9. Sari, A.B.T.; Fahrurrozi; Marwati, T.; Djaafar, T.F.; Hatmi, R.U.; Purwaningsih; Wanita, Y.P.; Lisdiyanti, P.; Perwitasari, U.; Juanssilfero, A.B.; et al. Chemical Composition and Sensory Profiles of Fermented Cocoa Beans Obtained from Various Regions of Indonesia. *Int. J. Food Sci.* **2023**, *2023*, 5639081. [CrossRef]
10. Subroto, E.; Andoyo, R.; Indiarto, R.; Lembong, E. Physicochemical properties, sensory acceptability, and antioxidant activity of chocolate bar fortified by solid lipid nanoparticles of gallic acid. *Int. J. Food Prop.* **2022**, *25*, 1907–1919. [CrossRef]
11. Purbaningrum, K.; Hidayat, C.; Witasari, L.D.; Utami, T. Flavor Precursors and Volatile Compounds Improvement of Unfermented Cocoa Beans by Hydrolysis Using Bromelain. *Foods* **2023**, *12*, 820. [CrossRef]
12. Apriyanto, M.; Sutardi; Supriyanto; Harmayani, E. Study on effect of fermentation to the quality parameter of cocoa bean in Indonesia. *Asian J. Dairy Food Res.* **2016**, *35*, 160–163. [CrossRef]
13. Agyirifo, D.S.; Wamalwa, M.; Otwe, E.P.; Galyuon, I.; Runo, S.; Takrama, J.; Ngeranwa, J. Metagenomics analysis of cocoa bean fermentation microbiome identifying species diversity and putative functional capabilities. *Heliyon* **2019**, *5*, e02170. [CrossRef] [PubMed]
14. Rahayu, E.S.; Triyadi, R.; Khusna, R.N.B.; Djaafar, T.F.; Utami, T.; Marwati, T.; Hatmi, R.U. Indigenous Yeast, Lactic Acid Bacteria, and Acetic Acid Bacteria from Cocoa Bean Fermentation in Indonesia Can Inhibit Fungal-Growth-Producing Mycotoxins. *Fermentation* **2021**, *7*, 192. [CrossRef]
15. Iacumin, L.; Pellegrini, M.; Colautti, A.; Orecchia, E.; Comi, G. Microbial Characterization of Retail Cocoa Powders and Chocolate Bars of Five Brands Sold in Italian Supermarkets. *Foods* **2022**, *11*, 2753. [CrossRef]
16. Schwan, R.F.; Wheals, A.E. The Microbiology of Cocoa Fermentation and its Role in Chocolate Quality. *Crit. Rev. Food Sci. Nutr.* **2004**, *44*, 205–221. [CrossRef]
17. Afoakwa, E.O.; Kongor, J.E.; Takrama, J.; Budu, A.S. Changes in nib acidification and biochemical composition during fermentation of pulp pre-conditioned cocoa (*Theobroma cacao*) beans. *Int. Food Res. J.* **2013**, *20*, 1843–1853.
18. Apriyanto, M.; Umanailo, M.C.B. Decrease polyphenols, ethanol, lactic acid, and acetic acid during fermentation with addition of cocoa beans innoculum. *Int. J. Sci. Technol. Res.* **2019**, *8*, 461–465.

19. De Vuyst, L.; Weckx, S. The cocoa bean fermentation process: From ecosystem analysis to starter culture development. *J. Appl. Microbiol.* **2016**, *121*, 5–17. [CrossRef]
20. De Vuyst, L.; Leroy, F. Functional role of yeasts, lactic acid bacteria and acetic acid bacteria in cocoa fermentation processes. *FEMS Microbiol. Rev.* **2020**, *44*, 432–453. [CrossRef] [PubMed]
21. Camu, N.; De Winter, T.; Addo, S.K.; Takrama, J.S.; Bernaert, H.; De Vuyst, L. Fermentation of cocoa beans: Influence of microbial activities and polyphenol concentrations on the flavour of chocolate. *J. Sci. Food Agric.* **2008**, *88*, 2288–2297. [CrossRef]
22. Wollgast, J.; Anklam, E. Review on polyphenols in Theobroma cacao: Changes in composition during the manufacture of chocolate and methodology for identification and quantification. *Food Res. Int.* **2000**, *33*, 423–447. [CrossRef]
23. Díaz-Muñoz, C.; De Vuyst, L. Functional yeast starter cultures for cocoa fermentation. *J. Appl. Microbiol.* **2022**, *133*, 39–66. [CrossRef]
24. Timothy, L.; Maarten, J.; Nicholas, C.; Luc, D.V. Kinetic Analysis of Strains of Lactic Acid Bacteria and Acetic Acid Bacteria in Cocoa Pulp Simulation Media toward Development of a Starter Culture for Cocoa Bean Fermentation. *Appl. Environ. Microbiol.* **2010**, *76*, 7708–7716. [CrossRef]
25. Ardhana, M.M.; Fleet, G.H. The microbial ecology of cocoa bean fermentations in Indonesia. *Int. J. Food Microbiol.* **2003**, *86*, 87–99. [CrossRef]
26. Delgado-Ospina, J.; Molina-Hernández, J.B.; Chaves-López, C.; Romanazzi, G.; Paparella, A. The Role of Fungi in the Cocoa Production Chain and the Challenge of Climate Change. *J. Fungi* **2021**, *7*, 202. [CrossRef] [PubMed]
27. Ackah, E.; Dompey, E. Effects of fermentation and drying durations on the quality of cocoa (*Theobroma cacao* L.) beans during the rainy season in the Juaboso District of the Western-North Region, Ghana. *Bull. Natl. Res. Cent.* **2021**, *45*, 175. [CrossRef]
28. Atmaja, M.I.P.; Haryadi, H.; Supriyanto, S. Quality Improvement of Non Fermented Cocoa Bean Through Pre-Incubation Treatments. *J. Ind. Beverage Crop.* **2016**, *3*, 11–20. [CrossRef]
29. Kamphuis, H.J. Production and Quality Standards of Cocoa Mass, Cocoa Butter and Cocoa Powder. In *Industrial Chocolate Manufacture and Use*; Beckett, S.T., Ed.; Blackwell Publishing Ltd.: Hoboken, NJ, USA, 2008; pp. 121–141, ISBN 9781444301588.
30. Delgado-Ospina, J.; Molina-Hernandez, J.B.; Viteritti, E.; Maggio, F.; Fernández-Daza, F.F.; Sciarra, P.; Serio, A.; Rossi, C.; Paparella, A.; Chaves-López, C. Advances in understanding the enzymatic potential and production of ochratoxin A of filamentous fungi isolated from cocoa fermented beans. *Food Microbiol.* **2022**, *104*, 103990. [CrossRef]
31. Cho, W.-I.; Chung, M.-S. Bacillus spores: A review of their properties and inactivation processing technologies. *Food Sci. Biotechnol.* **2020**, *29*, 1447–1461. [CrossRef]
32. Pereira, A.P.M.; Sant'Ana, A.S. Diversity and fate of spore forming bacteria in cocoa powder, milk powder, starch and sugar during processing: A review. *Trends Food Sci. Technol.* **2018**, *76*, 101–118. [CrossRef]
33. Copetti, M.V.; Iamanaka, B.T.; Frisvad, J.C.; Pereira, J.L.; Taniwaki, M.H. Mycobiota of cocoa: From farm to chocolate. *Food Microbiol.* **2011**, *28*, 1499–1504. [CrossRef]
34. Copetti, M.V.; Iamanaka, B.T.; Pitt, J.I.; Taniwaki, M.H. Fungi and mycotoxins in cocoa: From farm to chocolate. *Int. J. Food Microbiol.* **2014**, *178*, 13–20. [CrossRef]
35. Krapf, T.; Gantenbein-Demarchi, C. Thermal inactivation of *Salmonella* spp. during conching. *LWT—Food Sci. Technol.* **2010**, *43*, 720–723. [CrossRef]
36. Sharma, R.; Garg, P.; Kumar, P.; Bhatia, S.K.; Kulshrestha, S. Microbial Fermentation and Its Role in Quality Improvement of Fermented Foods. *Fermentation* **2020**, *6*, 106. [CrossRef]
37. Phoku, J.Z.; Barnard, T.G.; Potgieter, N.; Dutton, M.F. Mycotoxigenic potentials of the genera: Aspergillus, Fusarium and Penicillium isolated from houseflies (*Musca domestica* L.). *Acta Trop.* **2017**, *168*, 29–36. [CrossRef] [PubMed]
38. Pires, P.N.; Vargas, E.A.; Gomes, M.B.; Vieira, C.B.M.; Dos Santos, E.A.; Bicalho, A.A.C.; Silva, S.D.C.; Rezende, R.P.; De Oliveira, I.S.; Luz, E.D.; et al. Aflatoxins and ochratoxin A: Occurrence and contamination levels in cocoa beans from Brazil. *Food Addit. Contam. Part A* **2019**, *36*, 815–824. [CrossRef] [PubMed]
39. do Nascimento, M.D.S.; Pena, P.O.; Brum, D.M.; Imazaki, F.T.; Tucci, M.L.S.; Efraim, P. Behavior of Salmonella during fermentation, drying and storage of cocoa beans. *Int. J. Food Microbiol.* **2013**, *167*, 363–368. [CrossRef]
40. Purnamasari, L.; Ali, A.; Cuk, T.N. Assessment of Crude Aflatoksin B1 Production Based on Local *Aspergillus flavus* Mold Isolate in Corn and Corn + Ground Peanut Media. *Bul. Peternak.* **2016**, *40*, 133–137. [CrossRef]
41. Janssen, S.; Pankoke, I.; Klus, K.; Schmitt, K.; Stephan, U.; Wöllenstein, J. Two underestimated threats in food transportation: Mould and acceleration. *Philos. Trans. R. Soc. A Math. Phys. Eng. Sci.* **2014**, *372*, 20130312. [CrossRef]
42. Uygun, Y.; Jafri, S.A.I. Controlling risks in sea transportation of cocoa beans. *Cogent Bus. Manag.* **2020**, *7*, 1778894. [CrossRef]
43. Dewayani, W.; Syamsuri, R.; Septianti, E.; Suriany; Halil, W.; Nurlaila; Rahman, A. Good Handling Practice Study to Reduce The Level of Contamination in Cocoa Beans in East Luwu. *IOP Conf. Ser. Earth Environ. Sci.* **2022**, *1024*, 12067. [CrossRef]
44. Leandro-Muñoz, M.E.; Tixier, P.; Germon, A.; Rakotobe, V.; Phillips-Mora, W.; Maximova, S.; Avelino, J. Effects of microclimatic variables on the symptoms and signs onset of Moniliophthora roreri, causal agent of Moniliophthora pod rot in cacao. *PLoS ONE* **2017**, *12*, e0184638. [CrossRef]
45. Ndoumbe-Nkeng, M.; Cilas, C.; Nyemb, E.; Nyasse, S.; Bieysse, D.; Flori, A.; Sache, I. Impact of removing diseased pods on cocoa black pod caused by *Phytophthora megakarya* and on cocoa production in Cameroon. *Crop Prot.* **2004**, *23*, 415–424. [CrossRef]
46. Bariah, S.K.; Tajul, A.Y. Effect of Cocoa Pods Storage on the Temperature and Physicochemical Changes during Shallow Box Fermentation. *IJISET-Int. J. Innov. Sci. Eng. Technol.* **2017**, *4*, 197–203.

47. Septianti, E.; Salengke; Langkong, J.; Sukendar, N.K.; Hanifa, A.P. Characteristic Quality of Pinrang's Cocoa Beans During Fermentation Used Styrofoam Containers. *Canrea J. Food Technol. Nutr. Culin. J.* **2020**, *3*, 10–25. [CrossRef]
48. Djaafar, T.F.; Elghina, L.; Widodo, S.; Marwati, T.; Utami, T.; Rahayu, E.S. Study of Good Handling Practices and Critical Control Point Determination of Dried Fermented Cocoa Bean in Gunung Kidul Regency, Yogyakarta. *IOP Conf. Ser. Earth Environ. Sci.* **2019**, *309*, 12015. [CrossRef]
49. Rinaldo, R.; Chozin, M.A. Management of Handling Cocoa Pod (*Theobroma cacao* L.) in Central Java. *Bul. Agrohorti* **2016**, *4*, 210–214. [CrossRef]
50. Meulemans, C.C.; Consultancy, O.R.; Surapati, T.U.; Tjatjo, A. Colorimetric Measurements of Cocoa Beans (*Theobroma cacao*). *Indones. J. Agric. Sci.* **2013**, *3*, 52. [CrossRef]
51. Bachtiar, W.F.; Aji, G.K.; Norsita, D.I. Value Chain Performance of Cocoa Beans in Banjaroya, Kulon Progo. *Gontor AGROTECH Sci. J.* **2019**, *5*, 1. [CrossRef]
52. Tee, Y.-K.; Bariah, K.; Hisyam Zainudin, B.; Samuel Yap, K.-C.; Ong, N.-G. Impacts of cocoa pod maturity at harvest and bean fermentation period on the production of chocolate with potential health benefits. *J. Sci. Food Agric.* **2022**, *102*, 1576–1585. [CrossRef] [PubMed]
53. Afoakwa, E.O.; Quao, J.; Budu, A.S.; Takrama, J.; Saalia, F.K. Effect of pulp preconditioning on acidification, proteolysis, sugars and free fatty acids concentration during fermentation of cocoa (*Theobroma cacao*) beans. *Int. J. Food Sci. Nutr.* **2011**, *62*, 755–764. [CrossRef]
54. Hartuti, S.; Juanda, J.; Khatir, R. Efforts to Improve Cocoa Bean Quality (*Theobroma cacao* L.) through the Postharvest Handling Stage (Review). *Indones. J. Ind. Res.* **2020**, *15*, 38–52. [CrossRef]
55. Afoakwa, E.O.; Quao, J.; Budu, A.S.; Takrama, J.; Saalia, F.K. Influence of pulp-preconditioning and fermentation on fermentative quality and appearance of Ghanaian cocoa (*Theobroma cacao*) beans. *Int. Food Res. J.* **2012**, *19*, 127–133.
56. Hamdouche, Y.; Meile, J.C.; Lebrun, M.; Guehi, T.; Boulanger, R.; Teyssier, C.; Montet, D. Impact of turning, pod storage and fermentation time on microbial ecology and volatile composition of cocoa beans. *Food Res. Int.* **2019**, *119*, 477–491. [CrossRef] [PubMed]
57. Almasaudi, S.B. Acinetobacter spp. as nosocomial pathogens: Epidemiology and resistance features. *Saudi J. Biol. Sci.* **2018**, *25*, 586–596. [CrossRef]
58. Wong, D.; Nielsen, T.B.; Bonomo, R.A.; Pantapalangkoor, P.; Luna, B.; Spellberg, B. Clinical and Pathophysiological Overview of Acinetobacter Infections: A Century of Challenges. *Clin. Microbiol. Rev.* **2017**, *30*, 409–447. [CrossRef]
59. Brisse, S.; Fevre, C.; Passet, V.; Issenhuth-Jeanjean, S.; Tournebize, R.; Diancourt, L.; Grimont, P. Virulent Clones of *Klebsiella pneumoniae*: Identification and Evolutionary Scenario Based on Genomic and Phenotypic Characterization. *PLoS ONE* **2009**, *4*, e4982. [CrossRef]
60. Pérez Portuondo, I. Bacillus cereus y su papel en las intoxicaciones alimentarias. *Rev. Cuba. Salud Pública* **2012**, *38*, 98–108. [CrossRef]
61. Khan, A.R.; Mustafa, A.; Hyder, S.; Valipour, M.; Rizvi, Z.F.; Gondal, A.S.; Yousuf, Z.; Iqbal, R.; Daraz, U. *Bacillus* spp. as Bioagents: Uses and Application for Sustainable Agriculture. *Biology* **2022**, *11*, 1763. [CrossRef]
62. Miljaković, D.; Marinković, J.; Balešević-Tubić, S. The Significance of *Bacillus* spp. in Disease Suppression and Growth Promotion of Field and Vegetable Crops. *Microorganisms* **2020**, *8*, 1037. [CrossRef] [PubMed]
63. Iyanda, M.O.; Alhassan, E.A.; Adekanye, T.A. Design, Fabrication and Testing of a Cocoa Depodding Machine. *Mindanao J. Sci. Technol.* **2018**, *16*, 11–24.
64. Josué, D.; Bienvenu, K.; Fonsso Josué, D.; Mbetmi Guy-De-Patience, F.; Bonaventure, D. Design and development of cocoa pod breaking and beans extraction machine. *Int. J. Eng. Technol.* **2019**, *8*, 357–366.
65. Lozano Tovar, M.D.; Tibasosa, G.; González, C.M.; Ballestas Alvarez, K.; Lopez Hernandez, M.D.P.; Rodríguez Villamizar, F. Isolation and identification of microbial species found in cocoa fermentation as microbial starter culture candidates for cocoa bean fermentation in Colombia. *Pelita Perkeb.* **2020**, *36*, 236–248. [CrossRef]
66. Mota-Gutierrez, J.; Botta, C.; Ferrocino, I.; Giordano, M.; Bertolino, M.; Dolci, P.; Cannoni, M.; Cocolin, L. Dynamics and Biodiversity of Bacterial and Yeast Communities during Fermentation of Cocoa Beans. *Appl. Environ. Microbiol.* **2018**, *84*, e01164-18. [CrossRef]
67. Steensels, J.; Snoek, T.; Meersman, E.; Nicolino, M.P.; Voordeckers, K.; Verstrepen, K.J. Improving industrial yeast strains: Exploiting natural and artificial diversity. *FEMS Microbiol. Rev.* **2014**, *38*, 947–995. [CrossRef] [PubMed]
68. Djaafar, T.F.; Monika, D.C.; Marwati, T.; Triwitono, P.; Rahayu, E.S. Microbiology, Chemical, and Sensory Characteristics of Cocoa Powder: The Effect of *Lactobacillus plantarum* HL-15 as Culture Starter and Fermentation Box Variation. *Digit. Press Life Sci.* **2020**, 2, 8. [CrossRef]
69. Indarti, E.; Widayat, H.P.; Zuhri, N. Effect of fermentation container and thickness of bean mass during fermentation process of cocoa bean (*Theobroma cocoa* L). *Proc. Annu. Int. Conf. Syiah Kuala Univ.* **2011**, *1*, 64–69.
70. Nielsen, D.S.; Teniola, O.D.; Ban-Koffi, L.; Owusu, M.; Andersson, T.S.; Holzapfel, W.H. The microbiology of Ghanaian cocoa fermentations analysed using culture-dependent and culture-independent methods. *Int. J. Food Microbiol.* **2007**, *114*, 168–186. [CrossRef]
71. Rahardjo, Y.P.; Rahardja, S.; Samsudin; Saidah; Dalapati, A.; Amalia, A.F.; Purwaningsih, H.; Syamsu, K. A literature review on cocoa fermentation techniques to shorten fermentation time. *IOP Conf. Ser. Earth Environ. Sci.* **2022**, *974*, 12111. [CrossRef]

72. Mozzi, F.; Ortiz, M.E.; Bleckwedel, J.; De Vuyst, L.; Pescuma, M. Metabolomics as a tool for the comprehensive understanding of fermented and functional foods with lactic acid bacteria. *Food Res. Int.* **2013**, *54*, 1152–1161. [CrossRef]
73. Ordoñez-Araque, R.; Landines Vera, E.; Urresto-Villegas, J.; Caicedo-Jaramillo, C. Microorganisms during cocoa fermentation: Systematic review. *Foods Raw Mater.* **2020**, *8*, 155–162. [CrossRef]
74. Kustyawati, M.E.; Setyani, S. Effect of the Addition of Mixed Inoculums on the Chemical and Microbiological Changes During Cocoa Fermentation. *J. Teknol. Ind. dan Has. Pertan.* **2008**, *13*, 73–84.
75. Ouattara, D.H.; Ouattara, H.G.; Goualie, B.G.; Kouame, L.M.; Niamke, S.L. Biochemical and functional properties of lactic acid bacteria isolated from Ivorian cocoa fermenting beans. *J. Appl. Biosci.* **2014**, *77*, 6489–6499. [CrossRef]
76. Miguel MG DC, P.; de Castro Reis, L.V.; Efraim, P.; Santos, C.; Lima, N.; Schwan, R.F. Cocoa fermentation: Microbial identification by MALDI-TOF MS, and sensory evaluation of produced chocolate. *LWT* **2017**, *77*, 362–369. [CrossRef]
77. Hatmi, R.U.; Kobarsih, M.; Cahyaningrum, N. Fungi Level Analysis of Cocoa Beans Based on Fermentation Box Type and Duration. *Procedia Food Sci.* **2015**, *3*, 371–382. [CrossRef]
78. Guehi, T.S.; Dadie, A.T.; Koffi, K.P.B.; Dabonne, S.; Ban-Koffi, L.; Kedjebo, K.D.; Nemlin, G.J. Performance of different fermentation methods and the effect of their duration on the quality of raw cocoa beans. *Int. J. Food Sci. Technol.* **2010**, *45*, 2508–2514. [CrossRef]
79. Moreno-Zambrano, M.; Ullrich, M.S.; Hütt, M.-T. Exploring cocoa bean fermentation mechanisms by kinetic modelling. *R. Soc. Open Sci.* **2022**, *9*, 210274. [CrossRef]
80. Azzolini, M.; Fedrizzi, B.; Tosi, E.; Finato, F.; Vagnoli, P.; Scrinzi, C.; Zapparoli, G. Effects of *Torulaspora delbrueckii* and *Saccharomyces cerevisiae* mixed cultures on fermentation and aroma of Amarone wine. *Eur. Food Res. Technol.* **2012**, *235*, 303–313. [CrossRef]
81. De Vuyst, L.; Lefeber, T.; Papalexandratou, Z.; Camu, N. The Functional Role of Lactic Acid Bacteria in Cocoa Bean Fermentation. In *Biotechnology of Lactic Acid Bacteria*; Wiley: Hoboken, NJ, USA, 2010; pp. 301–325, ISBN 9780813820866.
82. Almeida, O.G.G.; Pinto, U.M.; Matos, C.B.; Frazilio, D.A.; Braga, V.F.; von Zeska-Kress, M.R.; De Martinis, E.C.P. Does Quorum Sensing play a role in microbial shifts along spontaneous fermentation of cocoa beans? An in silico perspective. *Food Res. Int.* **2020**, *131*, 109034. [CrossRef]
83. Freitas, S.R. Cocoa Fermentations Conducted with a Defined Microbial Cocktail Inoculum. *Appl. Environ. Microbiol.* **1998**, *64*, 1477–1483. [CrossRef]
84. Farrera, L.; Colas de la Noue, A.; Strub, C.; Guibert, B.; Kouame, C.; Grabulos, J.; Montet, D.; Teyssier, C. Towards a Starter Culture for Cocoa Fermentation by the Selection of Acetic Acid Bacteria. *Fermentation* **2021**, *7*, 42. [CrossRef]
85. Calvo, A.M.; Botina, B.L.; García, M.C.; Cardona, W.A.; Montenegro, A.C.; Criollo, J. Dynamics of cocoa fermentation and its effect on quality. *Sci. Rep.* **2021**, *11*, 16746. [CrossRef] [PubMed]
86. Zoi, P.; Gwen, F.; Edwina, R.; Carlos, J.J.; Freddy, A.; Heide-Marie, D.; Luc, D.V. Species Diversity, Community Dynamics, and Metabolite Kinetics of the Microbiota Associated with Traditional Ecuadorian Spontaneous Cocoa Bean Fermentations. *Appl. Environ. Microbiol.* **2011**, *77*, 7698–7714. [CrossRef]
87. Pereira, G.V.d.M.; Miguel, M.G.d.C.P.; Ramos, C.L.; Schwan, R.F. Microbiological and Physicochemical Characterization of Small-Scale Cocoa Fermentations and Screening of Yeast and Bacterial Strains to Develop a Defined Starter Culture. *Appl. Environ. Microbiol.* **2012**, *78*, 5395–5405. [CrossRef] [PubMed]
88. Fernández Maura, Y.; Balzarini, T.; Clapé Borges, P.; Evrard, P.; De Vuyst, L.; Daniel, H.-M. The environmental and intrinsic yeast diversity of Cuban cocoa bean heap fermentations. *Int. J. Food Microbiol.* **2016**, *233*, 34–43. [CrossRef]
89. Lagunes Gálvez, S.; Loiseau, G.; Paredes, J.L.; Barel, M.; Guiraud, J.-P. Study on the microflora and biochemistry of cocoa fermentation in the Dominican Republic. *Int. J. Food Microbiol.* **2007**, *114*, 124–130. [CrossRef]
90. Jamili; Yanti, N.A.; Susilowati, P.E. Diversity and the role of yeast in spontaneous cocoa bean fermentation from Southeast Sulawesi, Indonesia. *Biodiversitas* **2016**, *17*, 90–95. [CrossRef]
91. Koff, O.; Samagaci, L.; Goualie, B.; Niamke, S. Diversity of Yeasts Involved in Cocoa Fermentation of Six Major Cocoa-Producing Regions in Ivory Coast. *Eur. Sci. J. ESJ* **2017**, *13*, 496–516. [CrossRef]
92. Koné, M.K.; Guéhi, S.T.; Durand, N.; Ban-Koffi, L.; Berthiot, L.; Tachon, A.F.; Brou, K.; Boulanger, R.; Montet, D. Contribution of predominant yeasts to the occurrence of aroma compounds during cocoa bean fermentation. *Food Res. Int.* **2016**, *89*, 910–917. [CrossRef]
93. Daniel, H.-M.; Vrancken, G.; Takrama, J.F.; Camu, N.; De Vos, P.; De Vuyst, L. Yeast diversity of Ghanaian cocoa bean heap fermentations. *FEMS Yeast Res.* **2009**, *9*, 774–783. [CrossRef]
94. Hernández-Hernández, C.; López-Andrade, P.A.; Ramírez-Guillermo, M.A.; Guerra Ramírez, D.; Caballero Pérez, J.F. Evaluation of different fermentation processes for use by small cocoa growers in Mexico. *Food Sci. Nutr.* **2016**, *4*, 690–695. [CrossRef] [PubMed]
95. Adiko, E.; Ouattara, H.; Doué, G.; Niamké, S. Assessment of the Diversity of Lactic Acid Bacteria Involved in Cocoa Fermentation of Six Main Cocoa Producing Regions of Côte d'Ivoire. *Annu. Res. Rev. Biol.* **2018**, *27*, 1–16. [CrossRef]
96. Soumahoro, S.; Ouattara, H.G.; Droux, M.; Nasser, W.; Niamke, S.L.; Reverchon, S. Acetic acid bacteria (AAB) involved in cocoa fermentation from Ivory Coast: Species diversity and performance in acetic acid production. *J. Food Sci. Technol.* **2020**, *57*, 1904–1916. [CrossRef] [PubMed]
97. Ouattara, H.D.; Ouattara, H.G.; Droux, M.; Reverchon, S.; Nasser, W.; Niamke, S.L. Lactic acid bacteria involved in cocoa beans fermentation from Ivory Coast: Species diversity and citrate lyase production. *Int. J. Food Microbiol.* **2017**, *256*, 11–19. [CrossRef] [PubMed]

98. Chagas Junior, G.C.; Ferreira, N.R.; Gloria, M.B.A.; Gobira, R.M.; Maia, F.D.; Lopes, A.S. Identification of Lactic Acid Bacteria on Raw Material for Cocoa Bean Fermentation in the Brazilian Amazon. *Fermentation* **2022**, *8*, 199. [CrossRef]
99. De Bruyne, K.; Camu, N.; De Vuyst, L.; Vandamme, P. *Weissella fabaria* sp. nov., from a Ghanaian cocoa fermentation. *Int. J. Syst. Evol. Microbiol.* **2010**, *60*, 1999–2005. [CrossRef] [PubMed]
100. Almeida, O.G.G.; De Martinis, E.C.P. Metagenome-Assembled Genomes Contribute to Unraveling of the Microbiome of Cocoa Fermentation. *Appl. Environ. Microbiol.* **2021**, *87*, e00584-21. [CrossRef]
101. Magalhães da Veiga Moreira, I.; De Figueiredo Vilela, L.; Da Cruz Pedroso Miguel, M.G.; Santos, C.; Lima, N.; Freitas Schwan, R. Impact of a Microbial Cocktail Used as a Starter Culture on Cocoa Fermentation and Chocolate Flavor. *Molecules* **2017**, *22*, 766. [CrossRef]
102. Illeghems, K.; De Vuyst, L.; Weckx, S. Comparative genome analysis of the candidate functional starter culture strains *Lactobacillus fermentum* 222 and *Lactobacillus plantarum* 80 for controlled cocoa bean fermentation processes. *BMC Genomics* **2015**, *16*, 766. [CrossRef]
103. Nadia, A.B.; Jannah, S.N.; Purwantisari, S. Isolation and Characterization of Lactic Acid Bacteria from *Apis mellifera* Stomach and Their Potential as Antibacterial Using In Vitro Test Against Growth of *Staphylococcus aureus* and *Salmonella typhimurium*. *NICHE J. Trop. Biol.* **2020**, *3*, 35–44.
104. Korcari, D.; Fanton, A.; Ricci, G.; Rabitti, N.S.; Laureati, M.; Hogenboom, J.; Pellegrino, L.; Emide, D.; Barbiroli, A.; Fortina, M.G. Fine Cocoa Fermentation with Selected Lactic Acid Bacteria: Fermentation Performance and Impact on Chocolate Composition and Sensory Properties. *Foods* **2023**, *12*, 340. [CrossRef]
105. Lefeber, T.; Papalexandratou, Z.; Gobert, W.; Camu, N.; De Vuyst, L. On-farm implementation of a starter culture for improved cocoa bean fermentation and its influence on the flavour of chocolates produced thereof. *Food Microbiol.* **2012**, *30*, 379–392. [CrossRef]
106. Hill, C.; Guarner, F.; Reid, G.; Gibson, G.R.; Merenstein, D.J.; Pot, B.; Morelli, L.; Canani, R.B.; Flint, H.J.; Salminen, S.; et al. The International Scientific Association for Probiotics and Prebiotics consensus statement on the scope and appropriate use of the term probiotic. *Nat. Rev. Gastroenterol. Hepatol.* **2014**, *11*, 506–514. [CrossRef]
107. Foong, Y.J.; Lee, S.T.; Ramli, N.; Tan, Y.N.; Ayob, M.K. Incorporation of Potential Probiotic *Lactobacillus plantarum* Isolated from Fermented Cocoa Beans into Dark Chocolate: Bacterial Viability and Physicochemical Properties Analysis. *J. Food Qual.* **2013**, *36*, 164–171. [CrossRef]
108. Zarić, D.B.; Bulatović, M.L.; Rakin, M.B.; Krunić, T.Ž.; Lončarević, I.S.; Pajin, B.S. Functional, rheological and sensory properties of probiotic milk chocolate produced in a ball mill. *RSC Adv.* **2016**, *6*, 13934–13941. [CrossRef]
109. Faccinetto-Beltrán, P.; Gómez-Fernández, A.R.; Santacruz, A.; Jacobo-Velázquez, D.A. Chocolate as Carrier to Deliver Bioactive Ingredients: Current Advances and Future Perspectives. *Foods* **2021**, *10*, 2056. [CrossRef] [PubMed]
110. Chuah, L.-O.; Foo, H.L.; Loh, T.C.; Mohammed Alitheen, N.B.; Yeap, S.K.; Abdul Mutalib, N.E.; Abdul Rahim, R.; Yusoff, K. Postbiotic metabolites produced by *Lactobacillus plantarum* strains exert selective cytotoxicity effects on cancer cells. *BMC Complement. Altern. Med.* **2019**, *19*, 114. [CrossRef]
111. Gomes, R.J.; Borges, M.D.F.; Rosa, M.D.F.; Castro-Gómez, R.J.H.; Spinosa, W.A. Acetic acid bacteria in the food industry: Systematics, characteristics and applications. *Food Technol. Biotechnol.* **2018**, *56*, 139–151. [CrossRef]
112. Wang, Y.; Wu, J.; Lv, M.; Shao, Z.; Hungwe, M.; Wang, J.; Bai, X.; Xie, J.; Wang, Y.; Geng, W. Metabolism Characteristics of Lactic Acid Bacteria and the Expanding Applications in Food Industry. *Front. Bioeng. Biotechnol.* **2021**, *9*, 612285. [CrossRef]
113. Afrida, I.R. Detection of Acid Bacteria in the Process of Cocoa Production. *Proceeding Biol. Educ. Conf.* **2016**, *13*, 822–826.
114. Akinfala, T.O.; Houbraken, J.; Sulyok, M.; Adedeji, A.R.; Odebode, A.C.; Krska, R.; Ezekiel, C.N. Moulds and their secondary metabolites associated with the fermentation and storage of two cocoa bean hybrids in Nigeria. *Int. J. Food Microbiol.* **2020**, *316*, 108490. [CrossRef]
115. Dumadi, S.R. The Moisture Content Increase of Dried Cocoa Beans During Storage at Room Temperature. *J. Ilm. Teknol. Energi* **2011**, *1*, 45–54.
116. Marwati, T.; Purwaningsih; Djaafar, T.F.; Sari, A.B.T. Hernani Inhibition the growth of fungi and improving the quality of cocoa beans through fermentation using lactic acid bacteria. *IOP Conf. Ser. Earth Environ. Sci.* **2021**, *807*, 22048. [CrossRef]
117. Beg, M.S.; Ahmad, S.; Jan, K.; Bashir, K. Status, supply chain and processing of—A review. *Trends Food Sci. Technol.* **2017**, *66*, 108–116. [CrossRef]
118. Dina, S.F.; Limbong, H.P.; Rambe, S.M. Design and Performance Test of Solar Assisted Heat Pump Dryer for Cocoa Bean Drying. *J. Ris. Teknol. Ind.* **2018**, *12*, 21–33. [CrossRef]
119. Hayati, R.; Yusmanizar, Y.; Mustafril, M.; Fauzi, H. Study of Fermentation and Drying Temperature in Cacao Quality (*Theobroma cacao* L.). *J. Keteknikan Pertan.* **2012**, *26*, 129–135. [CrossRef]
120. Komolafe, C.A.; Waheed, M.A.; Kuye, S.I.; Adewumi, B.A.; Oluwaleye, I.O.; Olayanju, T.M.A. Sun drying of cocoa with firebrick thermal storage materials. *Int. J. Energy Res.* **2020**, *44*, 7015–7025. [CrossRef]
121. Guda, P.; Gadhe, S.; Jakkula, S. Drying of Cocoa Beans by Using Different Techniques. *Int. J. Agric. Innov. Res.* **2017**, *5*, 859–865.
122. Salazar, E.; Valenzuela, R.; Aguilar, M.; Aranda, N.; Sotelo, A.; Chire-Fajardo, G.C.; Ureña, M. Physicochemical properties and microbial group behavior of postharvest peruvian cocoa bean (*Theobroma cacao* L.). *Enfoque UTE* **2020**, *11*, 48–56. [CrossRef]
123. Sangronis, E.; Soto, M.J.; Valero, Y.; Buscema, I. Cascarilla de cacao venezolano como materia prima para infusiones. *Arch. Latinoam. Nutr.* **2014**, *64*, 123–130.

124. *ISO 2451:2017*; Cocoa Beans—Specification and Quality Requirements. 3rd ed. International Organization for Standardization (ISO): Geneva, Switzerland, 2017; Volume 2017.
125. Djaafar, T.F.; Utami, T.; Marwati, T.; Pramesi, P.C.; Wikandari, R.; Rahayu, E.S. The assessment of good manufacturing practices (GMP) implementation and critical control point (CCP) determination on the cocoa powder processing in Agricultural Techno Park Nglanggeran, Yogyakarta. *IOP Conf. Ser. Mater. Sci. Eng.* **2020**, *980*, 12034. [CrossRef]
126. Code of Practice by Codex Alimentarius Commission (CAC). *Code of Practice for the Prevention and Reduction of Ochratoxin A Contamination in Cocoa (CAC/RCP 72-2013)*; Food and Agriculture Organization (FAO): Rome, Italy, 2013.
127. Delgado-Ospina, J.; Di Mattia, C.D.; Paparella, A.; Mastrocola, D.; Martuscelli, M.; Chaves-Lopez, C. Effect of Fermentation, Drying and Roasting on Colombian Criollo Cocoa Beans and Shells. *Foods* **2020**, *9*, 520. [CrossRef] [PubMed]
128. Jeong, S.-M.; Kim, S.-Y.; Kim, D.-R.; Jo, S.-C.; Nam, K.C.; Ahn, D.U.; Lee, S.-C. Effect of Heat Treatment on the Antioxidant Activity of Extracts from Citrus Peels. *J. Agric. Food Chem.* **2004**, *52*, 3389–3393. [CrossRef]
129. Waldemar, S.; Julia, D.; André, L. Increased Biomass Production by Mesophilic Food-Associated Bacteria through Lowering the Growth Temperature from 30 °C to 10 °C. *Appl. Environ. Microbiol.* **2016**, *82*, 3754–3764. [CrossRef]
130. Anne, D.-R.; Jean-Philippe, L.; Jean-Marie, P. *Enterobacter* spp.: Update on Taxonomy, Clinical Aspects, and Emerging Antimicrobial Resistance. *Clin. Microbiol. Rev.* **2019**, *32*, e00002-19. [CrossRef]
131. Norlia, M.; Jinap, S.; Nor-Khaizura, M.A.R.; Radu, S.; Samsudin, N.I.P.; Azri, F.A. Aspergillus section Flavi and Aflatoxins: Occurrence, Detection, and Identification in Raw Peanuts and Peanut-Based Products Along the Supply Chain. *Front. Microbiol.* **2019**, *10*, 2602. [CrossRef]
132. Taniwaki, M.H.; Hocking, A.D.; Pitt, J.I.; Fleet, G.H. Growth and mycotoxin production by fungi in atmospheres containing 80% carbon dioxide and 20% oxygen. *Int. J. Food Microbiol.* **2010**, *143*, 218–225. [CrossRef]
133. Fagbohun, E.; Anibijuwon, I.; Egbebi, O.; Lawal, O. Fungi Associated with Spoilage of Dried Cocoa Beans During Storage in Ekti State of Nigeria. *J. Microbiol. Biotechnol. food Sci.* **2011**, *1*, 204–214.
134. Askun, T.; Elfem, R.; Taskin, E. Comparison of Rose Bengal Chloramphenicol Agar and Dichloran Glycerol Agar (DG18) for Enumeration and Isolation of Moulds from Raisins. *J. Appl. Biol. Sci.* **2007**, *1*, 71–74.
135. Koua, B.K.; Koffi, P.M.E.; Gbaha, P. Evolution of shrinkage, real density, porosity, heat and mass transfer coefficients during indirect solar drying of cocoa beans. *J. Saudi Soc. Agric. Sci.* **2019**, *18*, 72–82. [CrossRef]
136. Sánchez-Hervás, M.; Gil, J.V.; Bisbal, F.; Ramón, D.; Martínez-Culebras, P.V. Mycobiota and mycotoxin producing fungi from cocoa beans. *Int. J. Food Microbiol.* **2008**, *125*, 336–340. [CrossRef]
137. Dogan, A.; Erkan, M. Responses of High Carbon Dioxide Concentration on Postharvest Quality of Fresh Fig Fruit during Storage. *Horticulturae* **2023**, *9*, 293. [CrossRef]
138. Jin Choi, H.; Seuk Bae, Y.; Soo Lee, J.; Hea Park, M.; Gang Kim, J. Effects of Carbon Dioxide Treatment and Modified Atmosphere Packaging on the Quality of Long Distance Transporting "Maehyang" Strawberry. *Agric. Sci.* **2016**, *7*, 813–821. [CrossRef]
139. Mabbett, T. Quality control for stored coffee and cocoa. *Int. Pest Control. Mag.* **2013**, *55*, 91–94.
140. Jonfia-Essien, W.A.; Navarro, S.; Villers, P. Hermetic Storage: A Novel Approach to The Protection of Cocoa Beans. *African Crop Sci. J.* **2010**, *18*, 59–68. [CrossRef]
141. McClure, A.P.; Hopfer, H.; Grün, I.U. Optimizing consumer acceptability of 100% chocolate through roasting treatments and effects on bitterness and other important sensory characteristics. *Curr. Res. Food Sci.* **2022**, *5*, 167–174. [CrossRef] [PubMed]
142. Maïmouna Kouamé, L. Cocoa Fermentation from Agnéby-Tiassa: Biochemical Study of Microflora. *Am. J. Biosci.* **2015**, *3*, 203. [CrossRef]
143. Dulce, V.-R.; Anne, G.; Manuel, K.; Carlos, A.-A.; Jacobo, R.-C.; Sergio de Jesús, C.-E.; Eugenia, L.-C. Cocoa bean turning as a method for redirecting the aroma compound profile in artisanal cocoa fermentation. *Heliyon* **2021**, *7*, e07694. [CrossRef]
144. Anvoh, K.; Guéhi, T.; Beugré, G.; Kinimo, J.; Gnakri, D. Comparison of Biochemical Changes During Alcoholic Fermentation of Cocoa Juice Conducted by Spontaneous and Induced Processes for the Production of Ethanol. *Afr. J. Food Agric. Nutr. Dev.* **2010**, *10*, 2740–2754. [CrossRef]
145. Cempaka, L.; Aliwarga, L.; Purwo, S.; Penia Kresnowati, M.T.A. Dynamics of cocoa bean pulp degradation during cocoa bean fermentation: Effects of yeast starter culture addition. *J. Math. Fundam. Sci.* **2014**, *46*, 14–25. [CrossRef]
146. León-Roque, N.; Abderrahim, M.; Nuñez-Alejos, L.; Arribas, S.M.; Condezo-Hoyos, L. Prediction of fermentation index of cocoa beans (*Theobroma cacao* L.) based on color measurement and artificial neural networks. *Talanta* **2016**, *161*, 31–39. [CrossRef] [PubMed]
147. Romero-Cortes, T.; Salgado-Cervantes, M.A.; García-Alamilla, P.; García-Alvarado, M.A.; del C Rodríguez-Jimenes, G.; Hidalgo-Morales, M.; Robles-Olvera, V. Relationship between fermentation index and other biochemical changes evaluated during the fermentation of Mexican cocoa (*Theobroma cacao*) beans. *J. Sci. Food Agric.* **2013**, *93*, 2596–2604. [CrossRef]
148. Indiarto, R.; Pranoto, Y.; Santoso, U. Supriyanto In vitro antioxidant activity and profile of polyphenol compounds extracts and their fractions on cacao beans. *Pakistan J. Biol. Sci.* **2019**, *22*, 34–44. [CrossRef]
149. Indiarto, R.; Subroto, E.; Sukri, N.; Djali, M. Cocoa (*Theobroma cacao* L.) beans processing technology: A review of flavonoid changes. *Asian J. Plant Sci.* **2021**, *20*, 684–693. [CrossRef]
150. Payne, M.J.; Hurst, W.J.; Miller, K.B.; Rank, C.; Stuart, D.A. Impact of fermentation, drying, roasting, and Dutch processing on epicatechin and catechin content of cacao beans and cocoa ingredients. *J. Agric. Food Chem.* **2010**, *58*, 10518–10527. [CrossRef] [PubMed]

151. Kresnowati, M.T.A.P.; Suryani, L.; Affifah, M. Improvement of Cocoa Beans Fermentation by LAB Starter Addition. *J. Med. Bioeng.* **2013**, *2*, 274–278. [CrossRef]
152. Ho, V.T.T.; Zhao, J.; Fleet, G. The effect of lactic acid bacteria on cocoa bean fermentation. *Int. J. Food Microbiol.* **2015**, *205*, 54–67. [CrossRef]
153. Romero Vergel, A.P.; Camargo Rodriguez, A.V.; Ramirez, O.D.; Arenas Velilla, P.A.; Gallego, A.M. A Crop Modelling Strategy to Improve Cacao Quality and Productivity. *Plants* **2022**, *11*, 157. [CrossRef]
154. Tardzenyuy, M.E.; Jianguo, Z.; Akyene, T.; Mbuwel, M.P. Improving cocoa beans value chain using a local convection dryer: A case study of Fako division Cameroon. *Sci. Afr.* **2020**, *8*, e00343. [CrossRef]
155. Davit, J.M.; Yusuf, R.P.; Yudari, D.A.S. Effect of Fermentation Cocoa Processing and Non Fermentation on The Quality and Pricing Products on Tunjung Sari Unit Produktive in Tabanan District. *J. Agribus. Agritour.* **2013**, *2*, 191–203.
156. Romanens, E.; Pedan, V.; Meile, L.; Miescher Schwenninger, S. Influence of two anti-fungal Lactobacillus fermentum-Saccharomyces cerevisiae co-cultures on cocoa bean fermentation and final bean quality. *PLoS ONE* **2020**, *15*, e0239365. [CrossRef]
157. Santhanam Menon, A.; Hii, C.L.; Law, C.L.; Shariff, S.; Djaeni, M. Effects of drying on the production of polyphenol-rich cocoa beans. *Dry. Technol.* **2017**, *35*, 1799–1806. [CrossRef]
158. Ariyanti, M. Quality Characteristics of Cocoabeans (*Theobroma Cacao* L.) With Time Fermentation treatment Based on Iso 2323-2008. *Indones. J. Ind. Res.* **2017**, *12*, 34–42. [CrossRef]
159. Chire, G.C.; Valdivia, R.A.; Milber, O.U. Ochratoxin A in Cacao and Derivatives. *Prev. Meas.* **2014**, *17*, 9–15. [CrossRef]
160. Dano, S.D.; Manda, P.; Dembélé, A.; Kouassi Abla, A.M.; Bibaud, J.H.; Gouet, J.Z.; Ze Maria Sika, C.B. Influence of Fermentation and Drying Materials on the Contamination of Cocoa Beans by Ochratoxin A. *Toxins* **2013**, *5*, 2310–2323. [CrossRef] [PubMed]
161. Sukmawati, D.; Arman, Z.; Hasanah, R.; Balqis, M.; Setiawan, A.; Tafrijiyyah, F.; Sari, R.; Supiyani, A.; Prihantini, N.B.; Husna, S.N.A.; et al. Application of yeasts isolated from fermented cocoa beans for biocontrol of pathogenic mold in chocolate fruit. *J. Phys. Conf. Ser.* **2021**, *1869*, 012042. [CrossRef]
162. Zainuri; Sjah, T.; Prameswari, N.; Werdiningsih, W.; Tarmizi. Good agricultural and postharvest handling practices of Cocoa pods in Lombok to meet Cocoa bean quality for the global market. *IOP Conf. Ser. Earth Environ. Sci.* **2021**, *712*, 12028. [CrossRef]
163. Praseptiangga, D.; Guevara Zambrano, J.M.; Sanjaya, A.P.; Muhammad, D.R.A. Challenges in the development of the cocoa and chocolate industry in Indonesia: A case study in Madiun, East Java. *AIMS Agric. Food* **2020**, *5*, 920–937. [CrossRef]

horticulturae

MDPI

Article

Ripening Process of Tomato Fruits Postharvest: Impact of Environmental Conditions on Quality and Chlorophyll *a* Fluorescence Characteristics

Hyo-Gil Choi [1,2,*] and Kyoung-Sub Park [3,*]

1 Department of Horticulture, Kongju National University, Yesan 32439, Republic of Korea
2 Agriculture and Fisheries Life Science Research Institute, Kongju National University, Yesan 32439, Republic of Korea
3 Department of Horticultural Science, Mokpo National University, Muan 58554, Republic of Korea
* Correspondence: hg1208@kongju.ac.kr (H.-G.C.); unicos75@mokpo.ac.kr (K.-S.P.)

Abstract: This study aimed to investigate the combined effects of temperature and light conditions on tomato maturation. Tomato fruits that had completed volumetric growth at the mature green stage were harvested and matured in growth chambers composed of two temperature conditions (daytime–nighttime: 30–20 °C or 20–15 °C) and two light conditions (0 $\mu mol \cdot m^{-2} \cdot s^{-1}$ or 400 $\mu mol \cdot m^{-2} \cdot s^{-1}$), which were set for 12 h each day and night. Our findings indicate that tomato ripening was significantly influenced by both light and temperature. Tomatoes that matured under low-temperature conditions in the absence of light took more than three times longer to transition from the green stage to the breaker stage compared to those matured under high-temperature conditions with light exposure. Notably, tomato fruit maturation occurred at a faster rate under low-temperature and light conditions than under high-temperature and dark conditions. Changes in chlorophyll *a* fluorescence parameters were observed throughout the ripening process of tomato fruits. Tomato fruits ripened under low-temperature and dark conditions exhibited significantly lower NPQ (non-photochemical quenching) and R_{FD} (relative fluorescence decrease) values compared to other treatments, while their F_O (initial fluorescence) and F_M (maximum fluorescence) values were higher. The accumulation of sugar in tomato fruits was observed to be more influenced by light than temperature. On the other hand, the highest levels of phenolic content and lycopene were observed in tomato fruits matured under high-temperature and light conditions. Antioxidant activities, as measured by ABTS and DPPH assays, were highest in mature tomato fruits under high-temperature and light conditions, while they were lowest in fruits under low-temperature and dark conditions. In conclusion, this study highlights the critical role of temperature and light as crucial environmental factors influencing tomato maturation. Understanding these factors can contribute to optimizing postharvest conditions and enhancing fruit quality in the tomato industry.

Keywords: light; lycopene; maturity stage; sugar content; temperature

Citation: Choi, H.-G.; Park, K.-S. Ripening Process of Tomato Fruits Postharvest: Impact of Environmental Conditions on Quality and Chlorophyll *a* Fluorescence Characteristics. *Horticulturae* **2023**, 9, 812. https://doi.org/10.3390/horticulturae9070812

Academic Editors: Yang Bi, Yongcai Li and Di Gong

Received: 13 June 2023
Revised: 11 July 2023
Accepted: 13 July 2023
Published: 14 July 2023

1. Introduction

Tomato (*Solanum lycopersicum* L.) is one of the most widely grown horticultural crops worldwide [1], and is used both for raw consumption and for processing. In Korea, where the four seasons are distinct, tomato is mainly cultivated in a greenhouse as an annual plant and consumed raw [2]. Various metabolites of tomatoes are known to reduce the risk of diseases such as cancer, heart disease, and cardiovascular diseases [3,4]. Lycopene is the main carotenoid found in tomatoes, accounting for about 80% of all carotenoids produced [5]. It is well-known for its high antioxidant effects [6]. The color change in tomato fruit promoted by carotenoids, such as lycopene, is one of the major ripening characteristics [7]. As tomatoes are climacteric fruits, their synthesis is closely related to the rise in ethylene production. This process can even be triggered in detached fruit during storage periods [8].

Tomato fruits undergo significant changes in the content of sugars, organic acids, lycopene, phenolic compounds, and other phytochemicals as they ripen from the mature green stage to the red stage [9]. Furthermore, depending on the environmental conditions during the ripening process from the mature green stage, tomatoes exhibit significant differences in the accumulation of sugars, organic acids, and other components in the fruit [10]. The interaction of soluble sugars and organic acids in tomato fruits is an important quality factor due to its direct influence on the sweetness and sourness, which contribute to the overall taste intensity of tomatoes [11]. The antioxidant activity of tomato fruit is an important quality factor and is verified through analyses such as ABTS and DPPH radical quenching [12]. This antioxidant activity is influenced by conditions during the ripening process, and the ripening of tomatoes is accompanied by a gradual increase in various antioxidant compounds and a corresponding increase in antioxidant activity [13].

One of the most important environmental factors in the ripening process of tomatoes is temperature. The number of days required for tomato fruits to progress from the green stage to the breaker stage during ripening decreases as the temperature increases towards the optimum temperature for growth [14]. The content of lycopene increases during the ripening process of tomatoes, and it is know that this increase is greatly affected by temperature [15]. Ambient light is another important environmental factor affecting the color change of tomato fruit during the ripening process [1,16]. It is known that the phytochrome inherent in fruits plays a crucial role in regulating various aspects of tomato fruit ripening, including the accumulation of carotenoids such as a lycopene [17,18].

Horticultural crops, as living organisms, rely on photosynthesis to generate their own energy, making light an essential environmental factor for their growth, just like temperature. Light not only affects plant photosynthesis but also plays a vital role in signaling various metabolic processes [19,20]. One such process influenced by light conditions is the metabolism of ethylene, a hormone closely associated with plant maturation and aging [21]. In the case of postharvest peaches, exposure to blue light during ripening triggers the upregulation of ethylene biosynthesis genes, resulting in an accelerated fruit ripening process [22]. Similarly, when strawberries are exposed to light during ripening, the genes responsible for controlling pigmentation are upregulated, contributing to enhanced pigmentation [23]. Furthermore, treating harvested vegetables or fruits with LED light has been shown to have significant positive effects on quality and shelf life extension [24]. LED light treatment after harvesting can enhance the overall quality and extend the freshness of these produce items. Moreover, light has been found to impact the quality of tomato fruits by modulating the levels of soluble sugars, lycopene content, antioxidant activity, and organic acid content during the ripening process [25,26]. These findings highlight the multifaceted influence of light on the ripening of horticultural crops, showcasing its importance in horticultural crop postharvest practices.

The chlorophyll *a* fluorescence technique is a valuable tool for the non-destructive analysis of physiological responses, allowing confirmation of the plant's condition in response to various environmental conditions [27,28]. In addition to diagnosing the plant's stress state based on its cultivation environment [29], recent studies have also been conducted to classify the ripening stages of tomato fruits using chlorophyll *a* fluorescence [30,31]. In addition to representing plant conditions through the analysis of chlorophyll *a* fluorescence parameters, studies are being conducted to visually express it through images [32,33].

Therefore, this study was conducted to determine the extent to which tomato fruit is affected by temperature and light, which are the most sensitive environmental factors in the ripening process. Additionally, the study analyzed the chlorophyll fluorescence reaction to assess the differences in ripening stages based on each parameter's value and fluorescence image. The study also aims to detect fruits at the right time for harvesting through non-destructive chlorophyll *a* fluorescence imaging.

2. Materials and Methods

2.1. Plant Materials and Storage Conditions

The tomatoes (*Solanum lycopersicum* L. cv. Red 244) were grown hydroponically using a rock-wool medium in a Benno-type greenhouse located in Nonsan, Korea. The tomato fruits were harvested three times between December 2022 and March 2023 to repeat the experiment three times. On the morning of the experimental treatment, following volumetric growth, fruits in the mature green stage of the Red 244 cultivar were harvested, with an average weight of approximately 140 g. The harvested tomato fruits were transferred to growth chambers with different temperature and light conditions (Figure 1), where they were allowed to ripen fully until they reached the red stage.

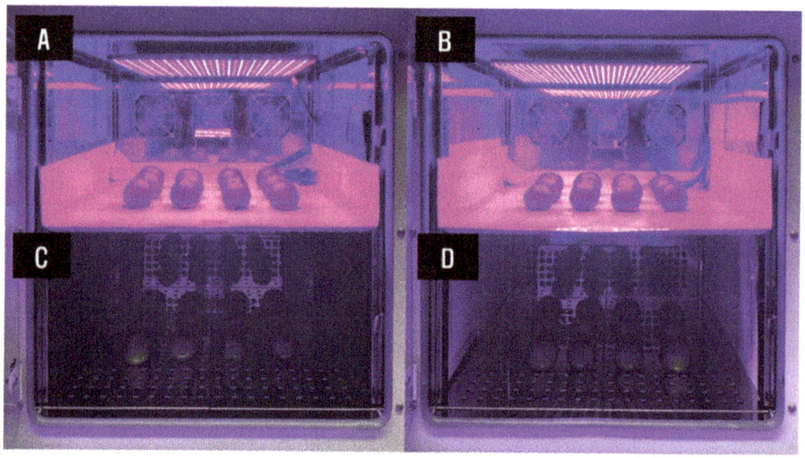

Figure 1. Tomato ripening under four different environmental conditions with varying temperatures and light levels. (**A**) The day and night temperatures were set at 30 °C and 20 °C, respectively, with a light intensity of 400 $\mu mol \cdot m^{-2} \cdot s^{-1}$ (STL). (**B**) The day and night temperatures were set at 20 °C and 15 °C, respectively, with a light intensity of 400 $\mu mol \cdot m^{-2} \cdot s^{-1}$ (WTL). (**C**) The day and night temperatures were set at 30 °C and 20 °C, respectively, with a light intensity of 0 $\mu mol \cdot m^{-2} \cdot s^{-1}$ (STD). (**D**) The day and night temperatures were set at 20 °C and 15 °C, respectively, with a light intensity of 0 $\mu mol \cdot m^{-2} \cdot s^{-1}$ (WTD).

In the management of tomato greenhouses in Korea, the cooling system is activated when the daytime and nighttime temperatures in summer exceed 30 °C and 20 °C, respectively. During winter, the average greenhouse temperature hovers around 20 °C, and heating is initiated when the night temperature drops below 15 °C. Furthermore, considering the performance of the artificial light used, an intensity of 400 $\mu mol \cdot m^{-2} \cdot s^{-1}$ is considered suitable for the efficient application of artificial light sources in greenhouses. Consequently, this study was designed to reflect these conditions. To create an environment in the growth chamber similar to that of a greenhouse, the day and night time were adjusted to 12 h each, and the relative humidity was maintained at 50–60%. As the light source for the experiment, an LED composed of blue and red in a ratio of 2:8 was used. The growth chambers with different temperature and light conditions were configured as follows:

- STL—the day and night temperatures were set at 30 °C and 20 °C, respectively, with a light intensity of 400 $\mu mol \cdot m^{-2} \cdot s^{-1}$;
- WTL—the day and night temperatures were set at 20 °C and 15 °C, respectively, with a light intensity of 400 $\mu mol \cdot m^{-2} \cdot s^{-1}$;
- STD—the day and night temperatures were set at 30 °C and 20 °C, respectively, with a light intensity of 0 $\mu mol \cdot m^{-2} \cdot s^{-1}$;

- WTD—the day and night temperatures were set at 20 °C and 15 °C, respectively, with a light intensity of 0 μmol·m^{-2}·s^{-1}.

2.2. Chlorophyll *a* Fluorescence

The chlorophyll *a* fluorescence induction kinetics of tomato fruits were measured after classifying the fruits into six ripening stages: mature green (where the surface of the tomato is completely green in color), breaker (with color transitioning from green to tannish-yellow, pink or red on no more than 10% of the surface), turning (where more than 10% but not more than 30% of the surface shows color ranging from green to tannish-yellow, pink, red, or a combination thereof), pink (where more than 30% but not more than 60% of the surface, in aggregate, displays pink or red coloration), light red (where more than 60% of the surface, in aggregate, shows pinkish-red or red), and red (where more than 90% of the surface, in aggregate, exhibits a red coloration). These ripening stages are defined according to the United States standards for grades of fresh tomatoes [34]. Three fruits were measured three times for each treatment. Prior to the chlorophyll *a* fluorescence analysis, the fruits were allowed to adapt to darkness for 30 min. The measurements were conducted using a fluorometer imaging system (FluorCam FC800; Photon Systems Instruments, Brno, Czech Republic) equipped with a progressive scan CCD camera (wavelength range: 400–1000 m; 720 × 560 pixels) and a prime lens (Fujinon HF8XA-1). The light panel consisted of four light-emitting diodes, including two red-orange light (620 nm) and two cool white light. The saturating super pulse light had an intensity of 4000 μmol·m^{-2}·s^{-1}, while the actinic light could reach up to 2000 μmol·m^{-2}·s^{-1}. The global light settings were set to 62% output for actinic light and 32% output for the super pulse light. The confirmation of chlorophyll *a* fluorescence images at different fruit ripening stages was performed using the protocol menu (quenching-Act1 analysis) of the FluorCam 7 software (FluorCam 7; Photon Systems Instruments, Brno, Czech Republic).

2.3. Analysis of Sugars, Acids, Phytochemicals, and Antioxidant Activity

To confirm the quality of phytochemicals in tomato fruits ripened under different environmental conditions of temperature and light, the red stage of the fruits was homogenized (Polytron, PT-MR3110D, Kinematica, Malters, Switzerland), and the extracts were centrifuged using a 64R Centrifuge (Beckman Coulter Inc., Brea, CA, USA) at 16,000× *g* for 30 min at 4 °C. The supernatant obtained after centrifugation was filtered through filter paper (Whatman No. 2, Sigma-Aldrich Co., St. Louis, MO, USA). The samples were stored in a cryogenic freezer at −70 °C, thawed, and then used for analysis.

After diluting 0.1 mL of the sample extracted from the fruit in 10 mL of distilled water, the sugar and acidity were measured using a fruit sugar–acidity meter (GMK-706R, G-WON Hitech Co., Ltd., Seoul, Republic of Korea). For the analysis of fructose, glucose, and sucrose in the fruit, fruit extracts were analyzed using an HPLC system (YL9100, Younglin Co., Anyang, Republic of Korea) equipped with a refractive index detector (YL9170 RI, Younglin Co., Anyang, Republic of Korea) and a Sugar-Pak column (4.6 mm × 250 mm, Supelco, Bellefonte, PA, USA) as previously described [35].

The lycopene analysis of extracts was conducted using an HPLC system (YL9100, Younglin Co., Anyang, Republic of Korea) equipped with a DB-C18 column (4.6 mm × 150 mm, Supelco, Bellefonte, PA, USA) and a DA detector (YL9120, Younglin Co., Anyang, Republic of Korea, as previously described [36].

The phenolic content in the tomato fruit extracts was determined using a UV–visible spectrophotometer (Evolution 300, Thermo Fisher Scientific., Waltham, MA, USA), with the gallic acid equivalents set as the standard. To accomplish this, aliquots of the extracts were sequentially treated with Folin–Ciocalteu reagent (50%) and Na_2CO_3 (20%). The resulting mixture was then incubated at 37 °C for 45 min. The absorbance of the test solutions was measured at 750 nm using the spectrophotometer, with the reagent used as a blank. The development of a blue color in each tube indicated the presence of phenolic compounds.

The antioxidant activity was assessed using methods described by [37], including the measurement of 2,2-diphenyl-1-picrylhydrazyl (DPPH) and 2,2′-azinobis-(3-ethylbenzothiazoline-6-sulfonate) (ABTS) radical scavenging abilities. The sample was mixed with the DPPH solution and allowed to react at room temperature for 30 min. The absorbance of the mixture was then measured at a wavelength of 517 nm to calculate the electron donating ability (EDA, %). EDA (%) was determined using the formula $[1 - ABS/ABC] \times 100$, where ABS represents the absorbance of the sample and ABC represents the absorbance of the control. For the ABTS assay, a mixture of 7.4 mM ABTS and 2.6 mM $K_2S_2O_8$ in a one-to-one ratio was prepared. The mixture was then filtered in the dark using Whatman No. 2 filter paper (Sigma-Aldrich Co., St. Louis, MO, USA). After overnight incubation, the mixture was adjusted to the desired absorbance at 734 nm by mixing with MeOH, and this value was used as the control. To measure the ABTS radical scavenging activity, the volume-adjusted ABTS solution and the sample were mixed together and allowed to react in a constant-temperature water bath at 37 °C for 1 min. The absorbance of the resulting mixture was then measured.

2.4. Experimental Design and Statistical Analysis

This experiment, aimed at confirming the changes in the ripening process of tomatoes under different environmental conditions such as temperature and light, was repeated three times using a random block design. The results of this experiment were analyzed using analysis of variance (ANOVA) with Duncan's multiple range test at a significance level of $p < 0.05$. Additionally, a two-way ANOVA was conducted using the SAS 9.4 (SAS Institute Inc., Cary, NC, USA).

3. Results

3.1. Changes of Color and Time to Maturity of Tomato Fruits

The ripening stages of tomatoes are generally classified into six stages: mature green, breaker, turning, pink, light red, and red. Figure 2A illustrates the color changes observed during these ripening stages in the growth chamber under different environmental conditions in this experiment. Tomatoes ripened in STL and WTL, which are subjected to light treatment, exhibited distinct difference across all ripening stages. However, in the cases of STD and WTD, which were not exposed to light irradiation, the progression through the six ripening stages was not clearly discernible. Notably, tomatoes ripened without light irradiation did not change to red as a whole, and the fruit surface displayed partial discoloration.

Figure 2B depicts the number of days required for tomato fruits to ripen under four distinct environmental conditions, featuring varying temperatures and light levels. Among the six stages of tomato ripening, a significant difference was observed in the number of days required to transition from the mature green stage to the breaker stage due to difference environmental conditions. However, during the ripening process of tomato fruit from the breaker stage to the red stage, there was no significant difference observed in the aging period based on environmental conditions. The results of the two-way ANOVA analysis reveal that both temperature and light significantly influence the ripening of tomato fruits. When comparing STL, which was subjected to high temperature and light irradiation, with WTD, which did not receive light irradiation and low temperature, the duration required for ripening was 2.5 times longer in WTD.

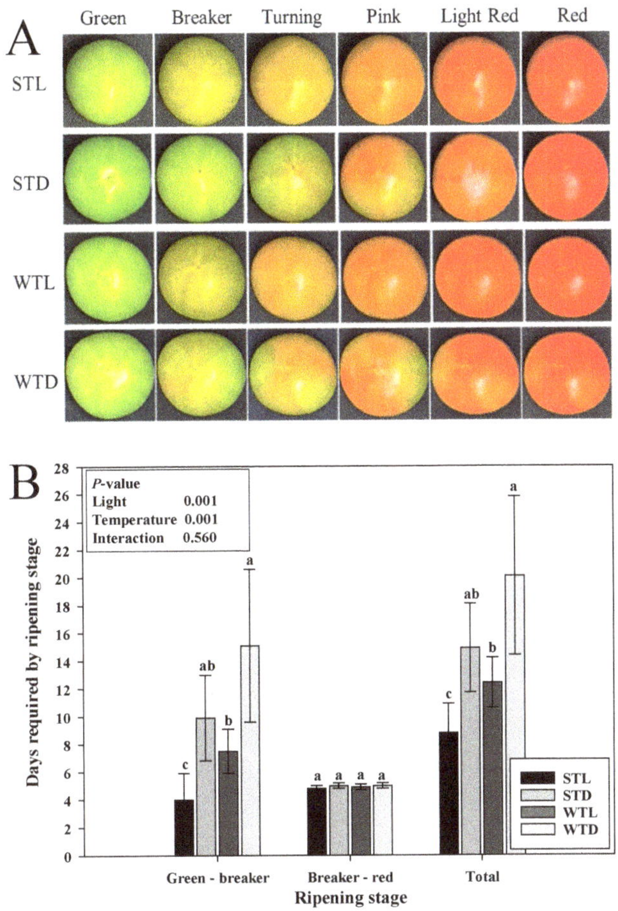

Figure 2. Changes of color (**A**) and time (**B**) to maturity of tomato fruit at different ripening stages under four distinct environmental conditions with varying temperatures and light levels. Vertical bars represent standard deviations (n = 30). Small letters at the data points indicate mean separation between the values using DMRT (p = 0.05). p values were determined through two-way ANOVA.

3.2. Chlorophyll a Fluorescence in Tomato Fruits

Upon examining the parameter-specific image of chlorophyll *a* fluorescence based on the ripening stages of tomato fruits, distinct differences in fluorescence were observed among the various ripening stages (Figure 3). Notably, consistent patterns of fluorescence images were observed throughout the measurements of various fruit samples at different ripening stages. In the fluorescence image analysis of tomato fruit ripening stages, no significant difference was observed in the minimum chlorophyll fluorescence in the dark-adapted state (F_O) or the maximum chlorophyll fluorescence in the dark-adapted state (F_M) values, assessed by ripening stage. However, the maximum PSII quantum yield (Q_Y) images showed distinct groupings across four stages: mature green stage, breaker stage, turning and pink stages, and light red and red stages. Similarly, the fluorescence decline ration in light (R_{FD}) images also exhibited distinct groupings across the mature green stage, the breaker and turning stages, and the pink stage, as well as the light red and red stages. Furthermore, it can be observed in the non-photochemical quenching (NPQ) image that significant changes in fluorescence occurred throughout the ripening stages. In particular, the transition from the mature green stage to the breaker stage, which is significantly influenced by environmental factors during the ripening process of tomatoes, could be distinctly distinguished using the chlorophyll *a* fluorescence image of the fruit.

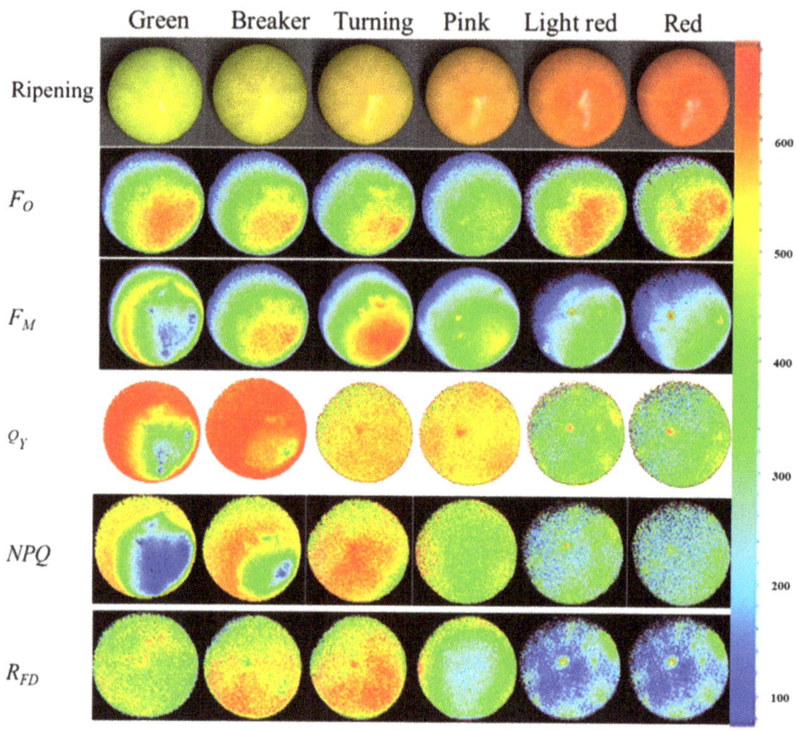

Figure 3. Fluorescence images of tomato fruit based on parameter values of chlorophyll *a* fluorescence under four different environmental conditions with varying temperatures and light levels. F_O: the minimum chlorophyll fluorescence in dark-adapted state; F_M: the maximum chlorophyll fluorescence in dark-adapted state; Q_Y: the maximum PSII quantum yield; NPQ: non-photochemical quenching; R_{FD}: the fluorescence decline ration in light.

Figure 4 shows the variations in chlorophyll *a* fluorescence parameter values at each stage of ripening for tomato fruits ripened in growth chambers under different environmental conditions. The F_O values in tomato fruits ripened under low-temperature conditions (WTL and WTD) exhibited an increasing trend until the pink stage, followed by a decrease. In contrast, fruits ripened under high temperature conditions (STL and STD) demonstrated a consistently decreasing pattern as ripening progressed. While the chlorophyll *a* fluorescence parameter values (except the F_O parameter) for each ripening stage of tomato fruits ripened under WTD conditions showed a gradual increase or decrease, the fluorescence parameter values of tomatoes ripened under STL conditions exhibited rapid increase or decrease. This indicates that the physiological changes in tomato fruits can be significantly determined by environmental conditions.

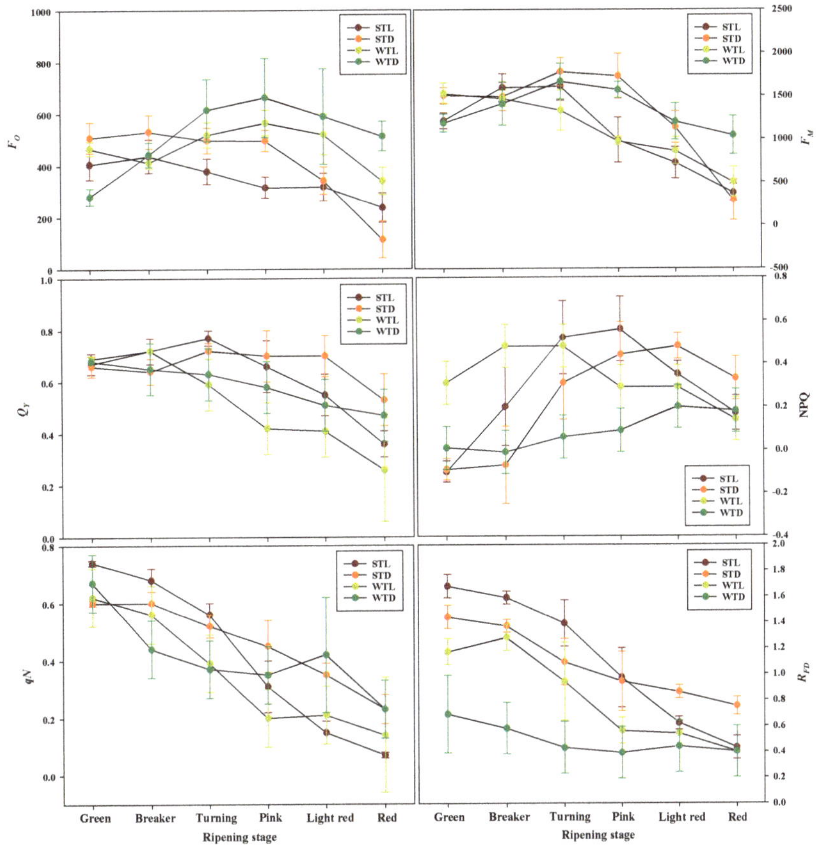

Figure 4. Changes in chlorophyll *a* fluorescence parameter values during the ripening of tomato fruit under four different environmental conditions with varying temperatures and light levels. Vertical bars are standard deviations (n = 9). F_O: the minimum chlorophyll fluorescence in the dark-adapted state; F_M: the maximum chlorophyll fluorescence in the dark-adapted state; Q_Y: the maximum PSII quantum yield; NPQ: non-photochemical quenching; q_N: non-photochemical quenching of variable fluorescence; R_{FD}: the fluorescence decline ration in light.

3.3. Phytochemicals and Antioxidant Activity in Tomato Fruits

The effects of temperature and light conditions during the ripening process of tomato fruits on the accumulation of various phytochemicals in the fruit were analyzed (Figure 5). The accumulation of soluble sugar content in tomato fruits was significantly influenced by light, with the highest accumulation observed when exposed to light under low temperature conditions. On the other hand, in terms of the acidity of tomato fruits, those ripened under high-temperature and -light conditions (STL) or low-temperature and dark conditions (WTD) exhibited significantly higher acidity levels. The phenol content in tomato fruits was found to increase under the STL condition, while no significant difference was observed in the other treatments. The lycopene content was highest in tomato fruits ripened under STL conditions, and lowest under WTD conditions. Furthermore, there was no significant difference in lycopene content between tomato fruits ripened under STD and WTL conditions.

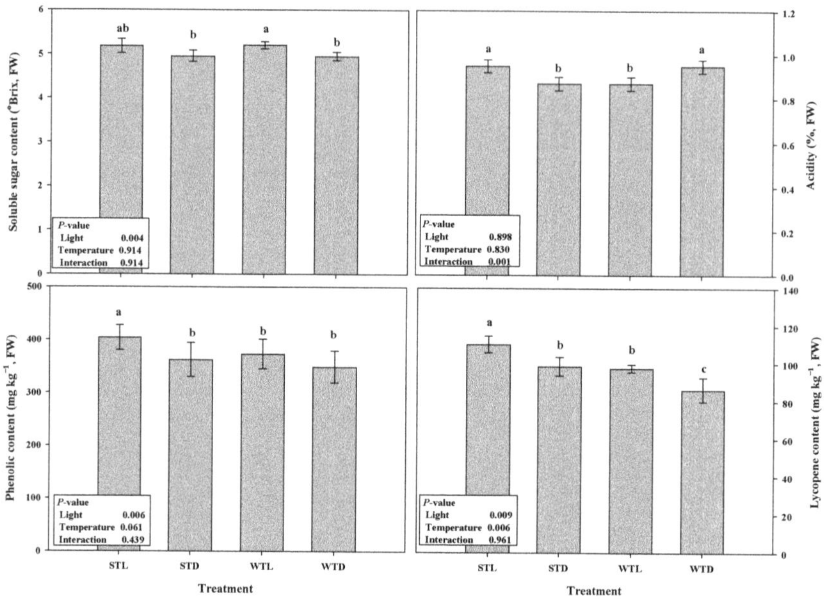

Figure 5. Phytochemical content of red stage matured tomato fruits under four different environmental conditions with varying temperatures and light levels. Vertical bars are standard deviations ($n = 3$). Small letters at the data points indicate mean separation between the values by DMRT ($p = 0.05$). p values were determined by two-way ANOVA.

Table 1 shows the variations in fructose, glucose, and sucrose levels in tomato fruits ripened under different environmental conditions. Irrespective of the treatment, fructose and glucose were identified as the primary sugars in tomato fruits. The content of sucrose, which is a glycoside formed by glucose and fructose, was relatively low. No significant effect of light on the sucrose content of tomato fruit was observed, whereas a substantial effect of light on fructose and glucose content was evident. Furthermore, within the temperature range examined in this experiment, no significant effect of temperature on the sugar content was observed.

Table 1. Soluble sugars content of red stage matured tomato fruits under four different environmental conditions with varying temperatures and light levels.

Treatment	Fructose	Glucose	Sucrose
		$mg \cdot g^{-1}$ FW	
STL	24.49 ± 1.31 a [z]	21.36 ± 1.11 ab	1.19 ± 0.26 a
STD	20.59 ± 1.12 b	19.25 ± 1.45 bc	0.97 ± 0.29 a
WTL	26.71 ± 1.34 a	22.78 ± 1.34 a	1.25 ± 0.11 a
WTD	19.87 ± 1.23 b	17.89 ± 1.36 c	0.98 ± 0.08 a
Effect (p value) *			
Temperature	0.4212	0.9217	0.8417
Light	0.0003	0.0054	0.1283
Interaction	0.1371	0.1566	0.8765

* p values were determined by two-way ANOVA. [z] Average values and standard deviation are presented, and values followed by different lower case letters within a column are different (DMRT at $p \leq 0.05$, $n = 3$).

The antioxidant activity of tomato fruits was confirmed by ABTS and DPPH assays when they were subjected to ripening under different environmental conditions of temperature and light (Figure 6). The antioxidant activity of tomato fruits varied based on the ripening environmental conditions. Particularly, they exhibited high activity under high-temperature and -light irradiation conditions (STL), while the activity significantly decreased under low-temperature and dark conditions. When considering the ABTS activity, the impact of light was found to be significant, while there was no significant difference in the effect of temperature. However, in the case of DPPH, both light and temperature had a significant influence on the activity.

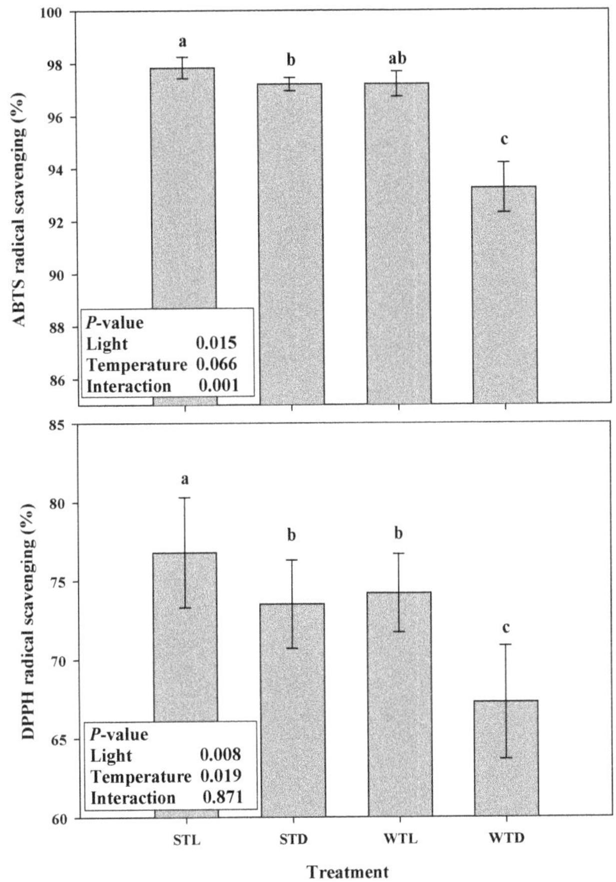

Figure 6. Antioxidant activity shown as ABTS and DPPH of red stage matured tomato fruits under four different environmental conditions with varying temperatures and light levels. Vertical bars are standard deviations (n = 3). Small letters at the data points indicate mean separation between the values by DMRT (p = 0.05). p values were determined by two-way ANOVA.

4. Discussion

Various environmental conditions, such as light, temperature, humidity, and CO_2, play an important role in the pre-harvest and post-harvest maturation of horticultural crops [38–40]. Tomato fruit, which is grown and consumed in various countries around the world, exhibits climacteric fruit characteristics and, like other horticultural crops, is influenced by a range of environmental factors [41]. In this study, ripening tomato fruits were subjected to different temperature and light conditions. Notably, tomatoes ripened in darkness without exposure to light exhibited a visible characteristic wherein the fruit's surface displayed mottled colors of red and green (Figure 2A). It is known that tomato pericarp cells regulate tomato carotenoid biosynthesis through light sensing [16]. Therefore, in this study, it was observed that the biosynthesis process does not occur smoothly when tomatoes mature in darkness, resulting in a mottled appearance of the tomato pericarp, indicating disrupted pigmentation patterns.

Tomatoes take the most time to transition from the mature green stage to the breaker stage compared to other ripening stages [14]. In particular, it is known that as the storage temperature increases up to 30 °C, the rate at which tomatoes transition from the mature green stage to the breaker stage is accelerated [14]. Tomatoes are primarily classified into ripening stages based on changes in skin color, and light plays a significant role in these color transformations [42]. It has been reported that the lycopene concentrations in tomato fruits exposed to LED light were higher compared to those exposed to dark conditions [42–44]. Consistent with the previous studies mentioned [14,42–44], the findings of this study demonstrate a pronounced influence of temperature and light on the ripening of tomato fruits (Figure 2). The time required for tomato fruits to transition from the green stage to the breaker stage varied up to threefold, depending on the environmental conditions during postharvest ripening. However, the period required for maturation from the breaker stage to the red stage of tomato fruit did not show a significant difference based on the treatment (Figure 2B). Based on these results, it is believed that the period when tomato fruit is most influenced by the environment is between the green stage and just before it reaches the breaker stage. Therefore, the environmental conditions during the transition period from the mature green stage to the breaker stage will greatly affect tomato ripening.

In a study utilizing spectrophotometric absorbance differences, the relative concentrations of PSII and PSI reaction centers were determined in the green fruit tissue of tomatoes [45]. According to the research report, the values of chlorophyll fluorescence parameters in tomato green fruits were found to be comparable to those measured in leaves [46]. Additionally, the RuBPCO activity, which was similar to that observed in the leaves during the immature stage of the fruit, decreased as the fruit ripened [46]. This finding suggests that chlorophyll fluorescence images can be effectively analyzed during tomato fruit ripening, given the demonstrated responsiveness of tomato fruit tissues to light. Recently, numerous studies have been conducted to utilize chlorophyll fluorescence induction for the classification of ripening stages in tomato fruits [28,47–49]. It has been reported that the non-photochemical quenching of photosystem (NPQ), one of the chlorophyll fluorescence parameters, is an effective method to classify tomato fruits according to their ripening stage [31]. The fluorescence decline ratio in light (R_{FD}), which is one of the chlorophyll fluorescence parameters, has also been reported as a reliable tool for classifying the ripening stage of tomatoes [47]. In this study, tomatoes were successfully classified by their ripening stage using fluorescence images obtained from chlorophyll fluorescence induction (Figure 3). Specifically, R_{FD}, NPQ, and Q_Y images were identified as highly effective tools for ripening stage classification. In the case of R_{FD} parameters, tomato fruit ripening in a location with favorable light and temperature conditions showed a steep decline as the ripening stage progressed, while it was observed to decline more gradually under unfavorable ripening conditions (Figure 4). Therefore, even without an image, the chlorophyll fluorescence induction parameter values such as R_{FD} can be used as essential tools to non-destructively assess the efficiency of tomato fruit maturation during the ripening process.

Tomatoes are recognized as an important horticultural crop due to their rich content of various phytochemicals [3,4]. It is known that environmental conditions, such as light and temperature, influence the metabolites of phytochemicals in the fruit during the tomato ripening process [44]. The formation of carotenoids in climacteric fruits does not require induction by light, but shaded fruits have a lower carotenoid content [50]. Furthermore, it is reported that red light affects chlorophyll degradation, whereas carotenoid synthesis is enhanced by blue light [50]. Also, the sugar content of horticultural crops varies depending on the light conditions. Tomatoes grown in shaded areas have lower sugar contents in their fruits compared to tomatoes grown in sunlight [1]. Similarly, strawberries grown in shaded areas, despite receiving some sunlight, exhibit reduced sugar contents compared to those grown under direct sunlight [51]. Consistent with previous studies [1,49], the present study found a significant increase in the sugar content of tomato fruits when they were irradiated with light during the ripening process, compared to those ripened in darkness (Table 1 and Figure 5). Ultimately, the biosynthesis of phytochemicals in tomatoes is regulated by the interaction between temperature and light [52].

As a result, the interaction between temperature and light is crucial in the ripening process of tomato fruits. Therefore, it is believed that by implementing LED lighting in close proximity to the fruit during the tomato cultivation process, it will be possible to enhance the quality of tomato fruits while promoting their ripening. This approach is expected to be effective not only during the winter season but also in the summer season.

5. Conclusions

Based on the results of this study, three key conclusions are presented. Firstly, when light is applied during the ripening stage of tomato fruits, it leads to uniform color changes and accelerates the ripening process. Secondly, chlorophyll fluorescence images and parameter values can be valuable tools for effectively classifying the ripening stages of tomato fruits. Lastly, employing LED illumination near the fruit proves to be an efficient cultivation method for growing tomatoes in both low- and high-temperature seasons.

Author Contributions: H.-G.C. contributed to the experimental design, conceptualization, data analysis, formal analysis, investigation, data curation, and original draft and writing; K.-S.P. contributed to the data analysis and writing. All authors have read and agreed to the published version of the manuscript.

Funding: This study received financial support from the Korea Institute of Planning and Evaluation for Technology in Food, Agriculture, and Forestry (IPET), Korea Smart Farm R&D Foundation (Project No. 421001-03) and Kongju National University (Project No. 2023-0223-01).

Data Availability Statement: The data presented in this study are available in the article.

Acknowledgments: We would like to express our thanks to the anonymous reviewers for their useful comments.

Conflicts of Interest: The authors declare no conflict of interest.

References

1. Pék, Z.; Szuvandzsiev, P.; Nemenyi, A.; Helyes, L.; Lugasi, A. The effect of natural light on changes in antioxidant content and color parameters of vine-ripened tomato (*Solanum lycoperisicum* L.) fruits. *HortScience* **2011**, *46*, 583–585. [CrossRef]
2. Park, H.; Kim, Y.-J.; Shin, Y. Estimation of daily intake of lycopene, antioxidant contents and activities from tomatoes, watermelons, and their processed products in Korea. *Appl. Biol. Chem.* **2020**, *63*, 50. [CrossRef]
3. Rowles, J.L.; Ranard, K.M.; Applegate, C.C.; Jeon, S.Y.; An, R.; Erdaman, J.W. Processed and raw tomato consumption and risk of prostate cancer: A systematic review and dose-response meta-anaylsis. *Prostate Cancer Prostatic Dis.* **2018**, *21*, 319–336. [CrossRef] [PubMed]
4. Salehi, B.; Sharifi-Rad, R.; Sharopov, F.; Namiesnik, J.; Roointan, A.; Kamle, M.; Kumar, P.; Martins, N.; Sharifi-Rad, J. Beneficial effects and potential risks of tomato consumption for human health: An overview. *Nutrition* **2019**, *62*, 201–208. [CrossRef] [PubMed]
5. Nagarajan, J.; Kay, H.P.; Krishnamurthy, N.P.; Ramakrishnan, N.R.; Aldawoud, T.M.S.; Galanakis, C.M.; Wei, O.C. Extraction of Carotenoids from Tomato Pomace via Water-Induced Hydrocolloidal Complexation. *Biomolecules* **2020**, *10*, 1019. [CrossRef]
6. Przybylska, S.; Tokarczyk, G. Lycopene in the Prevention of Cardiovascular Diseases. *Int. J. Mol. Sci.* **2022**, *23*, 1957. [CrossRef]
7. Li, S.; Zhu, B.; Pirrello, J.; Xu, C.; Zhang, B.; Bouzayen, M.; Chen, K.; Grierson, D. Roles of RIN and ethylene in tomato fruit ripening and ripening-associated traits. *New Phytol.* **2020**, *226*, 460–475. [CrossRef]
8. Orsi, B.; Sestari, I.; Preczenhak, A.P.; Tessmer, M.A.; Souza, M.A.D.S.; Hassimotto, N.M.A.; Kluge, R.A. Allelic variations in the tomato carotenoid pathway lead to pleiotropic effects on fruit ripening and nutritional quality. *Postharvest Biol. Technol.* **2021**, *181*, 111632. [CrossRef]
9. Helyes, L.; Pék, Z.; Lugasi, A. Tomato fruit quality and gontent depend on stage of maturity. *HortScience* **2006**, *41*, 1400–1401. [CrossRef]

10. Tigist, M.; Workneh, T.S.; Woldetsadik, K. Effects of variety on the quality of tomato stored under ambient conditions. *J. Food Sci. Technol.* **2013**, *50*, 477–486. [CrossRef]
11. Stevens, M.A. Relationships Between Components Contributing to Quality Variation Among Tomato Lines1. *J. Am. Soc. Hortic. Sci.* **1972**, *97*, 70–73. [CrossRef]
12. Tamasi, G.; Pardini, A.; Bonechi, C.; Donati, A.; Pessina, F.; Marcolongo, P.; Gamberucci, A.; Leone, G.; Consumi, M.; Magnani, A.; et al. Characterization of nutraceutical components in tomato pulp, skin and locular gel. *Eur. Food Res. Technol.* **2019**, *245*, 907–918. [CrossRef]
13. Bhandari, S.; Lee, J.G. Ripening-dependent changes in antioxidnats, color attributes, and antioxidant activity of seven tomato (*Solanum lycopersicum* L.) cultivars. *J. Anal. Methods Chem.* **2016**, *2016*, 5498618. [CrossRef] [PubMed]
14. Ciptaningtyas, D.; Benyakart, N.; Umehara, H.; Johkan, M.; Nakamura, N.; Nagata, M.; Orikasa, T.; Thammawong, M.; Shiina, T. Modeling the metachronous ripening pattern of mature green tomato as affected by cultivar and storage temperature. *Sci. Rep.* **2022**, *12*, 8241. [CrossRef]
15. JJarquín-Enríquez, L.; Mercado-Silva, E.M.; Maldonado, J.L.; Lopez-Baltazar, J. Lycopene content and color index of tomatoes are affected by the greenhouse cover. *Sci. Hortic.* **2013**, *155*, 43–48. [CrossRef]
16. Llorente, B.; D'Andrea, L.; Ruiz-Sola, M.A.; Botterweg, E.; Pulido, P.; Andilla, J.; Loza-Alvarez, P.; Rodriguez-Concepcion, M. Tomato fruit carotenoid biosynthesis is adjusted to actual ripening progression by a light-dependent mechanism. *Plant J.* **2016**, *85*, 107–119. [CrossRef]
17. Schofield, A.; Paliyath, G. Modulation of carotenoid biosynthesis during tomato fruit ripening through phytochrome regulation of phytoene synthase activity. *Plant Physiol. Biochem.* **2005**, *43*, 1052–1060. [CrossRef]
18. Gupta, S.K.; Sharma, S.; Santisree, P.; Kilambi, H.; Appenroth, K.; Sreelakshmi, Y.; Sharma, R. Complex and shifting interactions of phytochromes regulate fruit development in tomato. *Plant Cell Environ.* **2014**, *37*, 1688–1702. [CrossRef]
19. Artins, A.; Caldana, C. The metabolic homeostat TOR: The balance of holding on or letting grow. *Curr. Opin. Plant Biol.* **2022**, *66*, 102196. [CrossRef]
20. Lakshmanan, M.; Lim, S.-H.; Mohanty, B.; Kim, J.K.; Ha, S.-H.; Lee, D.-Y. Unraveling the light-specific metabolic and regulatory signatures of rice through combined in silico modeling and multi-omics analysis. *Plant Physiol.* **2015**, *169*, 3002–3020. [CrossRef]
21. Rodrigues, M.A.; Bianchetti, R.E.; Freschi, L. Shedding light on ethylene metabolism in higher plants. *Front. Plant Sci.* **2014**, *5*, 665. [CrossRef] [PubMed]
22. Kadomura-Ishikawa, Y.; Miyawaki, K.; Takahashi, A.; Masuda, T.; Noji, S. Light and abscisic acid independently regulated FaMYB10 in Fragaria × ananassa fruit. *Planta* **2015**, *241*, 953–965. [CrossRef] [PubMed]
23. Gong, D.; Cao, S.; Sheng, T.; Shao, J.; Song, C.; Wo, F.; Chen, W.; Yang, Z. Effect of blue light on ethylene biosynthesis, signalling and fruit ripening in postharvest peaches. *Sci. Hortic.* **2015**, *197*, 657–664. [CrossRef]
24. Nassarawa, S.S.; Abdelshafy, A.M.; Xu, Y.; Li, L.; Luo, Z. Effect of Light-Emitting Diodes (LEDs) on the Quality of Fruits and Vegetables During Postharvest Period: A Review. *Food Bioprocess Technol.* **2021**, *14*, 388–414. [CrossRef]
25. Helyes, L.; Lugasi, A.; Pék, Z. Effect of natural light on surface temperature and lycopene content of vine ripened tomato fruit. *Can. J. Plant Sci.* **2011**, *87*, 927–929. [CrossRef]
26. Xie, B.; Song, S.; Liu, H.; Sun, G.; Chen, R. Effects of Light Quality on The Quality Formation of Tomato Fruits. *Adv. Biol. Sci. Res* **2016**, *3*, 11–15.
27. Choi, H.G. Correlation among phenotypic parameters related to the growth and photosynthesis of strawberry (*Fragaria × ananassa* Duch.) grown under various light intensity conditions. *Front. Plant Sci.* **2021**, *12*, 647585. [CrossRef]
28. Na, Y.; Jeong, J.H.; Lee, S.Y.; Choi, H.G.; Kim, S.H.; Rho, I.R. Chloroophyll fluorescence as a diagnostic tool for abiotic stress tolerance in wild and cultivated strawberry species. *Hortic. Environ. Biotechnol.* **2014**, *55*, 280–286. [CrossRef]
29. Guidi, L.; Piccolo, E.L.; Landi, M. Chlorophyll Fluorescence, Photoinhibition and Abiotic Stress: Does it Make Any Difference the Fact to Be a C3 or C4 Species? *Front. Plant Sci.* **2019**, *10*, 174. [CrossRef]
30. Abdelhamid, M.A.; Sudnik, Y.A.; Alshinayyin, H.J.; Shaaban, F. Chlorophyll fluorescence for classification of tomato fruits by their maturity stage. *E3S Web Conf.* **2020**, *193*, 01065. [CrossRef]
31. Kasampalis, D.S.; Tsouvaltzis, P.; Siomos, A.S. Chlorophyll fluorescence, non-photochemical quenching and light harvesting complex as alternatives to color measurement, in classifying tomato fruit according to their maturity stage at harvest and in monitoring postharvest ripening during storage. *Postharvest Biol. Technol.* **2020**, *161*, 111036. [CrossRef]
32. Dong, Z.; Men, Y.; Li, Z.; Zou, Q.; Ji, J. Chlorophyll fluorescence imaging as a tool for analyzing the effects of chilling injury on tomato seedlings. *Sci. Hortic.* **2019**, *246*, 490–497. [CrossRef]
33. Senesi, G.S.; De Pascale, O.; Marangoni, B.S.; Caires, A.R.L.; Nicolodelli, G.; Pantaleo, V.; Leonetti, P. Chlorophyll Fluorescence Imaging (CFI) and Laser-Induced Breakdown Spectroscopy (LIBS) Applied to Investigate Tomato Plants Infected by the Root Knot Nematode (RKN) *Meloidogyne incognita* and Tobacco Plants Infected by Cymbidium Ringspot Virus. *Photonics* **2022**, *9*, 627. [CrossRef]
34. USDA. *United States Standards for Grades of Fresh Tomatoes*; USAD, Agricultural Marketing Service: Washington, DC, USA, 1991; pp. 1–13. Available online: https://www.ams.usda.gov/sites/default/files/media/Tomato_Standard%5B1%5D.pdf (accessed on 12 July 2023).
35. Choi, H.G.; Moon, B.Y.; Kang, H.J. Effect of LED light on the production of strawberry during cultivation in a plastic greenhouse and in a growth chamber. *Sci. Hortic.* **2015**, *189*, 22–31. [CrossRef]

36. Kim, H.-K.; Chun, J.-H.; Kim, S.-J. Method Development and Analysis of Carotenoid Compositions in Various Tomatoes. *Korean J. Environ. Agric.* **2015**, *34*, 196–203. [CrossRef]
37. Gil Choi, H.; Kwon, J.K.; Moon, B.Y.; Kang, N.J.; Park, K.S.; Cho, M.W.; Kim, Y.C. Effect of Different Light Emitting Diode (LED) Lights on the Growth Characteristics and the Phytochemical Production of Strawberry Fruits during Cultivation. *Korean J. Hortic. Sci. Technol.* **2013**, *31*, 56–64. [CrossRef]
38. Lee, S.G.; Kim, S.K.; Lee, H.J.; Lee, H.S.; Lee, J.H. Impact of moderate and extreme climate change scenarios on growth, morphological features, photosynthesis, and fruit production of hot pepper. *Ecol. Evol.* **2018**, *8*, 197–206. [CrossRef]
39. Dhall, P.K. Ethylene in post-harvest quality management of horticultural crops: A review. *Res. Rev. A J. Crop Sci. Technol.* **2013**, *2*, 9–24.
40. Lee, S.K.; Kader, A.A. Preharvest and postharvest factors influencing vitamin C content of horticultural crops. *Postharvest Biol. Technol.* **2000**, *20*, 207–220. [CrossRef]
41. Dorais, M.; Ehret, D.L.; Papadopoulos, A.P. Tomato (*Solanum lycopersicum*) health components: From the seed to the consumer. *Phytochem. Rev.* **2008**, *7*, 231–250. [CrossRef]
42. Nájera, C.; Guil-Guerrero, J.L.; Enríquez, L.J.; Álvaro, J.E.; Urrestarazu, M. LED-enhanced dietary and organoleptic qualities in postharvest tomato fruit. *Postharvest Biol. Technol.* **2018**, *145*, 151–156. [CrossRef]
43. Colquhoun, T.A.; Schwieterman, M.L.; Gilbert, J.L.; Jaworski, E.A.; Langer, K.M.; Jones, C.R.; Rushing, G.V.; Hunter, T.M.; Olmstead, J.; Clark, D.G.; et al. Light modulation of volatile organic compounds from petunia flowers and select fruits. *Postharvest Biol. Technol.* **2013**, *86*, 37–44. [CrossRef]
44. Gautier, H.; Diakou-Verdin, V.; Bénard, C.; Reich, M.; Buret, M.; Bourgaud, F.; Poëssel, J.L.; Caris-Veyrat, C.; Génard, M. How Does Tomato Quality (Sugar, Acid, and Nutritional Quality) Vary with Ripening Stage, Temperature, and Irradiance? *J. Agric. Food Chem.* **2008**, *56*, 1241–1250. [CrossRef] [PubMed]
45. Piechulla, B.; Glick, R.E.; Bahl, H.; Melis, A.; Gruissem, W. Changes in Photosynthetic Capacity and Photosynthetic Protein Pattern during Tomato Fruit Ripening. *Plant Physiol.* **1987**, *84*, 911–917. [CrossRef] [PubMed]
46. Carrara, S.; Pardossi, A.; Soldatini, G.; Tognoni, F.; Guidi, L. Photosynthetic Activity of Ripening Tomato Fruit. *Photosynthetica* **2001**, *39*, 75–78. [CrossRef]
47. Abdelhamid, M.A.; Sudnik, Y.; Alshinayyin, H.J.; Shaaban, F. Non-destructive method for monitoring tomato ripening based on chlorophyll fluorescence induction. *J. Agric. Eng.* **2020**, *52*, 1098.
48. Hoffmann, A.M.; Noga, G.; Hunsche, M. Fluorescence indices for monitoring the ripening of tomatoes in pre- and postharvest phases. *Sci. Hortic.* **2015**, *191*, 74–81. [CrossRef]
49. Fatchurrahman, D.; Amodio, M.L.; Chiara, M.L.V.D.; Chaudhry, M.M.A.; Colelli, G. Early discrimination of mature and immature green tomatoes (*Solanum lycopersicum* L.) using fluorescen imaging method. *Postharvest Biol. Technol.* **2020**, *169*, 111287. [CrossRef]
50. Dorais, M. Effect of cultural management on tomato fruit health qualities. *Acta Hortic.* **2007**, *774*, 279–294. [CrossRef]
51. Gil Choi, H.; Moon, B.Y.; Kang, N.J.; Kwon, J.K.; Bekhzod, K.; Park, K.S.; Lee, S.Y. Yield loss and quality degradation of strawberry fruits cultivated under the deficient insolation conditions by shading. *Hortic. Environ. Biotechnol.* **2014**, *55*, 263–270. [CrossRef]
52. Massot, C.; Bancel, D.; Lauri, F.L.; Truffault, V.; Baldet, P.; Stevens, R.; Gautier, H. High temperatue inhibits ascorbate recycling and light stimulation of the ascorbate pool in tomato despite increased expression of biosynthesis genes. *PLoS ONE* **2013**, *8*, e84474. [CrossRef] [PubMed]

horticulturae

MDPI

Article

Preservation of Quality and Bioactive Compounds in Mangoes Using Chitosan-Graphene-Oxide-Based Biodegradable Packaging

João Claudio Vilvert [1,2,*], Sérgio Tonetto de Freitas [2], Maria Aparecida Rodrigues Ferreira [3], Cristina dos Santos Ribeiro Costa [2], Ricardo Henrique de Lima Leite [4], Francisco Klebson Gomes dos Santos [5] and Edna Maria Mendes Aroucha [1]

1 Department of Engineering and Ambiental Sciences, Federal Rural University of the Semi-Arid Region, Mossoro 59625-900, Rio Grande do Norte, Brazil; aroucha@ufersa.edu.br
2 Brazilian Agricultural Research Corporation, Tropical Semi-Arid Embrapa, Petrolina 56302-970, Pernambuco, Brazil; sergio.freitas@embrapa.br (S.T.d.F.); cristinasrcosta@gmail.com (C.d.S.R.C.)
3 Department of Biological Sciences, State University of Feira de Santana, Feira de Santana 44036-900, Bahia, Brazil; ferreiraaparecida.21@gmail.com
4 Department of Engineering and Technology, Federal Rural University of the Semi-Arid Region, Mossoro 59625-900, Rio Grande do Norte, Brazil; ricardoleite@ufersa.edu.br
5 Department of Natural Sciences, Mathematics and Statistics, Federal Rural University of the Semi-Arid Region, Mossoro 59625-900, Rio Grande do Norte, Brazil; klebson@ufersa.edu.br
* Correspondence: jcvilvert@gmail.com; Tel.: +55-49-99977-9482

Abstract: Mango is a climacteric fruit that requires efficient postharvest technologies to maintain quality during storage and transportation. This study aimed to investigate the effect of biodegradable packaging from chitosan (CS) incorporated with graphene oxide (GO) on the quality, bioactive compounds and antioxidant activity of cold-stored 'Tommy Atkins' mangoes. Mangoes harvested at physiological maturity were stored without packaging or in CS, CS-GO or non-biodegradable polyethylene (PE) packaging at 12.0 °C and 89% relative humidity for 42 days. The results show that GO improved the water barrier and mechanical properties of CS packaging. All packaging delayed fruit ripening by reducing the respiration rate, mass loss, softening and changes in color, soluble solids, titratable acidity and beta-carotene content, also preserving the mangoes' visual appearance. In addition, all packaging maintained higher ascorbic acid, yellow flavonoid, phenolic compounds and antioxidant activity levels in the fruit, compared to non-packed ones. Chitosan packaging is a promising, eco-friendly alternative for the preservation of quality, bioactive compounds and antioxidant activity of cold-stored 'Tommy Atkins' mangoes, extending their postharvest life by at least 14 days.

Keywords: *Mangifera indica* L.; postharvest quality; biopolymers; shelf life; nanoparticles; packaging; MAP; fruit ripening; phenolic compounds; carotenoids

Citation: Vilvert, J.C.; de Freitas, S.T.; Ferreira, M.A.R.; Costa, C.d.S.R.; Leite, R.H.d.L.; dos Santos, F.K.G.; Aroucha, E.M.M. Preservation of Quality and Bioactive Compounds in Mangoes Using Chitosan-Graphene-Oxide-Based Biodegradable Packaging. *Horticulturae* **2023**, *9*, 1145. https://doi.org/10.3390/horticulturae9101145

Academic Editors: Yang Bi, Yongcai Li and Di Gong

Received: 19 August 2023
Revised: 11 September 2023
Accepted: 13 September 2023
Published: 19 October 2023

1. Introduction

Mangoes (*Mangifera indica* L.) are a tropical fruit with an attractive color, delicious taste and exotic flavor, and are also a rich source of nutrients and phytochemicals, such as provitamin A carotenoids, ascorbic acid and phenolic compounds [1,2]. These bioactive compounds provide benefits for the human body by acting as natural antioxidants, inhibiting oxidation damage caused by reactive oxygen species [3].

Mango production takes place in more than 100 countries, with an increasing international trade and a high demand in the mainstream market outlets in most of the developed countries [4]. Brazil is among the largest of the mango-exporting countries, with harvests of high-quality fruit taking place all year long. Tommy Atkins is the leading exporting cultivar in Brazil, as a well-known mango genotype with great international appeal, especially due to its ability to withstand longer transportation and shelf life, compared to other genotypes [1,5,6]. These traits also make Tommy Atkins the most widely grown mango cultivar in the world.

Mango is a highly perishable climacteric fruit, characterized by rapid and intense ripening-related changes, including increased respiration and ethylene production, softening, conversion of starch to sugars, degradation of organic acids and chlorophylls and synthesis of carotenoids [4]. Several technologies have been investigated and applied to extend mangoes' postharvest life [5,7–11]. Refrigeration at 10–13 °C is one of the most efficient techniques to maintain quality and extend the postharvest life of 'Tommy Atkins' mangoes without leading to chilling injury [12]. However, refrigeration alone may not be enough to export mangoes to distant markets, whose transport can take more than 30 days, limiting the worldwide mango distribution from the production regions to consumers.

In addition to refrigeration, modified atmosphere (MA) can also be used to maintain quality and extend mango postharvest life [5]. MA delays fruit ripening and senescence and preserves quality by reducing oxygen concentration and increasing carbon dioxide levels around the fruit as a result of its respiration and metabolic activity [13]. Traditionally used MA packaging films involve non-degradable materials, such as polyethylene (PE), which contributes to many environmental problems.

The application of biopolymers as edible coatings is a simple, non-toxic and environmentally friendly alternative for preserving mango quality, through formation of a transparent film on the fruit surface that acts as a barrier to water, oxygen and carbon dioxide, providing MA conditions to the fruit. The interest in biopolymers as eco-friendly and sustainable substitutes for petroleum-based plastics is strongly increasing because of their high biodegradability, low cost and ease to obtain from renewable resources [14,15].

Chitosan (CS) is a low-cost polymer derived from animal and fungal sources that has been tested for application in food, agriculture and pharmaceutical industries. The use of CS for food packaging has been successfully gaining attention due to its excellent film-forming ability associated with antioxidant and antimicrobial properties [16–18]. Besides many advantages over synthetic polymers, CS may have weak mechanical and moisture-barrier properties [19], both of which could be improved via the incorporation of nanoparticles to the film-forming solutions.

Graphene oxide (GO) is derived from graphene via the chemical modification of graphite, resulting in a molecule with various oxygen functional groups [20]. Amide linkages tend to be formed between the CS's highly reactive amine groups and GO's carboxylic acid groups, being expected that GO incorporates well into the biopolymeric CS matrix [14]. A homogeneous distribution and incorporation of GO into the CS matrix can enhance the mechanical, thermal, structural, morphological and barrier properties of the CS films, as previously reported in other studies [15,21].

Recently, CS-GO-based biodegradable packaging has been developed for fruit conservation using the casting method, resulting in an eco-friendly and high-performance packaging material that effectively maintain the quality and extend the postharvest life of melons [22]. Although MA conditions can have a strong effect on maintaining the quality and extending the postharvest life of mangoes, studies must be undertaken to determine the potential use of biodegradable packaging to make exporting mangoes to distant markets, that are still limited by the use of regular refrigerated containers, possible. Therefore, this study aimed to investigate the effect of biodegradable packaging from chitosan (CS) incorporated with graphene oxide (GO) on the quality, bioactive compounds and antioxidant activity of cold-stored 'Tommy Atkins' mangoes.

2. Materials and Methods

2.1. Fruit Material

'Tommy Atkins' mangoes were obtained from a commercial orchard in Petrolina, PE, Brazil (9°03′04.6″ S latitude, 40°17′46.5″ W longitude, and altitude of 376 m). Mangoes were harvested at the point of physiological maturity, represented by full shoulders at the fruit stem end and a predominant light-green skin color [23]. Mangoes without diseases or injuries were selected for uniformity of size, color and shape. Then, the fruit was washed, sanitized with a 200 ppm (*v*/*v*) sodium hypochlorite solution for 15 min and dried at room temperature before analysis.

2.2. Chemicals

An 85% deacetylated chitosan (CS) was purchased from Polymar (Fortaleza, Brazil). Graphene oxide (GO), gallic acid, DPPH, ABTS, Trolox, Folin–Ciocalteu's reagent, acetic acid, methanol, acetone, hexane, potassium persulfate, hydrochloric acid and anhydrous sodium carbonate were provided by Sigma-Aldrich (San Luis, MO, USA). Polyethylene (PE) packaging was acquired from Nissan Steel Industry (Kyoto, Japan).

2.3. Preparation and Characterization of CS-GO Biodegradable Packaging

Biodegradable packaging was prepared according to Paiva et al. [22]. Briefly, the CS (2%, *w*/*v*) and the GO at 0% and 0.25% (*w*/*w*) were dissolved in 1% acetic acid (*v*/*v*), followed by moderate stirring for 12 h. The film-forming solutions were homogenized in an ultrasonic bath for 10 min. Then, 240 g of each solution were transferred to an acrylic plate of 32 × 32 cm and dried for 10 h at 50 °C. Each package consisted of two overlapped films, which were then sealed in the edges, with an opening to insert the fruit, using a sealing machine.

For each formulation, five films were used in the characterization of the water vapor transmission rate (WVTR) and mechanical properties. The WVTR was measured in quintuplicates, following the methodology of Sun et al. [24], in accordance with the American Society for Testing and Materials (ASTM) E96-00 method. Permeation measuring cells were filled with about 5 mL of distilled water, and square film samples (2 × 2 cm) were placed over the cells. The cell–film sets were weighed and placed in a desiccator containing silica gel, with internal temperature of 25 °C and 50% relative humidity. Cell-films were weighed every hour for eight hours, and the following equation was used to calculate the WVTR (Equation (1)):

$$\mathrm{WVTR} = \mathrm{M}/(\mathrm{At}) \tag{1}$$

where WVTR is the water vapor transmission rate ($g/m^2/s$); M is the mass of water permeating through the film (g); A is the permeation area (m^2); and t is the permeation time (s).

The mechanical properties tensile strength (σ, in MPa), elongation at break (ε, in %) and Young's modulus (€, in MPa) were measured in film samples of 50 mm × 5 mm using a mechanical testing machine (model DL5000/10000, EMIC, São José dos Pinhais, Brazil), which operated according to ASTM D882-8312, at a test speed of 5 mm/min and an application force of 5 kN.

2.4. Postharvest Treatments

Mangoes were divided in four treatments, being CS-based biodegradable packaging (with and without GO), PE-based non-degradable packaging and the control (unpacked fruit). Two mangoes were inserted in each package, which was sealed using a sealing machine, creating conditions of modified atmosphere (MA). Then, the mangoes were stored at 12.0 ± 0.5 °C and 89 ± 3% relative humidity, in the central position at the end of a cold chamber (GEFRIO, Fortaleza, Brazil), with dimensions of 3 × 2 × 3 m (length × width × height), and air flow of 2 m s^{-1} [12]. A total of 96 fruit were stored for 42 days, simulating conditions of long-distance shipping. Every 14 days, six fruit per treatment were randomly sampled for analysis. About 50 g of each sample was stored at −85 °C for later bioactive compounds and antioxidant activity analyses.

2.5. Determination of Quality Parameters, Bioactive Compounds and Antioxidant Activity

2.5.1. Respiration Rate

Respiration rate of mangoes was measured with a gas analyzer model PA 7.0 (WITT, Witten, Germany). Each fruit was placed for 1 h into an airtight container, and the final CO_2 concentrations were recorded. Respiration rate was expressed as mg/kg/h.

2.5.2. Mass Loss

Mass loss was calculated by multiplying the difference between the initial mass and the mass at the end of the storage by 100 and dividing by the initial mass. Mass loss was expressed as percentage.

2.5.3. Skin and Pulp Color

Color was determined in skin and pulp tissues with a colorimeter model CR-400 (Konica Minolta, Tokyo, Japan). The color was expressed as hue angle (°h), where 0/360° represents red, 90° represents yellowish green, 180° represents turquoise blue and 270° represents violet.

2.5.4. External Appearance

External appearance was evaluated with a 9-point visual scale according to Lima et al. [25], where higher values represent better appearance and lower incidence of injuries, spots or rot, and 5 is the limit of acceptability, expressing fruit with 10% spots.

2.5.5. Pulp Firmness

Pulp firmness was measured with a texture analyzer TA.XTPlus (Stable Micro Systems, Godalming, UK), equipped with a 6 mm stainless steel probe and set for a 10 mm penetration distance. Pulp firmness results were obtained from the average of two measurements performed in the equatorial region of each fruit without the skin, and were expressed in N.

2.5.6. Soluble Solid Content (SSC)

Soluble solid content (SSC) was determined in the juice using a digital refractometer model PAL-1 (Atago, Tokyo, Japan). The results were expressed in percentage.

2.5.7. Titratable Acidity (TA)

Titratable acidity (TA) was measured by titrating 5 mL of juice diluted in 45 mL of distilled water with 0.1 N sodium hydroxide solution to an end point of pH 8.1, using an automatic titrator model 848 (Metrohm, Herisau, Switzerland). The results were expressed in g of citric acid per 100 g.

2.5.8. Beta-Carotene Content

Beta-carotene content was determined according to the method of Nagata and Yamashita [26]. One gram of mango pulp was added to test tubes with 10 mL of acetone:hexane (4:6, *v*/*v*). The solution was stirred for 1 min using a homogenizer model T18 digital (IKA, Guangzhou, China). An aliquot of the supernatant was transferred to quartz cuvettes and was read on a spectrophotometer at 663, 645, 505 and 453 nm. Beta-carotene content was determined according to Equation (2):

$$\beta\text{-carotene (mg/100 g)} = (0.216\ A_{663} - 1.22\ A_{645} - 0.304\ A_{505} + 0.452\ A_{453}) \times 10 \quad (2)$$

where A_{663}, A_{645}, A_{505} and A_{453} refer to absorbances at 663, 645, 505 and 453 nm, respectively.

2.5.9. Ascorbic Acid Content

Ascorbic acid (AsA) content was determined according to the AOAC method [27]. A total of 5 mL of mango juice was diluted in 100 mL of 0.5% (*w*/*v*) oxalic acid, which was then titrated with 0.02% (*v*/*v*) 2,6-dichlorophenol indophenol 0.02% (*v*/*v*) until reaching light pink color. The results were expressed as mg of AsA/100 g.

2.5.10. Total Phenolic Compounds (TPC) Content

For the preparation of the phenolic extract, 15 g of mango pulp was weighed in centrifuge tubes and left to extract in 20 mL of 50% methanol (*v*/*v*) for 1 h. The samples were later centrifuged at 10,000× *g* for 15 min at 4 °C. The supernatant was filtered and transferred to a 50 mL volumetric flask. The residue was extracted in 20 mL of 70% acetone (*v*/*v*) for 1 h. Centrifugation was repeated and the supernatant was filtered and added to the volumetric flask containing the supernatant of the first extraction. The final volume was then completed to 50 mL with distilled water.

TPC content was determined via the Folin–Ciocalteu method [27]. An aliquot of the phenolic extract was mixed with 1 mL of Folin–Ciocalteu reagent (1:3, *v*/*v*), 2 mL of 20% (*w*/*v*) sodium carbonate solution and 2 mL of distilled water in test tubes, which were shaken in a Vortex mixer and left to rest for 30 min, protected from light. Absorbance at 700 nm was measured using a spectrophotometer. Gallic acid was used to develop a standard calibration curve, and the results were expressed as g of gallic acid equivalents (GAE)/100 g.

2.5.11. Yellow Flavonoid (YF) Content

For determination of YF content, one gram of mango pulp was extracted in 10 mL of a 95% (*v*/*v*) ethanol/1.5 N HCl (85:15) solution [28]. Samples were homogenized for 1 min in an ULTRA-TURRAX homogenizer (model T18 digital, IKA, Guangzhou, China). The extract was transferred to a 50 mL balloon protected from light, conserved for 16 h under refrigeration and then filtered (Whatman No. 1 filter paper).

YF content was quantified by reading the extract absorbance at 374 nm on a spectrophotometer, using an absorption coefficient of 76.6. Results were expressed in mg/100 g.

2.5.12. Antioxidant Activity (AOX)

AOX was determined using the methods based on the capacity of the phenolic extract to scavenge the $ABTS^{\bullet+}$ and the $DPPH^{\bullet}$ radicals [29].

In the $ABTS^{\bullet+}$ method [30], the radical was generated by reaction of the ABTS stock solution (7 mmol L^{-1}) with potassium persulfate (140 mmol L^{-1}), kept in the dark for 16 h at ambient temperature. Prior to the analysis, the mixture was diluted in ethanol until reaching an absorbance of 0.700 ± 0.005 at 734 nm. Three different dilutions for each fruit extract were prepared in triplicate. In the dark, 30 μL of each dilution were added to 3 mL of the ABTS radical, followed by homogenization in a Vortex mixer. The samples were read at 734 nm after 6 min of the addition of the radical. A standard Trolox curve was used, and the results were expressed as μM of Trolox per g of fresh mass.

In the $DPPH^{\bullet}$ method [31], three dilutions of each extract were prepared in triplicate. In the dark, 100 μL of each extract dilution was added to 3.9 mL of the $DPPH^{\bullet}$ radical (0.06 mM). The mixture was stirred in a Vortex mixer and then left to rest in the dark. As a control, 100 μL of the control solution (50% methanol, 70% acetone and water, in the 4:4:2 ratio) was used instead of phenolic extract. The readings were carried out at 515 nm, 45 min after the addition of the $DPPH^{\bullet}$ radical, considering absorbance stabilization. The AOX was calculated as the extract concentration required to reduce the initial concentration of the $DPPH^{\bullet}$ radical by 50% (EC_{50}), with values expressed as mg fresh mass/mL of $DPPH^{\bullet}$.

2.6. Statistical Analysis

The experiment was performed using a completely randomized design, in a split-plot arrangement (4 × 4), with three replications of two mangoes. The plots were represented by packaging conditions (CS, CS-GO, PE and control), while the subplots consisted of storage times (0, 14, 28 and 42 days). The data were submitted to two-way ANOVA, and the least significant difference (LSD) test ($p \leq 0.05$) was used to compare the means between packaging types.

A principal component analysis (PCA) based on the correlation matrix among the variables was applied to summarize the data into few principal components responsible for most of data variance.

Statistical analyses were performed in R 4.0.2 (R Development Core Team, Vienna, Austria) and are detailed in Supplementary Material S1.

3. Results

3.1. Water Vapor Transmission Rate and Mechanical Properties of the Packaging

The incorporation of GO reduced the water vapor transmission rate (WVTR) of CS-based films by 35% ($p \leq 0.05$). The tensile strength and the Young's modulus of the CS film were increased by 21% and 19% through the addition of GO ($p \leq 0.05$), respectively. The elongation break was not influenced by the presence of GO ($p > 0.05$), which averaged 5.70% and 5.81% in the packaging with 0% and 0.25% GO, respectively (Table 1).

Table 1. Water vapor transmission rate (WVTR), tensile strength (σ), elongation at break (ε) and Young's modulus (∈) of chitosan (CS) and graphene oxide (GO)-based biodegradable packaging.

CS-Based Package	Water-Ba Rrier Properties	Mechanical Properties		
	WVTR ($g/m^2/s$)	σ ($\times 10^{-3}$ MPa)	ε (%)	∈ ($\times 10^{-3}$ MPa)
0% GO	52.73 ± 3.49 a	31.2 ± 2.5 b	5.70 ± 0.27 a	5.47 ± 0.29 b
0.25% GO	35.65 ± 2.42 b	37.9 ± 4.5 a	5.81 ± 0.32 a	6.52 ± 0.77 a

The values represent the mean ± standard deviation of five independent measurements. The values followed by different letters in the column are statistically different ($p \leq 0.05$) according to the F-test.

3.2. Fruit Quality Parameters

All quality parameters were significantly ($p \leq 0.05$) influenced by packaging treatment and storage period, as well as by the interaction between the two factors.

Mangoes in all treatments exhibited increased respiration rates during storage. However, after the 14th day of storage, the unpacked fruit showed a higher respiration rate than packed ones ($p \leq 0.05$). At 42 days of storage, fruit in all packaging had about 40% lower respiration rate than unpacked fruit (Figure 1a).

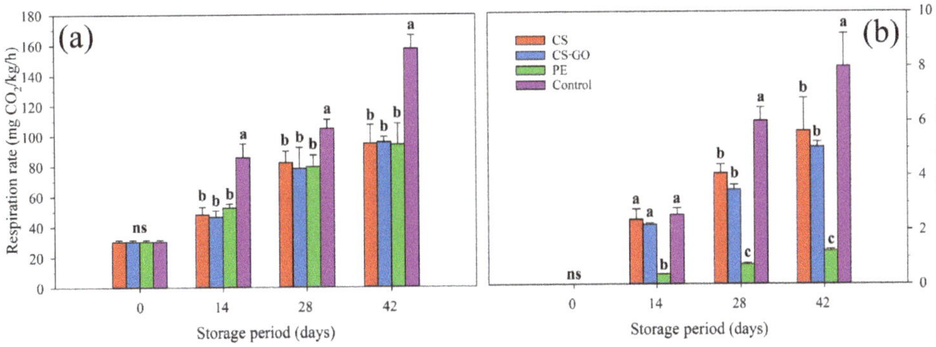

Figure 1. *Cont.*

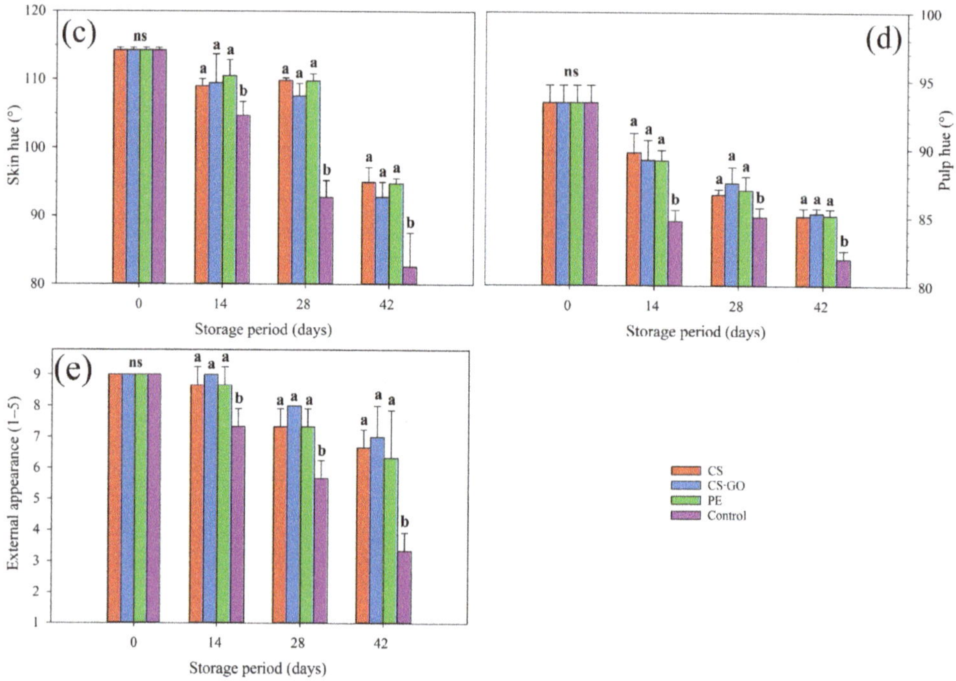

Figure 1. Effect of packaging conditions on respiration rate (**a**), mass loss (**b**), skin hue (**c**), pulp hue (**d**) and external appearance (**e**) of 'Tommy Atkins' mangoes stored at 12 °C for 42 days. The values followed by different letters, in each storage day, are statistically different ($p \leq 0.05$) according to the LSD test. ns indicates non-significant differences between packaging conditions ($p > 0.05$). Vertical bars represent the mean ± standard deviation of three replications per treatment ($n = 3$).

The packaging resulted in reduced mass loss ($p \leq 0.05$) compared with the control treatment. Synthetic packaging provided a much lower mass loss, compared to the other treatments ($p \leq 0.05$), with only 1.22% at 42 days. At the same day, mass losses of mangoes stored in CS and CS-GO-based packaging were 5.62% and 5.03%, respectively, values statistically lower ($p \leq 0.05$) than those observed in the control fruit (7.99%) (Figure 1b).

The hue angle of both the skin and pulp of 'Tommy Atkins' mangoes decreased during storage, which was more pronounced in unpacked fruit. At harvest, mangoes had a green skin color (h° = 114.3), which gradually changed to yellow (h° = 82.6) in unpacked mangoes after 42 days of storage; during the same period, packed fruit showed a higher ($p \leq 0.05$) average skin hue (h° = 94.2), indicating a yellowish-green color (Figure 1c). Pulp hue values of control mangoes decreased from 93.4° at harvest to 82.0° at the 42nd day of storage, which was statistically lower ($p \leq 0.05$) than the average of 85.2° found in packed fruit at the same day (Figure 1d).

At the 14th day of storage, the unpacked fruit already presented an external appearance inferior to the packed ones ($p \leq 0.05$), whose difference significantly remained until the end of storage. On the last day, the control fruit had an average score of 3.3 for external appearance, classifying them as unsuitable for commercialization due to the high severity of spots, injuries and wrinkles. On the same evaluation day, packed fruit had external appearance classified as regular/good, averaging 6.7 (Figure 1e).

Pulp firmness decreased over storage in all treatments, but this reduction was more intense in control fruit, which differed at 14 days of storage from packed ones ($p \leq 0.05$). Mangoes have shown a pulp firmness of 73.0 N at harvest, decreasing to 2.8 N and 5.0 N after 42 days of storage in unpacked and packed fruit, respectively (Figure 2a).

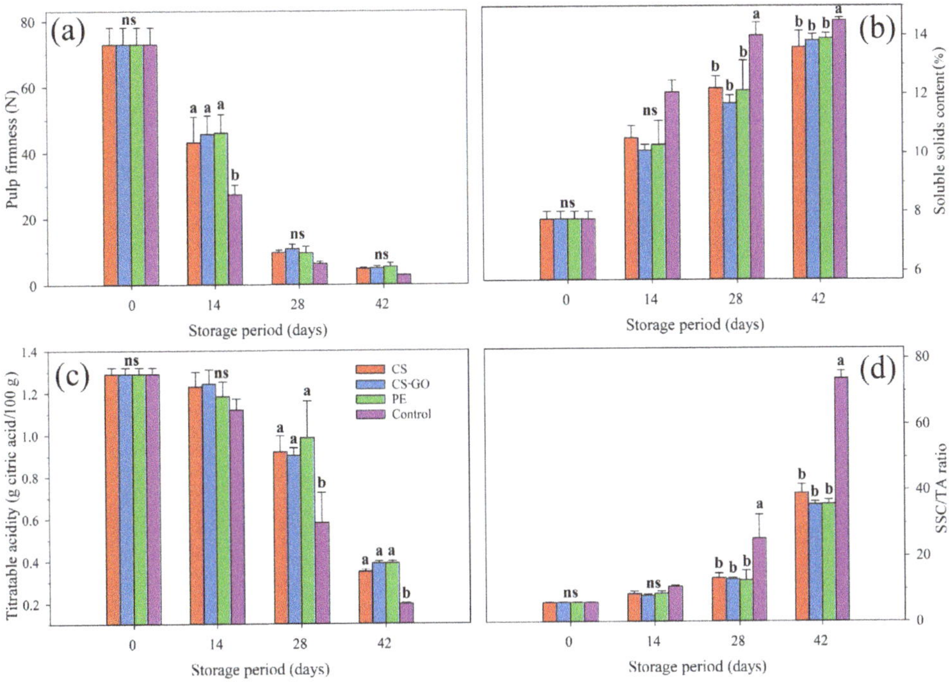

Figure 2. Effect of packaging conditions on pulp firmness (**a**), soluble solid content (SSC) (**b**), titratable acidity (TA) (**c**) and SSC/TA ratio (**d**) of 'Tommy Atkins' mangoes stored at 12 °C for 42 days. The values followed by different letters, on each storage day, are statistically different ($p \leq 0.05$) according to the LSD test. ns indicates non-significant differences between packaging conditions ($p > 0.05$). Vertical bars represent the mean ± standard deviation of three replications per treatment ($n = 3$).

Starting at 7.8% at harvest, the soluble solid content (SSC) increased until 14.5% and 13.8% in unpacked and packed mangoes, respectively (Figure 2b). Packed fruit have shown a lower SSC at 14 and 28 days of storage ($p \leq 0.05$), compared to control fruit.

Titratable acidity (TA) decreased in all treatments during storage. However, this reduction was more intense in unpacked fruit, differing from other treatments at 28 and 42 days of storage ($p \leq 0.05$) (Figure 2c). 'Tommy Atkins' mangoes were harvested with 1.29 g of citric acid/100 g, which decreased to 0.20 and 0.38 g of citric acid/100 g after 42 days of storage in unpacked and packed fruit, respectively.

The SSC/TA ratio increased in all treatments, but in unpacked fruit, the increase was 1126% over storage, while packed mangoes showed a statistically lower ($p \leq 0.05$) increase of 546%, 489%, and 492%, for fruit in CS, CS-GO and PE packaging, respectively (Figure 2d).

3.3. Beta-Carotene, Ascorbic Acid, Total Phenolic Compounds and Yellow Flavonoids

The interaction between storage time and packaging significantly ($p \leq 0.05$) influenced the β-carotene content in mangoes. Fruit showed an average β-carotene content of 0.23 mg/100 g at harvest, with an increase with storage time, regardless of packaging condition. However, unpacked fruit had statistically higher ($p \leq 0.05$) β-carotene content than those found in packed fruit at 28 and 42 days of storage. The highest mean β-carotene content was found in the control treatment at the 42nd day (115.12 mg/100 g) (Figure 3a).

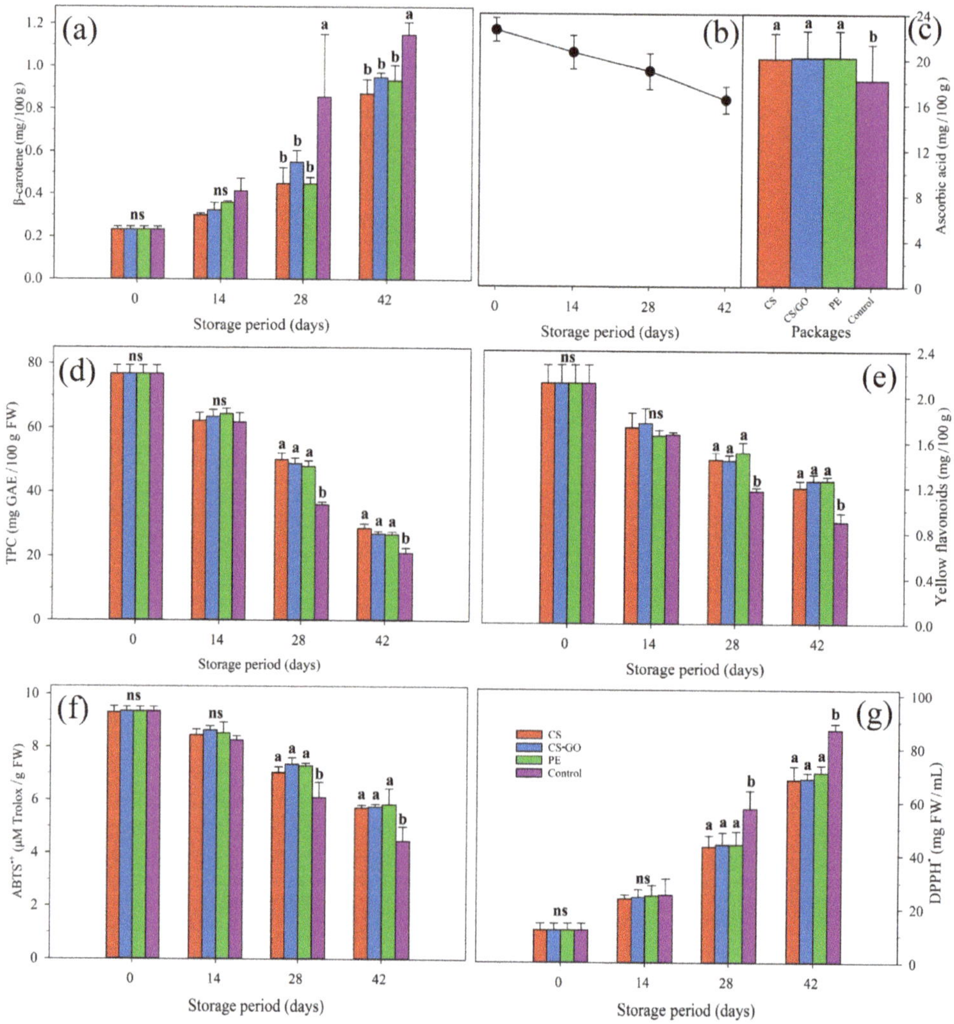

Figure 3. Effect of packaging conditions on β-carotene (**a**), ascorbic acid (AsA) (**b**,**c**), total phenolic compounds (TPC) (**d**) and yellow flavonoids (YF) (**e**) content and total antioxidant activity (AOX) via $ABTS^{\bullet+}$ (**f**) and $DPPH^{\bullet}$ (**g**) radical capture methods of 'Tommy Atkins' mangoes stored at 12 °C for 42 days. The values followed by different letters, in each storage day, are statistically different ($p \leq 0.05$) according to the LSD test. ns indicates non-significant differences between packaging conditions ($p > 0.05$). Vertical bars represent the mean ± standard deviation of three replications per treatment ($n = 3$).

Packaging treatment and storage time significantly ($p \leq 0.05$) affected the ascorbic acid (AsA) content in 'Tommy Atkins' mangoes, while the interaction between both was nonsignificant ($p > 0.05$). Mangoes showed, at the point of harvest, an average AsA content of 22.6 mg/100 g, which decreased over storage until averaging 16.5 mg/100 g at 42 days of storage (Figure 3b). The mean AsA content in packed fruit was 20.2 mg/100 g, which was 11% higher ($p \leq 0.05$) than that found in unpacked fruit (18.2 mg/100 g) (Figure 3c).

The total phenolic compound (TPC) content in mangoes was significantly ($p \leq 0.05$) influenced by storage time and packaging conditions and by the interaction between them.

At harvest, mangoes had a mean TPC content of 76.88 mg GAE/100 g FW, which decreased in all the samples during storage as a natural effect of fruit ripening. However, from 28 days onwards, packed fruit had significantly higher ($p \leq 0.05$) TPC content than unpacked ones (Figure 3d).

Yellow flavonoid (YF) content was influenced by storage time and packaging conditions, as well as by the interaction between both ($p \leq 0.05$). YF content in mangoes was 2.12 mg/100 g at the time of harvest, decreasing over storage, more markedly in control. At the end of the storage period, packed fruit averaged 1.24 mg/100 g for YF content, while unpacked fruit showed a 27% lower ($p \leq 0.05$) content (0.90 mg/100 g) (Figure 3e).

3.4. Antioxidant Activity

Antioxidant activity (AOX) of 'Tommy Atkins' mangoes showed a significant effect ($p \leq 0.05$) due to storage time and packaging conditions, as well as the interaction between them, which was observed in both the $ABTS^{\bullet+}$ and $DPPH^{\bullet}$ methods. A decrease was detected in fruit AOX in both $ABTS^{\bullet+}$ (Figure 3f) and $DPPH^{\bullet}$ (Figure 3g) methods during storage. However, a higher ($p \leq 0.05$) antioxidant activity was observed in packed fruit, compared to control fruit, which was reported from 28 days onwards in both methods (Figure 3f,g).

The correlation between $ABTS^{\bullet+}$ and $DPPH^{\bullet}$ values was significantly strong and negative (−0.970) (Figure 4). The $ABTS^{\bullet+}$ radical-scavenging activity is equivalent to that of Trolox; therefore, higher μM values of Trolox/g correspond to higher antioxidant activities. On the other hand, $DPPH^{\bullet}$ value is measured as the extract concentration required to reduce the initial concentration of the free radical by 50%, so lower values of mg/mL correspond to higher AOX.

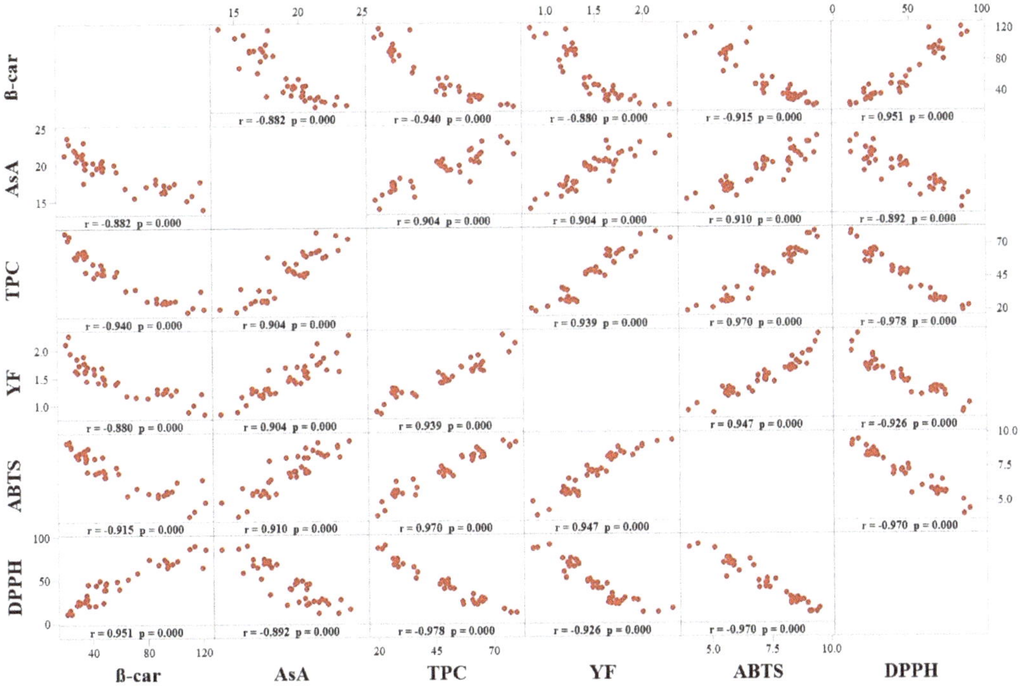

Figure 4. Pearson's correlation coefficient between bioactive compounds and antioxidant activities ($ABTS^{\bullet+}$ and $DPPH^{\bullet}$) of 'Tommy Atkins' mangoes stored at 12 °C for 42 days.

The contents of AsA (r = 0.910), TPC (r = 0.970) and YF (r = 0.947) had strong significant and positive ($p \leq 0.05$) correlations with the AOX, according to the ABTS•$^+$ assay. Similarly, these compounds also showed strong significant ($p \leq 0.05$) correlations with the antioxidant activity measured with the DPPH• method. Coefficients of correlation between DPPH• and the contents of AsA, TPC and YF were −0.892, −0.978 and −0.926, respectively.

In contrast, β-carotene content showed significant opposite correlations with both methods of antioxidant activity analyses, being negative with ABTS$^{•+}$ (−0.915) and positive with DPPH$^•$ (0.951) (Figure 4).

3.5. Principal Component Analysis

The data variability was explained in 92.02% by the first two principal components (PCs). PC1 was responsible for 88.00% of the total variance in the dataset and was positively correlated with the SSC, respiration rate, mass loss, SSC/TA ratio, β-carotene and DPPH$^•$. In contrast, the external appearance, skin and pulp hue, TA, pulp firmness, AsA, YF, TPC and ABTS$^{•+}$ were negatively correlated with PC1 (Figure 5).

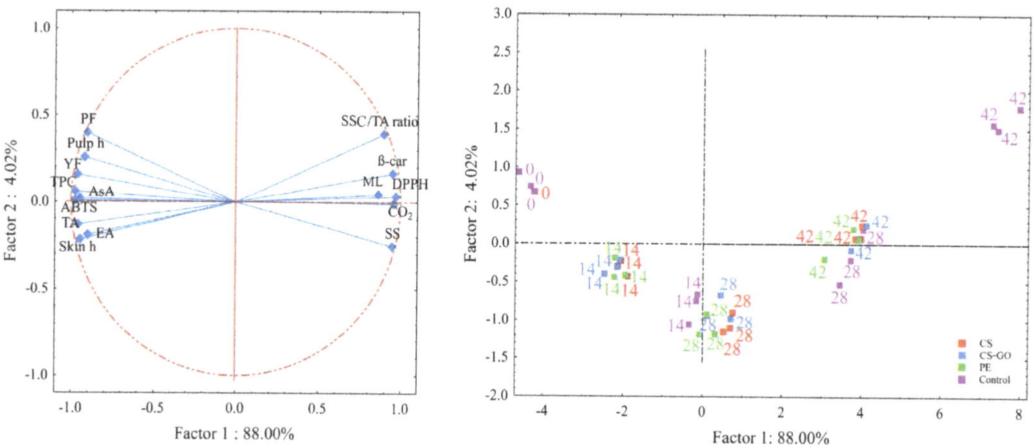

Figure 5. Principal component analysis plot of data from different quality attributes, bioactive compounds and antioxidant properties of 'Tommy Atkins' mangoes stored at 12 °C for 42 days.

The negative axes of PC1 were highly correlated with mangoes at harvest (0 day) and unpacked mangoes at 14 days of storage, which presented quality parameters typical of unripe fruit. Conversely, the positive axis of PC1 had a strong correlation with control fruit at 28 days and fruit of all treatments at 42 days, which indicates fruit of a more advanced ripening stage (Figure 5).

4. Discussion

'Tommy Atkins' is one of the mango varieties most valued by mango-exporting countries, due to its longer shelf life and ability to withstand long transportation compared to other varieties [1,6]. Even so, 'Tommy Atkins' mangoes require adoption of postharvest technologies to enable the export of high-quality fruit to the final destination, whose transport can take more than 30 days. The ideal storage temperature for 'Tommy Atkins' mangoes ranges from 10 to 13 °C, in order to reduce fruit metabolism and conserve fruit quality without lead to chilling injury [12]. However, due to the high perishability of mangos, other technologies are required in addition to cold storage to increase fruit postharvest life.

In our study, the adoption of a passive modified atmosphere (MA) through the use of biodegradable packaging was tested as an eco-friendly technology for mango export. In passive MAs, the packaging acts as a barrier against gas exchanges and moisture loss from the fruit to the environment [11], where the desired in-pack gas levels (low O_2 and high CO_2) are established naturally from the process of fruit respiration. The effectiveness of MA in preservation of fruit quality depends on factors related to fruit (species, variety, maturity stage and mass), packaging material (thickness and permeability to moisture, O_2 and CO_2) and storage conditions (temperature and moisture). In addition, the exposure of fruit to low oxygen concentrations increases the risk of stress-related physiological disorders that reduce fruit quality and acceptance by consumers [32].

CS has been extensively used in the creation of MA conditions for fruit conservation because its film-forming properties, which reduces gas exchanges between the fruit and the atmosphere [17,18]. However, CS has a significant water affinity, which can result in the weakening of the mechanical properties of materials based on this polymer [16]. According to our results, the incorporation of GO reduced the water vapor transmission rate (WVTR) of CS-based films. The WVTR represents the ability of a film to block the passage of water across its surface. This property is influenced by environmental factors such as temperature and relative humidity, film structure, including its thickness and area, difference in pressure and concentration gradient across the film [19]. Low WVTR values are favorable for the development of eco-friendly coatings or films to enhance the shelf life of food products, including fresh fruit [24]. Molecules of GO tend to fill the structural gaps when incorporated in the CS polymeric chain [33]. According to Ahmed et al. [21], the decrease in the diffusion of water vapor through the film is related to the creation of a tortuous path, due to the excellent dispersion of GO in the polymeric matrix.

The mechanical properties (tensile strength and Young's modulus) of the CS packaging were improved by the addition of GO, confirming the previous results of Paiva et al. [22]. The mechanical properties represent the resistance of the packaging to breakage and consequently reflect their potential to protect food from contact with the external environment [19]. For fresh fruit and vegetables, packaging is an important means of reducing gas exchanges with the external environment, thus reducing fruit respiration and metabolism and extending their shelf life. In this context, the CS-based biodegradable packaging is more suitable for use throughout refrigerated transportation, as a way to preserve the quality of the mangoes until they arrive to their final destination than for the transport of the fruit by the consumer. In this case, the biodegradable packaging does not require mechanical strength equivalent to commercial PE-based packaging for transport use [22].

Unpacked mangoes showed higher respiration rates than packed ones throughout the storage period. This result is in accordance with previous studies that also observed an effect of CS on reducing mango respiratory metabolism [34–36]. The decrease of fruit respiration rate is attributed to the gas barrier properties of the polymers, CS and PE, which reduce the availability of oxygen and concentrate the carbon dioxide surrounding the fruit, suppressing fruit metabolism.

Mass loss of mangoes was reduced via packaging treatments, especially PE-based packaging, which reduced the mass loss of fruit by 6.5-fold compared to the control at the end of storage. The positive effect of MA conditions on preventing mass loss has also been reported in mangoes coated with CS [8,37] or packed in plastic film [38]. The mass loss is related to fruit respiration and transpiration and results in quality losses that affect consumer acceptance. Packaging acts as a barrier with a protective function for the mango's surface, blocking the movement of moisture and solutes and improving mass retention [35].

Color is an important quality parameter for mango consumers, directly influencing their acceptance. 'Tommy Atkins' mangoes are characterized by a dark-red, blush skin color that covers much of the fruit's surface, with green and orange-yellow accents [23]. Mango ripening is accompanied by color changes from green to yellow-orange in the skin as a result of chlorophyll degradation concomitant with the accumulation of carotenoids in the mesocarp tissue [39]. However, the skin color may not have a good relationship with the internal quality of the fruit in some mango varieties [40]. In this context, pulp color changes are uniform during fruit ripening, so it is considered an adequate maturity index for mangoes [41]. Skin and pulp hue angles decreased as the duration of storage increased, which confirms color changes typical of mango ripening. In both fruit tissues, bagged fruit had higher hue values than control fruit, representing a delay in fruit ripening as a result of MA conditions. A previous study with lettuce revealed that MA regulated the expression of pheophorbide *a* oxygenase, a key enzyme in the chlorophyll degradation pathway [42].

The packaging reduced fruit softening, in agreement with Oliveira et al. [8], who reported the retention of firmness in chitosan-coated 'Tommy Atkins' mangoes compared to uncoated ones. Fruit softening is related to the solubilization of pectic substances, starch hydrolysis and transpiration [37]. Excessive softening is the major undesirable change that influences the marketability of mango [9]. Alterations in gaseous composition inside the packaging promoted by MA tends to reduce the activity of cell-wall-degrading enzymes [13], delaying the loss of firmness in packed mangoes.

Mango ripening is accompanied by the hydrolyzation of starch into simple sugars (glucose, fructose and sucrose), which increases fruit SSC and sweetness perception. Concomitantly, organic acids are used as respiratory substrates, decreasing fruit acidity [13]. The delay in these biochemical changes in packed fruit in relation to unpacked ones could be related to the reduced available oxygen for fruit upon MA conditions, which reduces fruit respiration [37].

β-carotene is a pro-vitamin A lipophilic molecule with important biological functions related to antioxidant activity, providing protection to human organism against cancer, cardiovascular diseases and diabetes by scavenging the oxygen free radicals [43]. As the major carotenoid in the pulp of 'Tommy Atkins' mangoes [44], β-carotene was quantified in our study to determine changes in fruit ripening. β-carotene content increased in all samples throughout the storage period, with a higher content in unpacked fruit. These results corroborate those presented in the literature, where mangoes stored under MA accumulated significantly less carotenoids over storage [45]. The changes in fruit metabolism promoted by packaging include the inhibition of ethylene synthesis, which in turn stimulates carotenoid biosynthesis by enhancing the transcription of phytoene synthase [46].

AsA (or vitamin C) is a natural water-soluble vitamin that acts as a potent antioxidant and free-radical scavenger [47]. AsA is the most abundant antioxidant compound in all plant tissues, and has a key role in crucial reactions throughout plant development, including its action as a promoter of enhanced pectin solubilization and depolymerization of polysaccharides during the ripening of climacteric fruit [48]. In our study, AsA content decreased in 'Tommy Atkins' mangoes during cold storage, as a result of the increased enzymatic degradation by ascorbic acid oxidase, which catalyzes the oxidation reaction of L-ascorbic acid to dehydroascorbic acid (DHA) [49].

Conditions of low oxygen availability promoted by packaging tend to delay the enzyme activity and oxidation reactions of AsA, conserving its content longer in packed fruit. Similar results were found in mangoes in MA conditions through the use of coatings based on guar-*Spirulina platensis-Aloe vera* extract [50], alginate [51] and CS-zein-cinnamaldehyde nano-cellulose [35]. The retention of AsA in mangoes after harvest is strictly related to high fruit quality, considering the high antioxidant and free-radical scavenging properties of AsA and its action as fruit preserver, through the inhibition of polyphenol oxidase activity, which oxidizes diphenols to quinones and leads to browning after wounding. AsA also acts as a promoter of enhanced pectin solubilization and depolymerization of polysaccharides during the ripening of climacteric fruit [52].

Phenolic compounds are a group of phytochemicals widely distributed in plants, well known for their beneficial biological effects, including the scavenging and neutralizing of reactive oxygen species, which are partially responsible for the incidence of chronic diseases [1,2]. Phenolic compounds also contribute to important organoleptic properties of fruits and vegetables, such as astringency, color, bitterness and flavor [50]. TPC content decreased throughout storage, as a natural effect of fruit ripening, so the preservation of these compounds is important in order to avoid a loss of fruit quality. By limiting the gas exchange between mangoes and the external environment, the packaging decreased polyphenol oxidase activity, which occurs in the presence of molecular oxygen and is responsible for undesirable enzymatic browning reactions and losses of nutraceutical value in fruit. The preservation of TPC in mangoes under MA was also observed in market conditions (>20 °C) up to four weeks in studies on fruit packed with polypropylene [50] and coated with guar-*Spirulina platensis-Aloe vera* extract [53].

Flavonoids are the major class of phenolic compounds, with the same biological functions, including antioxidant, antiviral, antifungal, anti-angiogenic, anti-tumorigenic, antidiabetic, and immunomodulatory bioproperties [54]. YF content decreased with the progression of the storage period, but with lower significant losses of these compounds in packaged mangoes. 'Banganapalli' and 'Totapuri' mangoes also exhibited higher flavonoid content when stored in a polypropylene-based packaging [55]. A similar effect was observed in 'Mahali' mangoes by Rastegar et al. [51], who applied an alginate-based coating and observed a 1.7-fold higher flavonoid content in coated fruit compared with the control fruit.

AOX represents the ability of phytochemicals to scavenge free radicals and inhibit oxidation, thus providing benefits to the immune system [29]. The AOX was measured by the capture of $ABTS^{\bullet+}$ and $DPPH^{\bullet}$ radicals from the pulp of 'Tommy Atkins' mangoes. In both methods, AOX decreased in all samples during storage, more markedly in unpacked mangoes. The preservation of the AOX of packed mangoes is a consequence of the delay in fruit ripening and the lower degradation of bioactive compounds during storage. Previous studies have also reported beneficial effects of MA in maintaining the antioxidant activity of mangoes during storage [45,50,51,53].

A principal component analyses (PCA) was carid out in order to summarize the data obtained from all quality parameters, bioactive compounds and AOX, evaluated on 'Tommy Atkins' mangoes, with the aim of assessing the effects of different packaging conditions and 42 days of storage at 12 °C on the quality of the fruit. The results demonstrate the delay in fruit ripening and the preservation of bioactive compounds provided by the packaging of 'Tommy Atkins' mangoes.

The postharvest quality of packed fruit was extended by at least 14 days compared to unpacked fruit stored under the same conditions. Moreover, due to the health-protective effects of mangoes which are of high interest in mangoes, postharvest treatments such as packaging stand out as an easy and effective method of preserving the phenolic compounds and antioxidant activity of the fruit, with the advantage that the CS-based packaging is eco-friendly and biodegradable.

5. Conclusions

The incorporation of GO improved the water-barrier and mechanical properties of CS packaging. The incorporation of all the types of packaging delayed the ripening of 'Tommy Atkins' mangoes by reducing their respiration rate, mass loss and softening, slowing changes in color, SSC, TA and beta-carotene and preserving the appearance of the fruit. The packaging also demonstrated a positive effect on preserving the bioactive compounds in the mangoes. The postharvest quality of the packed fruit was extended by at least 14 days compared to unpacked ones stored in the same conditions, with the advantage that the CS packaging is eco-friendly and biodegradable.

Supplementary Materials: The following supporting information can be downloaded at https://www.mdpi.com/article/10.3390/horticulturae9101145/s1, Table S1: F-values of analysis of variance for quality attributes of 'Tommy Atkins' mangoes stored in different packaging conditions for 42 days at 12 °C. One, two and three asterisks represent F-values at $p \leq 0.05$, $p \leq 0.01$ and $p \leq 0.001$, respectively, while ns represents non-significant F-value; Table S2: F-values of analysis of variance for bioactive compounds and antioxidant activity of 'Tommy At-kins' mangoes stored in different packaging conditions for 42 days at 12 °C. One, two and three asterisks represent F-values at $p \leq 0.05$, $p \leq 0.01$ and $p \leq 0.001$, respectively, while ns represents non-significant F-value; Table S3: Eigenvalues and variances of the principal components that represent 15 variables evaluated in 'Tommy Atkins' mangoes stored in different packaging conditions for 42 days at 12 °C; Table S4: Loadings in principal components 1 and 2 for 15 variables evaluated in 'Tommy Atkins' mangoes stored in different packaging conditions for 42 days at 12 °C.

Author Contributions: Conceptualization, J.C.V., S.T.d.F. and E.M.M.A.; methodology, J.C.V., S.T.d.F., R.H.d.L.L., F.K.G.d.S. and E.M.M.A.; formal analysis, J.C.V.; investigation, J.C.V., M.A.R.F. and C.d.S.R.C.; data curation, J.C.V.; writing—original draft preparation, J.C.V., M.A.R.F. and C.d.S.R.C.; writing—review and editing, S.T.d.F., R.H.d.L.L., F.K.G.d.S. and E.M.M.A.; supervision, S.T.d.F. and E.M.M.A. All authors have read and agreed to the published version of the manuscript.

Funding: This research was funded by the National Council for Scientific and Technological Development (CNPq, grant no. 131701/2019-3), the Coordination for the Improvement of Higher Education Personnel (CAPES, Financing Code 001), the Brazilian Agricultural Research Corporation (Embrapa, grant no. 22.13.06.025.00.04.007), the Federal Rural University of the Semi-Arid Region (UFERSA) and the Fundação de Amparo à Ciência e Tecnologia do Estado de Pernambuco (FACEPE, grant no. APQ-1046-5.01/22).

Data Availability Statement: The data that support the findings of this study are available from the corresponding author, upon reasonable request.

Acknowledgments: The authors thank José Gustavo Lima de Almeida and Cristiane Alves de Paiva for technical support in the preparation and characterization of the properties of the packaging. We also thank the company, Agrodan, for providing the mangoes.

Conflicts of Interest: The authors declare no conflict of interest.

References

1. Sousa, A.S.B.; Silva, M.C.A.; Lima, R.P.; Meireles, B.R.L.A.; Cordeiro, A.T.M.; Santos, E.F.S.; Amaro, A.L.; Pintado, M.M.E.; Silva, S.M. Phenolic Compounds and Antioxidant Activity as Discriminating Markers and Adding Value of Mango Varieties. *Sci. Hortic.* **2021**, *287*, 110259. [CrossRef]
2. Tacias-Pascacio, V.G.; Castañeda-Valbuena, D.; Fernandez-Lafuente, R.; Berenguer-Murcia, Á.; Meza-Gordillo, R.; Gutiérrez, L.F.; Pacheco, N.; Cuevas-Bernardino, J.; Ayora-Talavera, T. Phenolic Compounds in Mango Fruit: A Review. *J. Food Meas. Charact.* **2022**, *16*, 619–636. [CrossRef]
3. Win, S.T.; Setha, S. Enhancement of Anti-Inflammatory and Antioxidant Activities of Mango Fruit by Pre- and Postharvest Application of Salicylic Acid. *Horticulturae* **2022**, *8*, 555. [CrossRef]
4. Singh, Z.; Singh, R.K.; Sane, V.A.; Nath, P. Mango–Postharvest Biology and Biotechnology. *Crit. Rev. Plant Sci.* **2013**, *32*, 217–236. [CrossRef]
5. Santos, L.F.; Vilvert, J.C.; Souza, T.A.; Silva Alves, J.; Silva Ribeiro, T.; Neuwald, D.A.; Freitas, S.T. Minimum O_2 Levels During Storage to Inhibit Aerobic Respiration and Prolong the Postharvest Life of 'Tommy Atkins' Mangoes Produced in Different Growing Seasons. *Sci. Hortic.* **2023**, *318*, 112094. [CrossRef]
6. Costa, C.S.R.; Lima, M.A.C.; Lima Neto, F.P.; Silva Costa, A.E.; Vilvert, J.C.; Martins, L.S.S.; Santos Musser, R. Genetic Parameters and Selection of Mango Genotypes Using the FAI-BLUP Multitrait Index. *Sci. Hortic.* **2023**, *317*, 112049. [CrossRef]
7. Freitas, S.T.; Guimarães, Í.T.; Vilvert, J.C.; Amaral, M.H.P.; Brecht, J.K.; Marques, A.T.B. Mango Dry Matter Content at Harvest to Achieve High Consumer Quality of Different Cultivars in Different Growing Seasons. *Postharvest Biol. Technol.* **2022**, *189*, 111917. [CrossRef]
8. Ntsoane, M.L.; Zude-Sasse, M.; Mahajan, P.; Sivakumar, D. Quality Assesment and Postharvest Technology of Mango: A Review of its Current Status and Future Perspectives. *Sci. Hortic.* **2019**, *249*, 77–85. [CrossRef]
9. Oliveira, K.Á.R.; Conceição, M.L.; Oliveira, S.P.A.; Lima, M.S.; Galvão, M.S.; Madruga, M.S.; Magnani, M.; Souza, E.L. Postharvest Quality Improvements in Mango Cultivar Tommy Atkins by Chitosan Coating with *Mentha piperita* L. Essential Oil. *J. Hortic. Sci. Biotechnol.* **2020**, *95*, 260–272. [CrossRef]
10. Rastegar, S.; Hassanzadeh Khankahdani, H.; Rahimzadeh, M. Effects of Melatonin Treatment on the Biochemical Changes and Antioxidant Enzyme Activity of Mango Fruit during Storage. *Sci. Hortic.* **2020**, *259*, 108835. [CrossRef]
11. Nadeem, A.; Ahmed, Z.F.R.; Hussain, S.B.; Omar, A.E.-D.K.; Amin, M.; Javed, S.; Ali, A.; Ullah, S.; Razzaq, K.; Rajwana, I.A.; et al. On-Tree Fruit Bagging and Cold Storage Maintain the Postharvest Quality of Mango Fruit. *Horticulturae* **2022**, *8*, 814. [CrossRef]
12. Li, R.; Wang, Y.; Li, W.; Shao, Y. Comparative Analyses of Ripening, Texture Properties and Cell Wall Composition in Three Tropical Fruits Treated with 1-Methylcyclopropene during Cold Storage. *Horticulturae* **2023**, *9*, 126. [CrossRef]
13. Vasconcelos, O.C.M.; Duarte, D.; Silva, J.C.; Mesa, N.F.O.; Mederos, B.J.T.; Freitas, S.T. Modeling 'Tommy Atkins' Mango Cooling Time Based on Fruit Physicochemical Quality. *Sci. Hortic.* **2018**, *244*, 413–420. [CrossRef]
14. Seymour, G.; Poole, M.; Giovannoni, J.; Tucker, G. *The Molecular Biology and Biochemistry of Fruit Ripening*; John Wiley & Sons: Hoboken, NJ, USA, 2013.
15. Das, L.; Das, P.; Bhowal, A.; Bhattachariee, C. Synthesis of Hybrid Hydrogel Nano-Polymer Composite using Graphene Oxide, Chitosan and PVA and its Application in Waste Water Treatment. *Environ. Technol. Innov.* **2020**, *18*, 100664. [CrossRef]
16. Maraschin, T.G.; Correa, R.S.; Rodrigues, L.F.; Balzarettid, N.M.; Malmonge, J.A.; Galland, G.B.; Basso, N.R.S. Chitosan Nanocomposites with Graphene-Based Filler. *Mater. Res.* **2019**, *22*, e20180829. [CrossRef]

17. Han Lyn, F.; Tan, C.P.; Zawawi, R.M.; Nur Hanani, Z.A. Physicochemical Properties of Chitosan/ Graphene Oxide Composite Films and Their Effects on Storage Stability of Palm-Oil Based Margarine. *Food Hydrocoll.* **2021**, *117*, 106707. [CrossRef]
18. Parvin, N.; Rahman, A.; Roy, J.; Rashid, M.H.; Paul, N.C.; Mahamud, M.A.; Imran, S.; Sakil, M.A.; Uddin, F.M.J.; Molla, M.E.; et al. Chitosan Coating Improves Postharvest Shelf-Life of Mango (*Mangifera indica* L.). *Horticulturae* **2023**, *9*, 64. [CrossRef]
19. Silva, K.G.d.; Cavalcanti, M.T.; Martinsa, L.P.; Alves, R.d.C.; Lucena, F.A.d.; Santos, M.S.A.; Silva, S.X.d.; Costa, F.B.d.; Moreira, I.d.S.; Pereira, E.M. Coatings Based on Gelatin and Chitosan in the Conservation of Papaya (*Carica papaya* L.) Minimally Processed. *Horticulturae* **2023**, *9*, 729. [CrossRef]
20. Jakubowska, E.; Gierszewska, M.; Nowaczyk, J.; Olewnik-Kruszkowska, E. The Role of a Deep Eutectic Solvent in Changes of Physicochemical and Antioxidative Properties of Chitosan-Based Films. *Carbohydr. Polym.* **2021**, *255*, 117527. [CrossRef]
21. Manousi, N.; Rosenberg, E.; Deliyanni, E.; Zachariadis, G.A.; Samanidou, V. Magnetic Solid-Phase Extraction of Organic Compounds Based on Graphene Oxide Nanocomposites. *Molecules* **2020**, *25*, 1148. [CrossRef]
22. Ahmed, J.; Mulla, M.; Arfat, Y.A.; Thai, T.L.A. Mechanical, Thermal, Structural and Barrier Properties of Crab Shell Chitosan/Graphene Oxide Composite Films. *Food Hydrocoll.* **2017**, *71*, 141–148. [CrossRef]
23. Paiva, C.A.; Vilvert, J.C.; Menezes, F.L.G.; Leite, R.H.L.; Santos, F.K.G.; Medeiros, J.F.; Aroucha, E.M.M. Extended Shelf-Life of Melons using Chitosan and Graphene Oxide Based Biodegradable Bags. *J. Food Process. Preserv.* **2020**, *44*, e14871. [CrossRef]
24. National Mango Board. Available online: https://www.mango.org/ (accessed on 20 July 2023).
25. Sun, Q.; Sun, C.; Xiong, L. Mechanical, Barrier and Morphological Properties of Pea Starch and Peanut Protein Isolate Blend Films. *Carbohydr. Polym.* **2013**, *98*, 630–637. [CrossRef]
26. Lima, A.B.; Silva, S.M.; Rocha, A.; Nascimento, L.C.; Ramalho, F.S. Conservação Pós-Colheita de Manga "Tommy Atkins" Orgânica sob Recobrimentos Bio-Orgânicos. *Rev. Bras. Frutic.* **2012**, *34*, 704–710. [CrossRef]
27. Nagata, M.; Yamashita, I. Simple Method for Simultaneous Determination of Chlorophyll and Carotenoids in Tomato Fruit. *Nippon. Shokuhin Kogyo Gakkaishi* **1992**, *39*, 925–928. [CrossRef]
28. Association of Official Analytical Chemists (AOAC). *Official Methods of Analysis of the Association of Official Analytical Chemists*, 21st ed.; AOAC: Rockville, MD, USA, 2019.
29. Francis, F.J. Analysis of Anthocyanins. In *Anthocyanins as Food Colors*; Markakis, P., Ed.; Academic Press: New York, NY, USA, 1982; pp. 181–207.
30. Munteanu, I.G.; Apetrei, C. Analytical Methods Used in Determining Antioxidant Activity: A Review. *Int. J. Mol. Sci.* **2021**, *22*, 3380. [CrossRef]
31. Re, R.; Pellegrini, N.; Proteggente, A.; Pannala, A.; Yang, M.; Rice-Evans, C. Antioxidant Activity Applying an Improved ABTS Radical Cation Decolorization Assay. *Free Radic. Biol. Med.* **1999**, *26*, 1231–1237. [CrossRef] [PubMed]
32. Brand-Williams, W.; Cuvelier, M.E.; Berset, C.L.W.T. Use of a Free Radical Method to Evaluate Antioxidant Activity. *LWT Food Sci Technol.* **1995**, *28*, 25–30. [CrossRef]
33. Łysiak, G.P.; Rutkowski, K.; Walkowiak-Tomczak, D. Effect of Storage Conditions on Storability and Antioxidant Potential of Pears cv. 'Conference'. *Agriculture* **2021**, *11*, 545. [CrossRef]
34. Rambabu, K.; Bharath, G.; Banat, F.; Show, P.L.; Cocoletzi, H.H. Mango Leaf Extract Incorporated Chitosan Antioxidant Film for Active Food Packaging. *Int. J. Biol. Macromol.* **2019**, *126*, 1234–1243. [CrossRef]
35. Silva, G.M.C.; Silva, W.B.; Medeiros, D.B.; Salvador, A.R.; Cordeiro, M.H.M.; Silva, N.M.; Santana, D.B.; Mizobutsi, G.P. The Chitosan Affects Severely the Carbon Metabolism in Mango (*Mangifera indica* L. cv. Palmer) Fruit During Storage. *Food Chem.* **2017**, *237*, 372–378. [CrossRef] [PubMed]
36. Xiao, J.; Gu, C.; Zhu, D.; Huang, Y.; Luo, Y.; Zhou, Q. Development and Characterization of an Edible Chitosan/Zein-Cinnamaldehyde Nano-Cellulose Composite Film and its Effects on Mango Quality During Storage. *LWT* **2021**, *140*, 110809. [CrossRef]
37. Camatari, F.O.S.; Santana, L.C.L.A.; Carnelossi, M.A.G.; Alexandre, A.P.S.; Nunes, M.L.; Goulart, M.O.F.; Narain, N.; Silva, M.A.A.P. Impact of Edible Coatings Based on Cassava Starch and Chitosan on the Post-Harvest Shelf Life of Mango (*Mangifera indica*) 'Tommy Atkins' Fruits. *Food Sci. Technol.* **2018**, *38*, 86–95. [CrossRef]
38. Eshetu, A.; Ibrahim, A.M.; Forsido, S.F.; Kuyu, C.G. Effect of Beeswax and Chitosan Treatments on Quality and shelf life of Selected Mango (*Mangifera indica* L.) Cultivars. *Heliyon* **2019**, *5*, e01116. [CrossRef]
39. Costa, J.D.S.; Figueiredo Neto, A.; Almeida, F.A.C.; Costa, M.S. Conservation of 'Tommy Atkins' Mangoes Stored Under Passive Modified Atmosphere. *Rev. Caatinga* **2018**, *31*, 117–125. [CrossRef]
40. Yungyuen, W.; Vo, T.T.; Uthairatanakij, A.; Ma, G.; Zhang, L.; Tatmala, N.; Kaewsuksaeng, S.; Jitareerat, P.; Kato, M. Carotenoid Accumulation and the Expression of Carotenoid Metabolic Genes in Mango during Fruit Development and Ripening. *Appl. Sci.* **2021**, *11*, 4249. [CrossRef]
41. Vilvert, J.C.; Freitas, S.T.; Ferreira, M.A.R.; Costa, E.B.S.; Aroucha, E.M.M. Sample Size for Postharvest Quality Traits of 'Palmer' Mangoes. *Rev. Bras. Frutic.* **2021**, *43*, e014. [CrossRef]
42. Vanoli, M.; Rizzolo, A.; Grassi, M.; Spinelli, L.; Torricelli, A. Modeling Mango Ripening during Shelf Life Based on Pulp Color Nondestructively Measured by Time-Resolved Reflectance Spectroscopy. *Sci. Hortic.* **2023**, *310*, 111714. [CrossRef]
43. Chen, S.; Chen, X.; Fan, J.; Liu, H.; Zhou, J.; Chang, X. iTRAQ Proteomics Reveals Changes in the Lettuce (*Lactuca sativa* L. Grand Rapid) Proteome Related to Colour and Senescence under Modified Atmosphere Packaging. *J. Sci. Food Agric.* **2018**, *99*, 1908–1918. [CrossRef]

44. Bhatt, T.; Patel, K. Carotenoids: Potent to Prevent Diseases Review. *Nat. Prod. Bioprospecting* **2020**, *10*, 109–117. [CrossRef] [PubMed]
45. Ruales, J.; Baenas, N.; Moreno, D.A.; Stinco, C.M.; Meléndez-Martínez, A.J.; García-Ruíz, A. Biological Active Ecuadorian Mango 'Tommy Atkins' Ingredients–An Opportunity to Reduce Agrowaste. *Nutrients* **2018**, *10*, 1138. [CrossRef]
46. Sousa, F.F.; Pinsetta Junior, J.S.; Oliveira, K.T.E.F.; Rodrigues, E.C.N.; Andrade, J.P.; Mattiuz, B.-H. Conservation of 'Palmer' Mango with an Edible Coating of Hydroxypropyl Methylcellulose and Beeswax. *Food Chem.* **2021**, *346*, 128925. [CrossRef] [PubMed]
47. Dong, C.; Wang, Q.; Wang, Y.; Qin, L.; Shi, Y.; Wang, X.; Wang, R. NtDREB-1BL1 Enhances Carotenoid Biosynthesis by Regulating Phytoene Synthase in *Nicotiana tabacum*. *Genes* **2022**, *13*, 1134. [CrossRef] [PubMed]
48. Njus, D.; Kelley, P.M.; Tu, Y.J.; Schlegel, H.B. Ascorbic Acid: The Chemistry Underlying its Antioxidant Properties. *Free Radic. Biol. Med.* **2020**, *159*, 37–43. [CrossRef] [PubMed]
49. Mellidou, I.; Georgiadou, E.C.; Kaloudas, D.; Kalaitzis, P.; Fotopoulos, V.; Kanellis, A.K. Vitamins. In *Postharvest Physiology and Biochemistry of Fruits and Vegetables*; Yahia, E., Ed.; Academic Press: New York, NY, USA, 2019; pp. 359–384.
50. Mazurek, A.; Włodarczyk-Stasiak, M. A New Method for the Determination of Total Content of Vitamin C, Ascorbic and Dehydroascorbic Acid, in Food Products with the Voltammetric Technique with the Use of Tris(2-carboxyethyl)phosphine as a Reducing Reagent. *Molecules* **2023**, *28*, 812. [CrossRef]
51. Ebrahimi, F.; Rastegar, S. Preservation of Mango Fruit with Guar-Based Edible Coatings Enriched with *Spirulina platensis* and *Aloe vera* Extract During Storage at Ambient Temperature. *Sci. Hortic.* **2020**, *265*, 109258. [CrossRef]
52. Rastegar, S.; Hassanzadeh Khankahdani, H.; Rahimzadeh, M. Effectiveness of Alginate Coating on Antioxidant Enzymes and Biochemical Changes during Storage of Mango Fruit. *J. Food Biochem.* **2019**, *43*, e12990. [CrossRef]
53. Chylińska, M.; Szymańska-Chargot, M.; Deryło, K.; Tchórzewska, D.; Zdunek, A. Changing of Biochemical Parameters and Cell Wall Polysaccharides Distribution during Physiological Development of Tomato Fruit. *Plant Physiol. Biochem.* **2017**, *119*, 328–337. [CrossRef] [PubMed]
54. Rashid, M.I.; Fareed, M.I.; Rashid, H.; Aziz, H.; Ehsan, N.; Khalid, S.; Ghaffar, I.; Ali, R.; Gul, A.; Hakeem, K.R. Flavonoids and Their Biological Secrets. In *Plant and Human Health, Volume 2: Phytochemistry and Molecular Aspects*; Ozturk, M., Hakeem, K.R., Eds.; Springer: Cham, Switzerland, 2019; pp. 579–605. [CrossRef]
55. Perumal, A.B.; Nambiar, R.B.; Sellamuthu, P.S.; Emmanuel, R.S. Use of Modified Atmosphere Packaging Combined with Essential Oils for Prolonging Post-Harvest Shelf Life of Mango (cv. Banganapalli and cv. Totapuri). *LWT* **2021**, *148*, 111662. [CrossRef]

MDPI
St. Alban-Anlage 66
4052 Basel
Switzerland
www.mdpi.com

Horticulturae Editorial Office
E-mail: horticulturae@mdpi.com
www.mdpi.com/journal/horticulturae

www.ingramcontent.com/pod-product-compliance
Lightning Source LLC
Chambersburg PA
CBHW061446210926
51850CB00008B/1627